Artificial Intelligence
Frontiers in Statistics

Artificial Intelligence Frontiers in Statistics

AI and statistics III

Edited by

D.J. Hand

Professor of Statistics
The Open University, Milton Keynes, UK

CHAPMAN & HALL
London · Glasgow · New York · Tokyo · Melbourne · Madras

Published by Chapman & Hall, 2–6 Boundary Row, London SE1 8HN

Chapman & Hall, 2–6 Boundary Row, London SE1 8HN, UK

Blackie Academic & Professional, Wester Cleddens Road, Bishopbriggs, Glasgow G64 2NZ, UK

Van Nostrand Reinhold Inc., 115 5th Avenue, New York NY10003, USA

Chapman & Hall Japan, Thomson Publishing Japan, Hirakawacho Nemoto Building, 6F, 1-7-11 Hirakawa-cho, Chiyoda-ku, Tokyo 102, Japan

Chapman & Hall Australia, Thomas Nelson Australia, 102 Dodds Street, South Melbourne, Victoria 3205, Australia

Chapman & Hall India, R. Seshadri, 32 Second Main Road, CIT East, Madras 600 035, India

First edition 1993

© 1993 Chapman & Hall

Typeset in 10/12 Palatino by Thomson Press (India) Ltd, New Delhi.
Printed in Great Britain at the University Press, Cambridge

ISBN 0 412 40710 8

A catalogue record for this book is available from the British Library

Library of Congress Cataloging-in-Publication data
Artificial intelligence frontiers in statistics: AI and statistics
  III / edited by D.J. Hand. – 1st ed.
        p.   cm.
    Includes index.
    ISBN 0–412–40710–8
      1. Artificial intelligence. 2. Statistics. 3. Expert systems
(Computer science) I. Hand, D.J.
    Q335.5.A786   1993
    519.5′0285′63–dc20   92-34777
                                                      CIP

Contents

List of contributors   ix

Introduction D.J. Hand   xv

PART ONE   Statistical expert systems   1

1 **DEXPERT: an expert system for the design of experiments**   3
  T.J. Lorenzen, L.T. Truss, W.S. Spangler, W.T. Corpus
  and A.B. Parker

2 **Inside two commercially available statistical expert systems**   17
  J.F.M. Raes

3 **AMIA: Aide à la Modélisation par l'Intelligence Artificielle**
  (expert system for simulation modelling and sectoral forecasting)   31
  M. Ollivier, R. Arrus, M.-A. Durillon, S. Robert and B. Debord

4 **An architecture for knowledge-based statistical support systems**   39
  A. Prat, E. Edmonds, J.M. Catot, J. Lores, J. Galmes and P. Fletcher

5 **Enhancing explanation capabilities of statistical expert systems**
  through hypertext   46
  P. Hietala

6 **Measurement scales as metadata**   54
  D.J. Hand

PART TWO   Belief networks   65

7 **On the design of belief networks for knowledge-based systems**   67
  B. Abramson

8 **Lack-of-information based control in graphical belief systems**   82

9 Adaptive importance sampling for Bayesian networks applied
to filtering problems   90
A.R. Runnalls

10 Intelligent arc addition, belief propagation and utilization
of parallel processors by probabilistic inference engines   106
A. Ranjbar and M. McLeish

11 A new method for representing and solving Bayesian decision
problems   109
P.P. Shenoy

PART THREE   Learning   139

12 Inferring causal structure in mixed populations   141
C. Glymour, P. Spirtes and R. Scheines

13 A knowledge acquisition inductive system guided by
empirical interpretation of derived results   156
K. Tsujino and S. Nishida

14 Incorporating statistical techniques into empirical symbolic
learning systems   168
F. Esposito, D. Malerba and G. Semeraro

15 Learning classification trees   182
W. Buntine

16 An analysis of two probabilistic model induction techniques   202
S.L. Crawford and M. Fung

PART FOUR   Neural networks   215

17 A robust back propagation algorithm for function approximation 217
D.S. Chen and R.C. Jain

18 Maximum likelihood training of neural networks   241
H. Gish

19 A connectionist knowledge acquisition tool: CONKAT   256
A. Ultsch, R. Mantyk and G. Halmans

20 Connectionist, rule-based, and Bayesian decision aids:
an empirical comparison   264
S. Schwartz, J. Wiles, I. Gough and S. Phillips

PART FIVE   Text manipulation   279

21 Statistical approaches to aligning sentences and identifying word
correspondences in parallel texts: a report on work in progress   281
W.A. Gale and K.W. Church

22 Probabilistic text understanding   295
R.P. Goldman and E. Charniak

**23 The application of machine learning techniques in subject
classification**   312
I. Kavanagh, C. Ward and J. Dunnion

PART SIX   Other areas   325

24 A statistical semantics for causation   327
J. Pearl and T.S. Verma

**25 Admissible stochastic complexity models for classification
problems**   335
P. Smyth

**26 Combining the probability judgements of experts: statistical
and artificial intelligence approaches**   348
L.A. Cox

27 Randomness and independence in non-monotonic reasoning   362
E. Neufeld

**28 Consistent regions in probabilistic logic when using
different norms**   370
D. Bouchaffra

**29 A decision theoretic approach to controlling the cost of
planning**   387
L. Hartman

Index   401

Contributors

B. *Abramson*
Department of Computer Science
University of Southern California
Los Angeles
CA 90089-0782
USA

R. *Almond*
Department of Statistics GN-22
University of Washington
Seattle
WA 98195
USA

R. *Arrus*
CRISS
(Centre de Recherche en Informatique
  Appliquée aux Sciences Sociales)
Université des Sciences
38040 Grenoble Cedex
France

D. *Bouchaffra*
CRISS
(Centre de Recherche en Informatique
  Appliquée aux Sciences Sociales)
B.P. 47 X
38040 Grenoble Cedex
France

W. *Buntine*
RIACS & NASA Ames Research Center
Mail Stop 269-2
Moffet Field
CA 94035
USA

J.M. *Catot*
Departamento de Estadistica
E.T.S.E.I.B.
Diagonal 647
08028-Barcelona
Spain

E. *Charniak*
Department of Computer Science
Brown University
Box 1910
Providence
RI 02912
USA

D.S. *Chen*
Artificial Intelligence Laboratory
162 ATL
The University of Michigan
Ann Arbor
MI 48109-2110
USA

K.W. Church
AT&T Bell Laboratories
600 Mountain Avenue
Murray Hill
NJ 07079
USA

W.T. Corpus
Electronic Data Systems
5555 New King Street
MD-4029
Troy, MI 48098
USA

L.A. Cox Jr
US West Advanced Technologies
4001 Discovery Drive
Boulder
Colorado 80303
USA

S.L. Crawford
Advanced Decision Systems
1500 Plymouth Street
Mountain View
California
94043-1230
USA

B. Debord
CRISS
(Centre de Recherche en Informatique
   Appliquée aux Sciences Sociales)
Université des Sciences
38040 Grenoble Cedex
France

J. Dunnion
Department of Computer Science
University College
Dublin Belfield
Dublin 4
Ireland

M.A. Durillon
CRISS
(Centre de Recherche en Informatique
   Appliquée aux Sciences Sociales)
Université des Sciences
38040 Grenoble Cedex
France

E. Edmonds
LUTCHI Research Centre
Department of Computer Studies
Loughborough University of
   Technology
Loughborough
Leicestershire
LE11 3TU
UK

F. Esposito
Istituto di Scienze dell'Informazione
Università degli Studi
V. Amendola 173 (70126)
Bari
Italy

P. Fletcher
Departamento de Estadistica
E.T.S.E.I.B.
Diagonal 647
08028-Barcelona
Spain

R.M. Fung
Advanced Decision Systems
1500 Plymouth Street
Mountain View
California
94043-1230
USA

W.A. Gale
AT&T Bell Laboratories
600 Mountain Avenue
Murray Hill

NJ 07079
USA

J. Galmes
Departamento de Estadistica
E.T.S.E.I.B.
Diagonal 647
08028-Barcelona
Spain

H. Gish
BBN Systems and Technologies
10 Moulton Street
Cambridge
MA 01238
USA

R.P. Goldman
Department of Computer Science
Tulane University
301 Stanley Thomas Hall
New Orleans
LA 70115
USA

I. Gough
University of Queensland
Qld 4072
Australia

C. Glymour
Department of Philosophy
Carnegie Mellon University
Pittsburgh
PA 15213
USA

G. Halmans
Institute of Informatics
University of Dortmund
P.O. Box 500500
D-4600 Dortmund
Germany

D.J. Hand
Statistics Department
Faculty of Mathematics
The Open University
Walton Hall
Milton Keynes
Buckinghamshire
MK7 6AA
UK

L. Hartman
Computer Science
University of Rochester
Rochester
NY 14627
USA

P. Hietala
University of Tampere
Department of Mathematical Sciences
Statistics Unit
P.O. Box 607
SF-33101
Tampere
Finland

R.C. Jain
Artificial Intelligence Laboratory
162 ATL
The University of Michigan
Ann Arbor
MI 48109-2110
USA

I. Kavanagh
Department of Computer Science
University College
Dublin Belfield
Dublin 4
Ireland

T.J. Lorenzen
Mathematics Department
General Motors Research and

Environmental Staff
30500 Mound Road
Box 9055
Warren, MI 48090-9055
USA

J. Lores
Departamento de Estadistica
E.T.S.E.I.B.
Diagonal 647
08028-Barcelona
Spain

D. Malerba
Istituto di Scienze dell'Informazione
Università degli Studi
V. Amendola 173 (70126)
Bari
Italy

R. Mantyk
Institute of Informatics
University of Dortmund
P.O. Box 500500
D-4600 Dortmund
Germany

M. McLeish
Computing and Information Science
University of Guelph
Guelph
Ontario
N1G 2W1
Canada

E. Neufeld
Department of Computational Science
University of Saskatchewan
Saskatoon
Saskatchewan
S7N 0W0
Canada

K. Nishida
System 4G
Central Research Laboratory
Mitsubishi Electronic Corporation
8-1-1 Tsukaguchi-Honmachi
Amagasaki Hyogo 661
Japan

M. Ollivier
CRISS
(Centre de Recherche en Informatique
  Appliquée aux Sciences Sociales)
Université des Sciences
38040 Grenoble Cedex
France

A.B. Parker
Sun Microsystems, Inc.
MS PAL1-210
2550 Garcia Avenue
Mountain View, CA 94043-1100
USA

J. Pearl
Cognitive Systems Laboratory
Computer Science Department
University of California
Los Angeles
CA 90024
USA

S. Phillips
University of Queensland
Qld 4072
Australia

A. Prat
Departamento de Estadistica
E.T.S.E.I.B.
Diagonal 647
08028-Barcelona
Spain

J.F.M. Raes
University of Antwerp RUCA
Middleheimlaan 1
B 2020 Antwerp
Belgium

A. Ranjbar
Computing and Information Science
University of Guelph
Guelph
Ontario
N1G 2W1
Canada

S. Robert
CRISS
(Centre de Recherche en Informatique
    Appliquée aux Sciences Sociales)
Université des Sciences
38040 Grenoble Cedex
France

A.R. Runnalls
Computing Laboratory
University of Kent
Canterbury
UK

S. Schwartz
President's Office
J.D. Story Building
University of Queensland
Qld 4072
Australia

R. Scheines
Department of Philosophy
Carnegie Mellon University
Pittsburgh
PA 15213
USA

G. Semeraro
Istituto di Scienze dell'Informazione

Università degli Studi
V. Amendola 173 (70126)
Bari
Italy

P.P. Shenoy
School of Business
University of Kansas
Lawrence
KS 66045-2003
USA

P. Smyth
Jet Propulsion Laboratory 238 420
California Institute of Technology
4800 Oak Grove Drive
Pasadena
CA 91109
USA

W.S. Spangler
General Motors Advanced Engineering
Manufacturing B Building
MD-70
GM Technical Center
Warren, MI 48090-9040
USA

P. Spirtes
Department of Philosophy
Carnegie Mellon University
Pittsburgh
PA 15213
USA

L.T. Truss
Mathematics Department
General Motors Research and
    Environmental Staff
30500 Mound Road
Box 9055
Warren, MI 48090-9055
USA

A. Ultsch
Institute of Informatics
University of Dortmund
P.O. Box 500500
D-4600 Dortmund
Germany

C. Ward
Department of Computer Science
University College
Dublin Belfield
Dublin 4
Ireland

T.S. Verma
Cognitive Systems Laboratory
Computer Science Department
University of California
Los Angeles
CA 90024
USA

J. Wiles
University of Queensland
Qld 4072
Australia

Introduction

D.J. Hand

This volume contains selected refereed papers presented at the Third International Workshop on Artificial Intelligence and Statistics, Fort Lauderdale, Florida, in January 1991. These Workshops represent a showcase for state-of-the-art work in this area, which, in common with much computer technology, is progressing dramatically and across a broadening frontier. An ideal way to view developments in the area is to compare the contributions to the first (Gale, 1986) and second (Hand, 1990) Workshops with those in this volume.

In this volume I have attempted to partition the contributions into subdomains. This is difficult since many of the contributions naturally span such divisions. Two areas at the interface of statistics and AI have become particularly important: on the one hand, expert systems to assist researchers to do statistical analysis; and, on the other, statistical ideas applied within expert systems.

Part One of the present volume contains papers dealing with the first of these subdomains. Lorenzen *et al.* describe DEXPERT, an expert system for experimental design. Early work on statistical expert systems had to steer a line between attempting something so grandiose that it could never be accomplished in a reasonable time and producing something so specialized that, while it worked, it would find only very limited application. The work on DEXPERT illustrates perfectly the great progress that has been made, being a functioning system which nevertheless covers a wide domain. DEXPERT produces alternative designs, a layout sheet to help collect the data, and later analyses the data and interprets the results for the user. It is extremely flexible, and represents a true statistical expert system.

A further demonstration of progress is that some systems are now commercially available. In his chapter, Raes looks at two such systems, the SPRINGEX system and STATISTICAL NAVIGATOR.

The European Statistical Office, Eurostat, has a great interest in statistical expert systems as it has to deal with large quantities of data arising from many different sources and nationalities. Eurostat has therefore funded research projects in the area through its Development Of Statistical Expert Systems (DOSES) programme, and an example of such a project is given here in the paper by Ollivier *et al.* Their system, AMIA, assists forecasters using simulation models.

An earlier milestone system was the GLIMPSE system to assist researchers building linear models using the GLIM statistical package. Advantage of the experience gained through the construction of GLIMPSE, as well as other systems, is being taken in the FOCUS project described by Prat *et al.*

Paula Hietala describes the use of hypertext to provide explanation and help facilities in statistical expert systems, illustrating it through using the Apple HyperCard system with the ESTES expert system for time series analysis.

Metadata is information about data. Much metadata is currently implicit, but making such material explicit so that software can access it will assist expert systems to guide the researcher. In his chapter, Hand looks at one particular kind of metadata: measurement scales.

Part Two of this volume presents work on belief networks and their generalizations, one of the areas of AI research which is being substantially influenced by statistical concepts. A high-level discussion of the use of belief networks as the overall framework for knowledge-based systems is given by Abramson, and Almond describes how a successive model refinement process can lead to an expert system control algorithm.

Runnals describes a method for analysing a system of continuous variables represented as a Bayesian network stimulated by work on aircraft navigation systems. Ranjbar and McLeish describe several innovations arising from an attempt to improve the response time of an expert system based on a causal probabilistic network. Shenoy introduces 'valuation networks' to represent and solve Bayesian decision problems.

Part Three includes chapters primarily concerned with learning in some sense. That by Glymour *et al.* describes an algorithm for inferring causal relationships from statistical data and partial background knowledge. The chapter by Tsujino and Nishida constructs knowledge bases from examples given by an expert and then modifies the result using abstract domain knowledge provided by an expert. The chapter by Esposito *et al.* considers the effectiveness of integrating statistical data analysis with symbolic concept learning systems.

Learning through recursive construction of classification trees has become an established subdomain of both statistical classification and knowledge acquisition in AI. Buntine outlines a Bayesian tree learning algorithm and Crawford and Fung compare the CART recursive tree growing algorithm with CONSTRUCTOR, a system for generating probabilistic networks.

Neural networks also represent an important part of the interface between statistics and AI, being a fundamentally AI construct, the understanding of which requires advanced probabilistic and statistical ideas. Part Four of the book is concerned with this subdomain.

In their chapter, Chen and Jain describe a robust back propagation learning algorithm which is resistant to noise effects. Gish discusses the application of the important statistical criterion of maximum likelihood in the training of neural networks. Ultsch *et al.* integrate neural network technology with rule-based approaches in the CONKAT system. Finally in this section, Schwartz *et al.* explore an area which is very important but not yet sufficiently investigated: comparisons between the various approaches to classification. In particular, they compare a connectionist approach, a rule induction approach, and a Bayesian updating approach.

Part Five contains three chapters concerned in one way or another with text manipulation. That by Gale and Church describes a statistical method for matching the sentences in large bodies of text in different languages. That by Goldman and Charniak describes a probabilistic approach to text understanding. And that by Kavanagh *et al.* describes a system which uses machine learning techniques to determine the subject of a piece of text.

The final part, Part Six, contains further chapters illustrating the breadth of interaction between the two disciplines, and how AI and statistics can benefit each other. Pearl and Verma discuss how genuine causal influences can be distinguished from spurious covariances. Smyth investigates the application of stochastic complexity theory to classification problems (such problems, in one form or other, being a staple of both disciplines). Cox discusses the important practical problem of combining probability judgements of different experts. Neufeld discusses non-monotonic reasoning, in which many patterns of non-monotonic inference can be approximated within a probabilistic framework. Bouchaffra extends Nilsson's probabilistic logic. Finally, Hartman describes a decision-theoretic approach to controlling the cost of planning.

REFERENCES

Gale, W.A., ed. (1986) *Artificial Intelligence and Statistics*, Reading, MA: Addison-Wesley.

Hand, D.J., ed. (1990) Artificial intelligence and statistics II. *Annals of Mathematics and Artificial Intelligence*, **2**.

PART ONE
Statistical expert systems

DEXPERT: an expert system for the design of experiments

1

T.J. Lorenzen, L.T. Truss, W.S. Spangler, W.T. Corpus and A.B. Parker

INTRODUCTION

DEXPERT is an expert system within General Motors for the design and analysis of experiments. The fundamental basis for this program is a descriptive mathematical model of the experiment under consideration. From this mathematical model, expected mean squares are computed, tests are determined, and the power of the tests computed. This supplies the information needed to compare different designs and choose the best possible design. DEXPERT provides verbal interpretations to facilitate these comparisons. A layout sheet is then generated to aid in the collection of data. Once the data has been collected and entered into the system, DEXPERT analyses and interprets the results using a number of analytical and graphical methods.

DEXPERT's capabilities include fixed and random factors (even with fractional designs), arbitrary nestings, the ability to handle incomplete randomization (including the technique known as blocking), fractional and mixed fractional designs, crossed fractional designs (including the complete confounding pattern), standard response surface designs such as Plackett–Burman, central composite and Box–Behnken designs, and D-optimal designs. Analytic capabilities include ANOVA, variance component estimation, comparison of means, polynomial approximation, data transformation, and graphical displays of means (with confidence intervals), interactions, normal effects plots, residual plots, time plots, response surface plots, and contour plots.

Applications fall into five generic categories. **Screening** uses fractional designs to select the most important factors from a large set of candidate factors. The user specifies those interactions that need to be estimated and interactions that may have an effect but can be confounded among themselves. A design can be selected that can either minimize the total size of the experiment or minimize the number of expensive combinations of certain factors.

Complicated influence is used when a more comprehensive study on a smaller number of factors is desired. It is the default category and primarily uses full factorial designs.

Artificial Intelligence Frontiers in Statistics: AI and statistics III. Edited by D.J. Hand. Published in 1993 by Chapman & Hall, London. ISBN 0 412 40710 8

Robust design is used when factors can be put into either the control or noise category. Control factors must be controllable by the factory, and noise factors are either not controllable or not easily controllable by the factory. The purpose of these experiments is to find the settings of the control factors that minimize the influence of the noise factors on the response. Emphasis is placed on the control by noise interactions. Analysis takes place on the raw data (treating control and noise factors simply as factors), on means and variances calculated over the noise factors, and on the loss function (after adjustment if there is an adjustment factor with known properties). Both inner/outer arrays and combined arrays can be generated and their properties compared.

Model response uses certain preclassified designs (Plackett–Burman, central composite, Box–Behnken), and D-optimality to design for specified polynomial models. Arbitrary candidate regions and categorical variables can be used with D-optimality.

Finally, **sequential optimization** can be used to find the best operating conditions for the factors. We start out with a quadratic fit around the current conditions and continue to use quadratic fits to move toward the optimum. When the centre of the region moves twice in the same direction, the range in this direction doubles. When the centre of the region stays the same twice, the range halves. When the optimum is within a design region, the region contracts for a more accurate estimate.

The interface of the system is tuned to the level of sophistication of the user. For a novice, the system minimizes the presentation of statistical information while providing extensive help facilities, suggestions, and consistency checks. Default decisions are supplied that will produce a reasonable design. Appropriate warnings are given, if necessary. Verbal summaries and interpretations are generated and can be printed. For an expert, the system displays much more detailed information and makes no decisions for that user. An automatic help facility is also available to guide the first-time user through the system.

REQUIRED FACTOR INFORMATION

DEXPERT interacts with users through a series of screens. The initial screen gathers information about the user. The user's name is entered for the purpose of allocating files. The skill level of the user is selected to determine the proper interface. Interface issues will not be discussed in this chapter. Rather, we will be concerned with the technical workings of DEXPERT.

Having set up the underlying file structure, the user must next specify the goal of the experiment, the response variable(s), the factors, and their properties. There is general help available regarding the best selection of response variable(s) and factors, but the final selection is the responsibility of the user.

Next, the user enters the factors to be varied in this experiment. For each factor, the user must specify a name and certain characteristics of the factor. An abbreviation is generated based on each factor name, but may be changed. For each factor, the user must indicate the number of levels to be used, whether the factor is fixed or random, whether it is quantitative or qualitative, if any nesting or sliding (Taguchi, 1987) levels occur, and whether there is an implied run order for this factor. If desired, defaults can be set to minimize the input effort. For example, screening experiments typically involve fixed factors having two quantitative levels. Of course, defaults can be overridden.

The selection of the number of levels is not critical as these are often changed in the redesign stage. The user should simply make a best guess, knowing that it will probably be changed later.

The basic factor information for the robust designs application category is the same with one important difference: each factor must be classified as either a control factor or a noise factor. The classification is necessary in order to carry out the goal of finding the control factor settings that minimize variation caused by the noise factors.

A control factor is a factor whose levels can be controlled in a factory setting. A control factor cannot be random since, by definition, one cannot control a random factor. Noise factors cannot be controlled in a factory setting. A special type of noise factor is called **transmitted variation**. Transmitted variation is due to the variation of parts coming into the manufacturing process. For example, one may order steel at a thickness of 3 mil but the actual thickness may range from 2.5 to 3.5 mil. A transmitted variation factor is associated with a corresponding control factor and is always fixed and quantitative.

Several differences occur when the model response goal is selected. Since we are approximating a surface with a polynomial function, it no longer makes sense to talk about the number of levels of a factor. Rather, the user must enter the specific polynomial model that needs to be fitted. Both quantitative and qualitative factors can be considered but all factors must be fixed. For all quantitative factors, the user enters the range of interest; for all qualitative factors, the exact values used in the experiment are entered.

For sequential optimization of a process, all factors must be fixed and quantitative. Initial factor inputs consist solely of starting conditions and the original ranges for each factor. The ranges will be altered as necessary in the iterative process of finding the optimal.

GENERATING DESIGNS

DEXPERT has three different modes for generating and evaluating designs: the full factorial mode; the fractional factorial mode; and the response surface mode. A separate subsection will be devoted to each mode. However, the link from these modes to the areas of application will not be clear. The next section will make this link.

To simplify the discussion in this chapter, we will assume the factor abbreviations are given by the capital letters of the alphabet: A, B, C, etc. Of course, the system works with the abbreviations specified by the user.

FULL FACTORIAL DESIGNS

For full factorial designs, DEXPERT starts by generating a descriptive mathematical model. Start with the first factor and attach the lower-case subscript i yielding A_i. Take the next factor, attach the next subscript (alphabetically) and form all interactions with the previous terms. For example, add B_j, yielding the terms A_i, B_j, and AB_{ij}. Continue in this fashion for all crossed factors.

For nested factors, attach the next available subscript but put the subscripts associated with the nesting inside parentheses. For example, if C were nested within A, use the notation $C_{k(i)}$. If C were nested within A and B, use the notation $C_{k(ij)}$. Again form the interaction with all other terms separately combining subscripts outside parentheses and inside parentheses. Do not include any terms that have the same index inside and outside the parentheses. For example, if B were nested in A, and D within C, the complete set of terms would be A_i, $B_{j(i)}$, C_k, AC_{ik}, $BC_{jk(i)}$, $D_{l(k)}$, $AD_{il(k)}$, and $BD_{jl(ik)}$. Note, for example, that $AB_{ij(i)}$ is not included since i appears inside and outside the parentheses.

Restrictions on randomization also affect the mathematical model through the addition of one or more terms. Suppose, for example, that the user wishes to select a level of factor B, and run all combinations involving that level of B before changing levels of B. Then we say there is a **restriction error** on the B_j term or the **experiment is blocked** on the B_j. This is denoted in the model by the addition of a random term *restr* having one level and nested in B (Anderson, 1970; Lorenzen, 1984). For this particular example, we would denote the restriction error *restr*$_{k(j)}$. Restriction error terms do not interact with any other term in the model. If the restriction error were more complicated (say, the user wanted to change the combinations of A and B as seldom as possible), then the restriction would fall on the AB_{ij} interaction, and the system would denote this *restr*$_{k(ij)}$. This restriction error concept generalizes to all types of restriction on randomization.

The underlying mathematical model is the basis for generating and evaluating the properties of a design. These properties include the degrees of freedom (df) for each term, the type of test available for each term, the term(s) testing each term, the df for the test, the detectability of each term, a classification of detectability, and the expected mean squares (EMS) for each term. In addition, the required experiment size is computed.

Since all of these designs are balanced, the df are computed using well-known formulae (see, for example, Peng, 1967). The formulae depend on the number of levels of each factor and the nesting pattern. The total experiment size is simply the product of the number of levels of each factor and the number of repeats. The remaining calculations are based on the EMS.

The calculation of the EMS is based on an algorithm due to Lorenzen (1977). This algorithm is equivalent to the standard algorithm given by Bennett and Franklin (1954), except that it takes more paper to compute by hand. However, the Lorenzen algorithm is easier to program on a computer. The EMSs are represented as a square matrix with dimension equal to the number of terms in the mathematical model. Each row corresponds to a given term in the mathematical model. The matrix is filled with zeros and ones. A one in a column indicates the corresponding column term is in the EMS for that row term. Separate vectors store the coefficients associated with each column and whether the column terms are fixed or random. The square matrix, coefficient vector, and term-type vector completely determine the EMS for each term.

The next step is to determine the type of test available on each term. A DIRECT test on a term X exists if the EMS for another term Y equals the EMS for X with the X-component set equal to 0. With our EMS representation, Y directly tests X if the row corresponding to Y equals the row corresponding to X with the diagonal element of X set to 0.

An APPROXIMATE test exists if a linear combination of terms has an EMS which forms a DIRECT test. The approximate df associated with this test is the harmonic mean of the df of the terms involved in the linear combination. This approximation will be replaced by Satterthwaite's (1946) approximation after data is collected.

A CONCLUDE SIGNIFICANT ONLY test occurs when the denominator contains an extra term and is therefore too big. If the test with a denominator that is too big is significant, then the test with a proper-sized denominator would surely have been significant. If the test is insignificant, perhaps the term being tested is insignificant or perhaps the extra term in the denominator is significant.

A CONCLUDE INSIGNIFICANT ONLY test type has one extra term in the numerator (or one too few in the denominator). If the test is insignificant, then both terms in the numerator are insignificant; so, in particular, the tested term is insignificant. If the test is significant, we cannot tell which term is the cause. For both of these conservative tests, the extra term should be the highest-order term possible in order to maximize the likelihood that the extra term has little influence on the test.

Having computed all possible tests, the next step is to compute the minimal detectable difference. The minimal detectable difference is the magnitude of an effect that can be detected by this experiment. In particular, we assume an α error of 0.05 and a β error of 0.10 and express the minimal detectable difference as the number of standard deviations of the denominator in the test. For example, a minimal detectable difference of 2.0 means the effect of that term must be twice as big as the denominator in the test in order to be 90% certain the experiment will declare it significant. The minimal detectable difference depends on four quantities: the df for the numerator; the df for the denominator; the coefficient preceding the term in the EMS; and whether the term is random or fixed. In particular, the minimal detectable difference Δ for a fixed term is given by

$$\Delta = \sqrt{\lambda/(C \times \mathrm{df}_{num})}$$

where λ is the non-centrality parameter of the non-central F-distribution achieving an α error of 0.05 and a β error of 0.10, and C is the coefficient from the EMS. For a random term, Δ is given by

$$\Delta = \sqrt{(F_{0.05} \times F_{0.90} - 1)/C}$$

It has been our experience that users have difficulty understanding the minimal detectable difference, Δ. To help the user, we have five categories for Δ. A Δ less than 0.5 has a detectable difference classified as 'extremely small'. A Δ between 0.5 and 1.5 is classified as 'small'. A Δ between 1.5 and 3.0 is classified as 'medium'. A Δ between 3.0 and 5.0 is classified as 'large'. Finally, a Δ larger than 5.0 is classified as 'extremely large'. The detectable difference is the size of an effect that must exist before this experiment

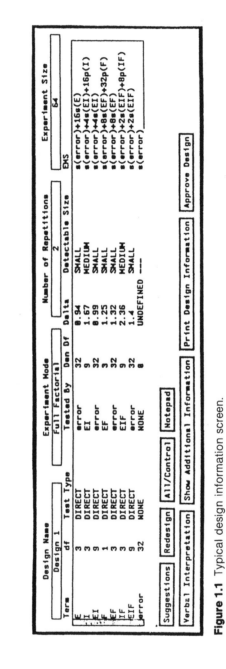

Figure 1.1 Typical design information screen.

will find it. In many cases, an 'extremely small' difference is wasteful of experimental effort. An 'extremely large' difference is generally detected without the aid of statistics so the experiment is generally useless. The remaining categories are useful to the experimenter who must make a trade-off between the detectable difference and the overall size of the experiment. DEXPERT cannot make this trade-off. It is one of the many decisions made by the user throughout a session. However, DEXPERT can provide useful information to the user to help make a better decision.

A design information screen is displayed in Fig. 1.1. Experts would see all of the columns displayed in the figure. Novices would only see the term, test type, and detectable size columns.

FRACTIONAL FACTORIAL DESIGNS

Prior to displaying the design information screen for fractional designs, DEXPERT needs to gather information about the nature of the fractionation. The user must specify the interactions of interest, interactions that may be important (allowed to be confounded with themselves but not with the main effects or interactions of interest), and interactions that can be assumed negligible.

To generate a design, DEXPERT first decomposes each factor into its prime components. For example, if A has six levels, A is decomposed into A_1 having two levels and A_2 having three levels. Next, collect all similar prime components, that is, all two-level components, all three-level components, all five-level components, etc., and separately fractionate each component. The SAS® procedure FACTEX is called for each individual component, crossing each design to obtain the final design. Some care must be taken with prime components. For example, if A has four levels, A is decomposed into A_1 and A_2 and we must be sure to estimate A_1, A_2, and A_1A_2. However, if A has six levels, we only need to estimate A_1 within the two-level block and A_2 within the three-level block because the A_1A_2 interaction is automatically estimated when the two-level block is crossed with the three-level block.

An added complexity with fractional designs is the determination of the confounding pattern. DEXPERT provides the confounding pattern for all cases, whether the factors have the same number of levels or not, and even if there is nesting.

Restrictions on randomization work for fractional designs as they worked for full factorial designs. However, because terms are confounded, the effects of restrictions are much less predictable.

Computation of the degrees of freedom is also more difficult for fractional designs, especially for factors having three or more levels. In fact, for factors having three or more levels, the computation of df depends on the method of analysing the data. We analyse the data using the type I computational method, sorted by order and alphabetically within an order. Thus, computation of df corresponds to the type I computational method.

A slight alteration of the full factorial EMS algorithm handles the fractionation case.

SAS® is a registered trademark of SAS Institute Inc.

Start by computing the EMS for every term using the full factorial algorithm. Next, alter every coefficient by multiplying by the fraction in the experiment, e.g., $\frac{1}{2}, \frac{1}{3}, \frac{1}{6}$. Any coefficient less than 1 gets replaced with 1. Finally, work out the EMS for every primary term by forming the union of all confounded terms. The method is illustrated in a technical paper by Lorenzen and Truss (1989) and will handle arbitrary fractionation and combinations of random, fixed and nested terms.

The search for the tests is exactly like the search used for full factorial designs with one exception: we look for a test on the primary term before looking for a test on the confounded string. That a test on a primary term could occur despite the confounding caused by fractionation came as a surprise to the authors of this chapter. See the appendix of the technical paper by Lorenzen and Truss (1989) for an example.

Computation of the detectable difference Δ is exactly the same so will not be further discussed. The output screen for fractional designs looks like the screen for full factorial designs.

RESPONSE SURFACE DESIGNS

Response surface designs are not generated as are full and fractional designs. Rather, a response surface design is selected from certain pre-specified designs. These include Plackett–Burman, central composite, Box–Behnken, and D-optimal designs. Only fixed factors can be considered. For the Plackett–Burman, central composite with star point equal to 1, Box–Behnken, and D-optimal designs, factors can be either qualitative or quantitative. For general central composite designs, all factors must be quantitative.

Plackett–Burman designs are for linear models when all factors are at two levels. The size of the design is the next multiple of 4 greater than the number of parameters.

Both central composite and Box–Behnken designs are for full quadratic models. Central composite designs consist of a factorial portion, star points and centre points. The factorial portion is, by default, the smallest-resolution V design. The star points are located a distance α from the centre. The parameter α must either be specified by the user or automatically chosen to obtain rotatability, orthogonality, or both. The number of centre points must be specified by the user unless both the rotatability and orthogonality options are specified. Box–Behnken designs are predetermined based on the number of factors. Box–Behnken designs are used when the corners of the region are difficult to attain or are unusual in some sense.

The D-optimality portion is used for general situations when the polynomial model is not linear or a full quadratic, or when the working region is not hyper-rectangular. The user is responsible for specifying the polynomial model, defining the working region, selecting candidate points, and selecting the size of the design.

The power of the tests, Δ, can be computed using algorithms appropriate for the factorial case if the proper coefficient C can be determined. The first step is to rescale the problem so the lowest value corresponds to -1 and the highest value corresponds to 1. Then, for any quantitative term X (like A, A^2, or AB^2C), the coefficient is computed from the design's scaled X-values and is defined (Odeh and Fox, 1975) as

$$C = \sum X^2/2 \qquad (X \text{ quantitative})$$

For any qualitative term, or interaction of qualitative terms, we approximate the coefficient using the harmonic average of the number of times each level occurs.

$$C = \frac{I}{\Sigma_{i=1}^{I}(1/n_i)} \qquad (X \text{ qualitative})$$

where I is the number of levels of X in the design and n_i is the number of times the ith level of X occurs in the design. If the term X is the product of a quantitative term Y and a qualitative term Z, the coefficient is again approximated using a harmonic average:

$$C = \frac{I}{\Sigma_{i=1}^{I}(1/\Sigma Y_i^2/2)} \qquad (Y \text{ quantitative}, Z \text{ qualitative})$$

where I is the number of levels of Z in the design and Y_i is summed over the ith level of Z. For quantitative factors, the computed Δ is defined as the magnitude of the parameter estimate expressed in terms of its standard deviation.

RELATING APPLICATION AREAS TO SYSTEM CAPABILITIES

The system capabilities fall into three main categories: full factorial designs; fractional factorial designs; and response surface designs. Yet there are five areas of application: screening; complicated influence; robust design; model response; and sequential optimization. This section relates the five areas of application to the three system capabilities.

By definition, screening primarily uses fractional factorial concepts. However, when there are mixed level designs, nestings, or inexpensive factors, part of the design may be a full factorial design. DEXPERT handles all of these complexities.

The area of application called complicated influence primarily uses full factorial designs. However, this area can also handle fractional designs, exactly as in screening.

Robust designs require more discussion. The purpose of these designs is to find the control factor combinations that minimize the influence of noise factors. If a control factor has an effect on the influence of a noise factor, there will be a control by noise interaction. In order to detect a control by noise interaction, the design must be constructed in such a way that the control by noise interactions are estimable. So, by default, all control by noise interactions must be estimable, as well as the interaction of a control factor with its associated transmitted variation factor. Any other control by transmitted variation interaction must be added by the user. Of course, these default interactions can be changed by the user, adding or deleting interactions as knowledge permits. Robust designs typically use fractional factorial concepts, but may also include some full factorial portions.

The model response goal uses response surface methodologies. The only addition to make is that response surfaces can be fitted to factorial and fractional factorial designs.

The sequential optimization goal also uses response surface techniques. In particular, each iteration uses a central composite design with star points at $\alpha = 1$ in order locally to fit a quadratic model. By the sequential nature of the search, some points for the next central composite design are the same as the previous design and need not be run again.

SUGGESTIONS FOR REDESIGN

The methodology is now set in place to generate and evaluate any design for any goal. However, different designs will have properties that are better suited to different needs. Since the design generation is very fast, many designs should be generated and compared for suitability.

The mathematical model approach, combined with the object-oriented representation scheme in DEXPERT, allows for easy translation of the mathematical designs into English summaries. For example, by grouping the terms in a design according to their test type and their minimal detectable difference categories, DEXPERT generates a concise description of the properties of a particular design. A similar strategy is used to produce text which compares two alternative designs. Non-statisticians have found the DEXPERT verbal interpretations to be extremely valuable.

A novice user may have difficulty knowing what things to try during redesign. To help the novice (and even the intermediate-level user), redesign suggestions have been implemented in DEXPERT. These do not guarantee the best design but will guide the novice in the search selection. An expert user, based on experience, will find good designs faster than following DEXPERT's suggestions and will probably know more options to try.

There are many options available for redesign of experiments. The user can add or remove factors, change levels of any factors, change repeats, change restrictions on the randomization process, assume certain terms are negligible, fractionate, change fractionation requirements, run a full factorial, change the number of centre points and/or the location of the star points (if central composite), change the size of the design, unmix the fractional design, or search through designs to achieve certain detectable differences. For the last option, the user specifies the desired detectable differences and factors whose levels may be changed. DEXPERT then performs a search using branch-and-bound with dynamic programming (Winston, 1984) to find the smallest design that attains the specified differences.

The redesign suggestions in DEXPERT come from a series of forward chaining rules. In general, there is a tiered approach to suggestions. We first try to make mild suggestions. If none of these rules applies, we follow through with rules that suggest more drastic changes. The system automatically searches every term looking for a DIRECT test. If the test type is not DIRECT, specific rules are fired. In addition, the user can have rules fired by indicating the experiment is too big, that complete randomization is not desirable, or that more detectability is desired on some terms. If the user indicates that complete randomization is not desirable or more detectability is desired, DEXPERT simply invokes the appropriate redesign tools. All other rules make specific suggestions about the particular design at hand.

As an example of a rule, suppose the experimenter wishes to decrease the size of the experiment (by far the most popular request). First try reducing the levels of repeats or of the random factors. Next, reduce the levels of any fixed quantitative factors greater than three to three levels. Next, reduce three-level quantitative factors to two levels. After that, reduce the levels of qualitative factors. Finally, fractionate the design.

Many such rules have been built into DEXPERT. These rules are likely to evolve over time as we learn more about the interaction between DEXPERT and its users.

DATA COLLECTION SHEET

At this stage, the user has reviewed many alternatives and approved a particular design. The approved design has been judged 'best' for the user with respect to the resources required and the information to be obtained. The design has a corresponding set of experimental conditions at which data must be collected, and those combinations are found on the data collection sheet. The data must be collected exactly as specified.

DATA ANALYSIS

Once all of the data has been collected, the experimenter returns to DEXPERT to enter the collected data either interactively or via a UNIX file. DEXPERT analyses the collected data and interprets the results using a number of analytical and graphical methods. Analysis is performed separately on each response variable. Analytical capabilities in DEXPERT include the analysis of variance, variance component estimation, percent contribution, comparison of means, polynomial approximation, predictions, transformation of response variables, pooling, and, if possible, checks on the assumptions made by the user. Graphical displays include main effect and interaction plots (with confidence intervals), normal effects plots, residual plots, time plots, response surface plots, and contour plots, all of which can be shown on the screen as well as printed. The analyses are performed using the appropriate SAS procedures. DEXPERT writes SAS code specific to the task at hand, captures the output, and displays the information using a form, a graph, or a verbal interpretation. PROC GLM is called for all factorial and fractional factorial designs. PROC REG is used for all response surface designs. Data sets are generated for graphs and other analyses.

Graphs are all interactive in nature. The user can include confidence intervals, rewrite the headings, label points, draw reference lines, and attach notes. After the graphs have been customized, hard copies are available. Figure 1.2 shows some typical main effect and interaction plots with confidence bands.

Robust designs require special analyses. Analyses will be performed on the raw data and also on the data summarized over the noise factors. These summaries are of three types: means calculated over the noise factors; the log of the standard deviation calculated over the noise factors; and the quadratic loss function combining closeness to target and variability caused by the noise factors. If there is an external adjustment factor with known properties that can be used to bring the process back to its target, then the loss function is computed after proper adjustment. The user must provide the target value for the response variable and, optionally, a cost value to scale the loss

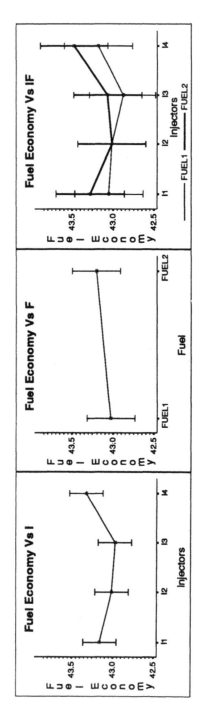

Figure 1.2 Typical main effect and interaction plots.

function properly. More details are given in a technical paper by Lorenzen and Villalobos (1990).

KNOWLEDGE REPRESENTATION

DEXPERT has been developed using KEE (Knowledge Engineering Environment from IntelliCorp, Inc.). DEXPERT uses a frame-based, object-oriented representation to store user inputs and information about experiment designs. Each factor and each term in the mathematical model is represented as a unit in the knowledge base. The slots on the factors represent information given by the user (e.g., number of levels, fixed or random). The slots on the terms represent information generated by DEXPERT (e.g., test type, minimal detectable differences). A separate unit contains information about the experiment itself, such as the list of terms in the mathematical model or the number of repetitions.

Each design in DEXPERT is represented as a separate context world. Context worlds allow the information in different designs to share the same unit/slot storage space simultaneously. The set of designs generated by the user is represented as a tree structure where the parent of each new design is the design it was created from, thereby preserving the history of the design session. A new design inherits all information from its parent except for the user requested modifications and any consequences of these modifications. All alternative designs remain available to the user for retrieval and comparison in order to select the one best suited to the user's needs.

SUMMARY

DEXPERT is an extremely sophisticated program for the design and analysis of experiments. DEXPERT generates designs according to the user's needs rather than selecting a design from some pre-specified catalogue. As such, many different designs are presented and the user must decide which is best suited for the particular needs of the experiment. Upon selection of a design, DEXPERT generates a layout sheet describing the exact way the experiment is to be run. After the data has been collected and entered into the system, DEXPERT analyses the data and helps interpret the results.

In addition to technical capabilities, DEXPERT contains a rule base to guide the user in the selection of a design and the analysis of the data. DEXPERT also interprets complicated results in useful English summaries. These are understandable by users lacking a strong statistical background.

To the best of our knowledge, no other system can match the design and analysis capabilities of DEXPERT.

REFERENCES

Anderson, V.L. (1970) Restriction errors for linear models (an aid to develop models for designed experiments). *Biometrics*, **26**, 255–268.

Bennett, C.A. and Franklin, N.L. (1954) *Statistical Analysis in Chemistry and the Chemical Industries*, New York: Wiley.

Lorenzen, T.J. (1977) Derivation of expected mean squares and F-tests in statistical experimental design. Research Publication GMR-2442, Mathematics Department, General Motors Research Laboratories, Warren, MI 48090-9055.

Lorenzen, T.J. (1984) Randomization and blocking in the design of experiments. *Communications in Statistics—Theory and Methods*, **13**, 2601–2623.

Lorenzen, T.J. and Truss, L.T. (1989) DEXPERT—Design of EXPeriments using Expert Reasoning Tools. Research Publication GMR-6778, Mathematics Department, General Motors Research Laboratories, Warren, MI 48090-9055.

Lorenzen, T.J. and Villalobos, M.A. (1990) Understanding robust design, loss functions, and signal to noise ratios. Research Publication GMR-7118, Mathematics Department, General Motors Research Laboratories, Warren, MI 48090-9055.

Odeh, R.E. and Fox, M. (1975) *Sample Size Choice*, New York: Marcel Dekker.

Peng, K.C. (1967) *The Design and Analysis of Scientific Experiments*, Reading, MA: Addison-Wesley.

Satterthwaite, F.E. (1946) An approximate distribution of estimates of variance components. *Biometrics Bulletin*, **2**, 110–114.

Taguchi, G. (1987) *System of Experimental Design*, Dearborn, MI: American Supplier Institute.

Winston, P.H. (1984) *Artificial Intelligence*, Reading, MA: Addison-Wesley.

Inside two commercially available statistical expert systems

2

J.F.M. Raes

INTRODUCTION

The research into applications of artificial intelligence in statistics carried out in the last decade has resulted in numerous papers describing **hypothetical** or **toy** statistical expert systems and a number of working demonstration prototypes. However, few of these systems have become available to the general public, among them, Hand's (1987) system KENS and also GLIMPSE (see, for example, Wolstenholme *et al.*, 1988). Nevertheless, interest in the development of more intelligent statistical software is still going strong and valuable contributions are being delivered at several academic and other research institutes.

MULREG is an example of a *less stupid* environment for multiple regression analysis which, although it does not use specific artificial intelligence or expert-system concepts or technology, achieves much of what typical statistical expert systems aim to do. For example, it will recommend certain techniques basing its advice on metadata (the measurement level) and on properties of the data themselves (parameters of their distribution). As a commercial product it is far beyond the experimental prototype stage. See DuMouchel (1990) for an extensive discussion of MULREG.

Two other statistical expert systems, SPRINGEX and the STATISTICAL NAVIGATOR, became commercially available in the late 1980s. Both differ significantly from MULREG in several ways. First, they only guide the user to an appropriate statistical analysis technique, while MULREG offers a complete data-analysis environment. Second, they do use expert system technology and claim to be statistical expert systems. Third, both systems are inexpensive products available for IBM-compatible microcomputers.

We will not try to validate, justify or contest whether these systems merit the label of 'statistical expert system'. Despite several efforts the term 'statistical expert system' and its alternatives such as 'consultation system' or 'statistical knowledge-based system' are still ill defined—see, for example Streitberg (1988) and the comments on that article

Artificial Intelligence Frontiers in Statistics: AI and statistics III. Edited by D.J. Hand. Published in 1993 by Chapman & Hall, London. ISBN 0 412 40710 8

(Haux, 1989; Wittkowski, 1990)—and we feel that this labelling issue is of minor importance. We prefer to focus our attention on more basic questions like what the systems do and how they do it and we will pay special attention to the knowledge incorporated in those systems.

SPRINGEX

SPRINGEX (Spring Systems, Chicago) is an optional add-on expert system module for the TURBO SPRING-STAT system, a statistical package. 'The purpose of the expert system is to help you match your data handling and analytical needs with the available TURBO Spring-Stat procedures', it says in a help file. SPRINGEX is written in Turbo PROLOG and is based on an example expert system from the Turbo PROLOG reference guide called GENI. Through a series of yes/no questions the user's problem is diagnosed and an appropriate action within the SPRING-STAT system is proposed. There is no direct interaction between the expert system module and the other modules of the SPRING-STAT system.

Figures 2.1–2.3 are screendumps from a SPRINGEX session which ends with the advice to run the non-parametric Friedman or Kendall procedures of SPRING-STAT. As illustrated by the example session, the questions SPRINGEX asks bear in mind the facilities of the SPRING-STAT system. For instance, the first question in the example checks the need for running SPRING-STAT's data-management module, while the fifth and seventh questions pertain to the usefulness of the graphics module. Other questions (9, 10 and 13) refer to specific SPRING-STAT data files.

SPRINGEX is knowledge-based and has a backward-chaining inference engine that tries to prove hypotheses or rules by querying the user about the fulfilment of the conditions associated with those rules. SPRINGEX also has a primitive 'why' explanation facility which is illustrated in Fig. 2.4. The explanation given is an answer to 'why do you want to know "is it true that you want to see if the variables/groups are DIFFERENT" ', which is the last question in Fig. 2.3. The explanation is a list of the rules that have succeeded together with the conditions that have been asserted.

The knowledge base of SPRINGEX is contained in an ASCII file which can be adapted freely by the user. We have stripped this knowledge base, removing all SPRING-STAT specific entries and the items relating to time-series analysis, retaining only the knowledge on bivariate, multivariate and non-parametric statistics. The original 48 rules and 92 conditions were thus reduced to 22 rules and 35 conditions. This knowledge is represented in Table 2.1. The first three columns are rules, while the conditions associated with those rules are in the last column.

The combination of the rules and the conditions not only represents SPRINGEX's knowledge but also its strategy. This is how SPRINGEX operates: First of all the system tries to prove the rule *statistic* in the first column (in our simplified strategy this seems trivial but originally there were other alternatives like *datamanagement* or *graphics*). Proving the rule implies satisfying its condition [*you want to run a statistical procedure*], and therefore the user is asked 'is it true that [*condition*]' with possible answers 'yes' or 'no'. If the condition is false, the session ends.

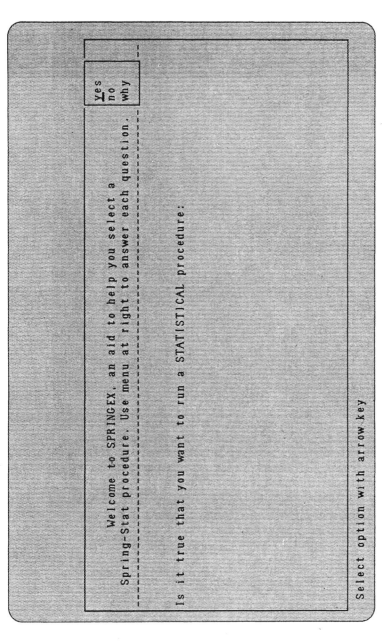

Figure 2.1 A screendump showing the layout of the user interface and SPRINGEX's first question. The answer is pasted in from the menu in the upper right-hand corner.

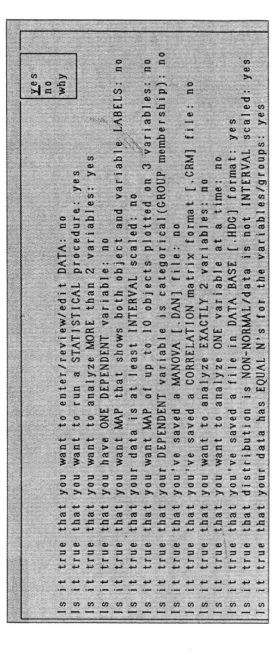

```
                                                        ┌─────┐
                                                        │ Yes │
                                                        │ no  │
                                                        │ why │
                                                        └─────┘

Is it true that you want to enter/review/edit DATA: no
Is it true that you want to run a STATISTICAL procedure: yes
Is it true that you want to analyze MORE than 2 variables: yes
Is it true that you have ONE DEPENDENT variable: no
Is it true that you want MAP that shows both object and variable LABELS: no
Is it true that your data is at least INTERVAL scaled: no
Is it true that you want MAP of up to 10 objects plotted on 3 variables: no
Is it true that your DEPENDENT variable is categorical(GROUP membership): no
Is it true that you've saved a MANOVA [.DAN] file: no
Is it true that you've saved a CORRELATION matrix format [.CRM] file: no
Is it true that you want to analyze EXACTLY 2 variables: no
Is it true that you want to analyze ONE variable at a time: no
Is it true that you've saved a file in DATA BASE [.HDG] format: yes
Is it true that distribution is NON-NORMAL/data is not INTERVAL scaled: yes
Is it true that your data has EQUAL N's for the variables/groups: yes
```

Figure 2.2 Further questions and answers. Suppose we have observed the scores of each of the same three persons under four different conditions and that we want to verify the conditions are indeed different (see, for example, Siegel and Castellan, 1988, pp. 174–183).

```
│ Is it true that you want to see if the variables/groups are DIFFERENT: yes │

│ I think you may want to run: [ Non-Par ] FRIEDMAN or KENDALL               │
```

Figure 2.3 Final question and SPRINGEX's advice.

```
I try to show that: nonparametric is a [ Non-Par ] FRIEDMAN or KENDALL
By using rule number 38
  Rule 38: nonparametric is a [ Non-Par ] FRIEDMAN or KENDALL
    if you want to analyze MORE than 2 variables
    and your data has EQUAL N's for the variables/groups
    and you want to see if the variables/groups are DIFFERENT
I have shown that: statistic is a nonparametric
By using rule number 32:

  Rule 32: statistic is a nonparametric
    if you've saved a file in DATA BASE [.HDG] format
    and distribution is NON-NORMAL/data is not INTERVAL scaled
I have shown that: Spring-Stat (c) is a statistic
By using rule number 21:

  Rule 21: Spring-Stat (c) is a statistic
    if you want to run a STATISTICAL procedure
```

Figure 2.4 Explanation of 'why do you want to know "is it true that you want to see if the variables/groups are DIFFERENT"'.

Table 2.1 SPRINGEX's stripped-down knowledge on bivariate, multivariate and non-parametric statistics

Intermediate rules	Conclusive rules	Conditions
Statistic		you want to run a STATISTICAL procedure
Multivariate		you want to analyse MORE than 2 variables
	STEP REGRESSION	you have ONE DEPENDENT variable you have TWO or more INDEPENDENT VARIABLES you want to do multiple linear REGRESSION you want to see variables added a STEP at a time
	MDS & CLUSTER	you want a hierarchical CLUSTERING of objects
	ANOVA	your data is at least INTERVAL scaled you have ONE DEPENDENT variable your INDEPENDENT VARIABLES are categorical your data is from a planned/controlled EXPERIMENT you want to see VARIANCE partitioned by treatment
	CORRELATION MAT	your data is at least INTERVAL scaled you want to see SSXP; DISPERSION or CORRELATION matrix you want to see Barlett's Test or MULT R^2
	MANOVA	your DEPENDENT variable is categorical (GROUP membership) you want to know if the GROUPS have different PROFILES
	STEP DISCRIM	you want to see which variables DISCRIMINATE best you want to see variables added a STEP at a time you want a DISCRIMINANT function for each GROUP
	MULT DISCRIM	you want to see which variables DISCRIMINATE best you want a DISCRIMINANT function for each GROUP you want GROUPS & variables reduced to 2-D space
	FACTOR ANALYSIS	you have TWO or more INDEPENDENT variables you want variables reduced to smaller # of FACTORS you want FACTORS rotated to SIMPLE STRUCTURE

CANONICAL CORRELATION	you have TWO or more INDEPENDENT variables you have multiple DEPENDENT AND INDEPENDENT variables you want to see RELATIONSHIP between variable GROUPINGS
Bivariate	
LOGIT MODEL	you want to analyse EXACTLY 2 variables your DEPENDENT variable is a proportion
CONTINGENCY TABS	BOTH variables are CATEGORICAL you want to see a CROSSTAB
t-TEST	BOTH of your variables are PROPORTIONS you want to see if the variables/groups are DIFFERENT
z-TEST	your data is at least INTERVAL scaled you want to see if the variables/groups are DIFFERENT
Non-parametric	distribution is NON-NORMAL/data is not INTERVAL scaled
WILCOXON TEST or SIGN TEST	you want to analyse EXACTLY 2 variables your data has EQUAL N's for the variables/groups you want to see if the variables/groups are DIFFERENT
KENDALL'S TAU/SPEARMAN'S RHO	you want to analyse EXACTLY 2 variables your data has EQUAL N's for the variables/groups you want to see if the variables/groups are CORRELATED
FRIEDMAN or KENDAL	you want to analyse MORE than 2 variables your data has EQUAL N's for the variables/groups you want to see if the variables/groups are DIFFERENT
MANN–WHITNEY or KOLMOGOROV	you want to analyse EXACTLY 2 variables your data has UNEQUAL N's for the variables/groups you want to see if the variables/groups are DIFFERENT
KRUSKAL–WALLIS	you want to analyse MORE than 2 variables your data has UNEQUAL N's for the variables/groups you want to see if the variables/groups are DIFFERENT

If the condition is found to be true by the user the rule *statistic* succeeds. The next hypothesis that the system tries to prove is *multivariate* and therefore its associated condition has to be judged by the user. If the answer is negative the rule *bivariate* will be tried out and if this one also fails the last hypothesis is *nonparametric*.

When the rule *multivariate* succeeds, all multivariate hypotheses (STEP REGRESSION, MDS & CLUSTER, ANOVA, . . .) are tried out one after the other until one of them succeeds, which will end the session. These rules typically have more than one condition. As soon as one of the conditions is found to be false the rule fails. The conditions found to be true or false are remembered by the system so the user need not answer them again. In this way a rule can succeed or fail without explicitly testing all of its conditions. Although this may be convenient in a way, it can also be distracting as the order of the questions in different sessions can vary depending on the answers given to previous questions.

If none of the multivariate rules is appropriate the rules *bivariate* and *nonparametric* will be tested in the same way as the rule *multivariate*.

A closer inspection of the contents of Table 2.1 reveals that SPRINGEX's knowledge is unclear (and SPRINGEX does not provide explanation of technical terms), superficial and incomplete. Moreover, the rules concerning *Mann–Whitney or Kolmogorov* and *Kruskal–Wallis* are simply incorrect. The condition 'your data has UNEQUAL N's for the variable/groups' may be typical of these techniques but it is not a necessary condition. The strategy also has very little structure: the user must answer many questions, most of them irrelevant ones, in order to arrive (or even not arrive) at a certain technique.

In conclusion, we may say that, although SPRINGEX has taken into account the research into applications of artificial intelligence in statistics (references are made to Gale, 1986; Gale and Pregibon, 1982; and Hand, 1984), the result is rather meagre. The user interface is primitive and clumsy, the knowledge weak and the inference process questionable. We do not believe that SPRINGEX could be of any practical use in the sense that it could really help an inexperienced user in a non-trivial task. To illustrate this we suggest that the reader should try to follow the example session. On the other hand, SPRINGEX has the definite assets of being available and of running without problems, which distinguish it from most other efforts in this area.

STATISTICAL NAVIGATOR

'Statistical Navigator is an expert system program using artificial intelligence strategies to guide the researcher to the appropriate form of statistical analysis', says Brent (1988). It uses an expert system shell called EXSYS, which provides help and explanation facilities and which allows for uncertainty in its inference process.

STATISTICAL NAVIGATOR covers a wide range of statistical techniques: multivariate causal analysis; scaling and classification; measures of agreement and/or reliability; measures of process/change; measures of association; tests of significance (hypothesis tests); univariate description of a variable; and exploratory data analysis. We will only discuss hypothesis tests.

By asking a series of questions STATISTICAL NAVIGATOR diagnoses the user's problem. The first questions relate to the objectives of the analysis. Afterwards the assumptions the user is willing to make about the data are determined. Grossly simplified, this comes down to the determination of the sample design followed by the selection of an appropriate parametric or non-parametric technique. Four techniques are thus selected and ordered. STATISTICAL NAVIGATOR has extensive reporting capacities, including a short description of the technique and references to literature and statistical packages that implement the technique.

STATISTICAL NAVIGATOR's knowledge about statistical techniques and their appropriateness under certain conditions is hard-coded in the rule of the EXSYS expert-system shell. The rules also incorporate a simple mechanism for reasoning with certainty factors. The EXSYS system can show the user the rules it is using in its reasoning process. However, such an explanation is hard to read and understand and it only gives a partial view of what is going on.

In order to gain complete insight into the workings of STATISTICAL NAVIGATOR we have analysed the system files, and the knowledge contained therein pertaining to hypothesis tests has been represented in Table 2.2. The rows of Table 2.2 are the hypothesis tests among which STATISTICAL NAVIGATOR can differentiate. The columns contain the different properties of the techniques. These are the questions which the user has to answer. A + entry in the table means that the presence of this property is favourable for the selection of the corresponding hypothesis test. A − sign signifies that the presence of that property will contribute negatively to the selection of the corresponding technique. A blank entry in the table means that the presence or absence of a certain property is irrelevant for the selection of the corresponding hypothesis test.

The reasoning process of STATISTICAL NAVIGATOR can now be explained straightforwardly. The user scores the presence of each of the 19 properties on a scale from 0 to 10. Then the system calculates an overall score for each of the 42 hypothesis tests, taking into account whether a property score contributes positively, negatively (which is calculated as 10 minus the score), or not at all. This overall score is then standardized by dividing through the number of properties that have been taken into account for a certain technique.

Two observations regarding this mechanism can be formulated. First, when a property contributes (positively or negatively) to the selection of a technique, it does so with the same relative importance as every other contributing property. Second, the relative importance of a contributing property varies from technique to technique, depending on the number of other contributing properties.

In the light of these observations it is easy to see that the fewer contributing properties a technique has, the more clear-cut its selection (or rejection) will be, thus promoting the selection of 'simple' techniques. On the other hand, 'complex' techniques may also be selected even if one of the properties flagrantly contra-indicates the use of the technique, because the positive indications of the many other properties may out-weigh the objection. Clearly, the use of intermediate certainty factors will complicate the decision process dramatically, making it completely incomprehensible.

Table 2.2 STATISTICAL NAVIGATOR's knowledge on hypothesis tests

```
D I C H O T O M O U S   D E P E N D E N T   V A R I A B L E
2   B Y   2   T A B L E S
M O R E   T H A N   2   T A B L E S
H A N D L I N G   C O N T I N U O U S   D I S T R I B U T I O N
H O M O G E N E I T Y   O F   V A R I A N C E
P O P U L A T I O N   D I S T R I B U T I O N   O F   V A R I A B L E S
P O P U L A T I O N   S T A N D A R D   D E V I A T I O N S   K N O W N
L A R G E   S A M P L E   G R E A T E R   T H A N   5   P E R   C E L L
N O M I N A L   D E P E N D E N T   V A R I A B L E
O R D I N A L   D E P E N D E N T   V A R I A B L E
I N T E R V A L   D E P E N D E N T   V A R I A B L E
E M P H A S I S   O N   E X T R E M E   V A R I A B L E
R E L A T I V E L Y   P O W E R F U L   V A R I A B L E
S I N G L E   I N D E P E N D E N T   V A R I A B L E
R E P E A T E D   M E A S U R E D   O N   A L L   V A R I A B L E S
R E P E A T E D   M E A S U R E D   O N   A T   L E A S T   O N E   V A R I A B L E
T W O   O R   M O R E   T R E A T M E N T S
O N E   S A M P L E   T E S T S
S A M P L E S   T E S T
```

ONE-SAMPLE t-TEST

ONE-SAMPLE z-TEST

WILCOXON SIGNED-RANKS TEST FOR ORDINAL DATA

TWO-SAMPLE t-TEST

TWO-SAMPLE z-TEST

FISHER'S RANDOMIZATION TEST

t-TEST FOR MATCHED GROUPS

SINGLE-FACTOR REPEATED MEASURES ANOVA

FISHER'S MATCHED-PAIRS TEST

WILCOXON MATCHED-PAIRS TEST

ONE-WAY ANOVA

FACTORIAL ANOVA, COMPLETELY RANDOMIZED

FACTORIAL ANOVA, COMPLETELY WITHIN-SUBJECTS DESIGN

FACTORIAL ANOVA—MIXED DESIGN

COCHRAN'S Q-TEST

GOODMAN'S TWO PROCEDURES

CHI-SQUARE TEST FOR HOMOGENEITY

LIGHT & MARGOLIN'S ANOVA-LIKE PROCEDURE

McNEMAR TEST

FISHER'S EXACT TEST

PEARSON'S CHI-SQUARE HOMOGENEITY TEST

BINOMIAL SIGN TEST

CHI-SQUARE ONE-SAMPLE TEST

CHI-SQUARE TEST FOR INDEPENDENCE

FISHER'S EXACT TEST FOR INDEPENDENCE

CHI-SQUARE FOR MULTIDIMENSIONAL TABLES

CHI-SQUARE GOODNESS OF FIT TEST

NORMAL APPROXIMATION TO THE z-TEST

BINOMIAL ONE-SAMPLE TEST FOR MEDIAN

BINOMIAL ONE-SAMPLE TEST FOR QUANTILES

KOLMOGOROV–SMIRNOV TEST

MANN–WHITNEY U TEST

TWO-SAMPLE MEDIAN TEST

KOLMOGOROV–SMIRNOV TWO-SAMPLE TEST

WALD–WOLFOWITZ RUNS TEST

SIGN TEST

WILCOXON MATCHED-PAIRS TEST

BLOMQVIST'S DOUBLE-MEDIAN TEST

TUKEY & OLMSTEAD TEST

KRUSKAL–WALLIS TEST

MEDIAN TEST

HODGES–LEHMAN TEST

The selection of a technique cannot be seen independently of other techniques, as the award of a positive, negative, or no indication of a property to a technique depends on other techniques. Therefore, the combination of the 42 hypothesis tests and 19 properties must be judged as a whole. Only the properties of the techniques as being one, two, or three and more sample hypothesis tests are mutually exclusive (as one can see from the first three columns of Table 2.2), although this can be confounded by the use of certainty factors.

Apart from the properties that determine the number of the samples involved, none of the other properties is relevant for all the techniques. In fact the number of techniques for which a property is relevant varies between 1 and 28. The two most discriminating properties are 'repeated measures on all variables' (which indicates 28 techniques) and 'single independent variable' (25), followed by 'repeated measures on at least one variable' (19), the properties relating to the measurement scale (interval (11), ordinal (18) and nominal (15)), and 'large sample' (17). Together, these seven properties account for 133 of the total of 184 entries, leaving only 51 entries to the other nine properties.

A similar analysis from the point of view of the techniques reveals that (again not counting the properties related to the number of samples) a hypothesis test involves between 1 and 8 properties, more or less symmetrically distributed with a median and mode of 5.

The 42 × 16 matrix of the combinations of techniques and properties is rather sparse, being filled with non-blanks for only 27% of the cells.

It should be clear that STATISTICAL NAVIGATOR contains a wealth of information concerning hypothesis tests. Note also that this only one of the eight knowledge domains of STATISTICAL NAVIGATOR.

By concentrating on the knowledge involved we have not been able to do justice to a number of other features of STATISTICAL NAVIGATOR. Suffice it to say that STATISTICAL NAVIGATOR has a rich and rather effective environment for helping a user select an appropriate technique.

However, another, much more transparent, way of interacting with the user is proposed, which does not change the basic operating characteristics of the system. Basically, the idea is to put the knowledge matrix of STATISTICAL NAVIGATOR on screen, much like a spreadsheet. The user could then fill in the scores for the different properties after which the techniques would be automatically sorted in order of preference. This would make the knowledge and the operation of the system very transparent. It would also easily allow for 'what if' sensitivity analysis, a feature which is also available in STATISTICAL NAVIGATOR but which is very cumbersome. Changing the score of a property would immediately result in a different ordering of the techniques. This visual interactive exploration could be supplemented by other tools, such as a help system to explain concepts and terms.

DISCUSSION AND CONCLUSION

The appearance of commercially available statistical expert systems has taken the quest for more intelligent statistical software from an experimental stage, with a lot of attention going to concepts, tools and techniques, to a level where the overall result and effectiveness of a system are more important. These systems are ready for testing and experimentation and open to criticism. As such they are a genuine enrichment to other statistical expert systems.

The SPRINGEX system, with its simple yes/no question and answer paradigm, feels more like a random generator of questions than a guide for the inexperienced user. It lacks a clear structure in its approach. However, unlike STATISTICAL NAVIGATOR, it has a strategy, that is, its knowledge is structured. It first looks for global categories of techniques (multivariate, bivariate, non-parametric) and then tries to recommend the techniques it knows within a category in a certain order. STATISTICAL NAVIGATOR has no real strategy. It will let the user judge the presence or absence of all the properties it knows, always in the same order. The order of the questions is organized according to the determination of the objectives of the analysis followed by the determination of the assumptions the user is willing to make about the data. However, the knowledge gained in the first part is not put to use in the latter part, for example, to leave out a number of irrelevant questions. The proper techniques are selected only after all the information has been entered, and would not be influenced if the questions were asked in any other order.

SPRINGEX's knowledge covers a broad range of statistical techniques, but it is incomplete and superficial. STATISTICAL NAVIGATOR is much stronger in this area. The knowledge with regard to hypothesis tests, as implemented in STATISTICAL NAVIGATOR, is the most comprehensive treatment the author is aware of. Evaluating the quality of this knowledge is very tricky and requires an enormous amount of expertise, because the different pieces of information, the techniques and their relevant properties, are interrelated and cannot be judged independently. Nevertheless, one may wonder whether all of the 42 different techniques and 19 properties are equally necessary and valuable. Certainly most authors of (non-parametric) statistics textbooks limit themselves to a much smaller selection of techniques, and use correspondingly fewer properties to classify the techniques. The inexperienced user may easily feel lost when confronted with this vast amount of knowledge and thus become frustrated.

Here is an example of how research into artificial intelligence and statistics can stimulate research in the field of statistics itself. Given that we have the technology to make all statistical knowledge on hypothesis tests readily available (at least partially: the techniques are only selected, for proper execution and interpretation the user is referred to other sources), what are the most worthwhile techniques that should be included, and which techniques are less important? What are the most important properties that can discriminate between alternative techniques, and which properties only come second as refiners?

In our opinion, statistical expert systems for the selection of techniques should not be immense and opaque collections of knowledge. On the contrary, such consultation systems should carefully select and structure relevant knowledge. In this way users will gain a better and deeper insight into the problem domain as they consult these systems, and this will make them more self-confident and raise their interest in the application of statistics.

REFERENCES

Brent, E.E. (1988) *STATISTICAL NAVIGATOR: An Expert System to Assist in Selecting Appropriate Statistical Analyses*, Colombia, MO: The Idea Works, Inc.

DuMouchel, W. (1989) The structure, design principles, and strategies of Mulreg. *Annals of Mathematics and Artificial Intelligence*, **2**, 117–134.

Gale, W.A. (1986) *Artificial Intelligence and Statistics*, Reading, MA. Addison-Wesley.

Gale, W.A. and Pregibon, D. (1982) An expert system for regression analysis, in *Computer Science and Statistics: Proceedings of the 14th Symposium on the Interface*, Heiner, K.W., Sacher, R.S. and Wilkinson, J.W. (eds), New York: Springer-Verlag, pp. 110–117.

Hand, D.J. (1984) Statistical expert systems: design. *The Statistician*, **33**, 351–369.

Hand, D.J. (1987) A statistical knowledge enhancement system. *The Journal of the Royal Statistical Society, A*, **150**, 334–345.

Haux, R. (1989) Statistische Expertensysteme (mit Kommentare). *Biometrie und Informatik in Medizin und Biologie*, **20**, 3–65.

Siegel, S. and Castellan, N.J. Jr. (1988) *Nonparametric Statistics for the Behavioral Sciences*, New York: McGraw-Hill.

Streitberg, B. (1988) On the nonexistence of expert systems: Critical remarks on artificial intelligence in statistics (with discussion). *Statistical Software Newsletter*, **14**(2), 55–62.

Wittkowski, K.M. (1990) Statistical knowledge-based systems—critical remarks and requirements for approval, in *COMPSTAT 1990*, Momirovic, K. and Mildner, V. (eds), Heidelberg: Physica, pp. 49–56.

Wolstenholme, D.E., O'Brien, C.M. and Nelder, J.A. (1988) GLIMPSE: a knowledge-based front end for statistical analysis. *Knowledge-Based Systems*, **1**, 173–178.

AMIA: Aide à la Modélisation par l'Intelligence Artificielle

3

(Expert system for simulation modelling and sectoral forecasting)

M. Ollivier, R. Arrus, M.-A. Durillon, S. Robert and B. Debord

INTRODUCTION

Within current forecasting practice, the computing facilities used to run macro- and micro-economic models which handle large quantities of statistical data, are programmed using more or less powerful procedural languages. These programs carry out only the calculations involved in the resolution of the models and in some cases provide a preprogrammed presentation of the results. As a consequence, numerous stages within the complex process of forecasting rely on human expertise and have to be carried out manually by users. These stages include the writing of the model itself and all the related operations; the definition of scenarios, necessary for carrying out the various simulations of future events; the analysis of the simulations results; and the comparison of these results with the conclusions which can be drawn from them.

It has been noticed that the automation of some of these stages is more a question of conceptual rules than of algorithmic operations. This is why, in our view, expert systems are well adapted to this type of problem. In addition, automation has to be accompanied by an interactive mode of operation, which is essential in the execution of the different phases.

In order to establish the foundations of the present project, we have chosen a well-defined and sufficiently large field of application which has already been the subject of reliable forecasting work, namely, long-term energy demand modelling. This field of activity has given rise to a family of well-established models, the MEDEE models, which are used by the Commission of the EC (DG XII) as well as by several countries in Europe and elsewhere (Lapillonne and Chateau, 1982). Most of these applications use a numerical resolution program as the basis of the computing operations. This not only provides a way of resolving the models, but also provides certain other functions: modelling support; organization of relations; syntax checking; identification of loops; etc.

Artificial Intelligence Frontiers in Statistics: AI and statistics III. Edited by D.J. Hand. Published in 1993 by Chapman & Hall, London. ISBN 0 412 40710 8

We can thus build an expert system type of environment around these models and the numerical resolution program, on the basis of real operational practices and needs and with the guarantee of immediate development of our results. We call it Assistance à la Modélisation par l'Intelligence Artificielle or AMIA (Debord *et al.*, 1990).

CURRENT SITUATION

In the statistical field, the best-known expert systems are mainly centred on the integration of data bases and statistical methods. A first approach is to develop the data management capacity using statistical programs (SPSS, GENSTAT, SAS, etc.), hence the introduction of new functions in line with developing needs. A second approach is based on a conventional data-base management system, whose capacity to handle statistics (mean, variance, frequency, etc.) is then developed. A third approach exploits the advantages of both statistical analysis and of data management for which specialized programs already exist, making possible, for example, linear regression analysis with REX (Hand, 1989) or analysis by principal components with ALADIN (Abdali and Prévôt, 1989). Finally, and without making any claim to exhaustivity, we could mention RX (Gottinger, 1988), which combines statistical and medical knowledge, GUHA (Gale and Pregibon, 1982), which generates hypotheses in exploratory analysis, and a computer-aided fatigue diagnosis expert system which uses survey data.

The field of forecasting proper has not yet been the object of full-scale expert systems applications, because expertise in this field (other than the handling of statistical data by econometric models or by simulation models) relies on qualitative factors whose relationships are too vague to be easily formalized. However, the work of the forecaster is increasingly necessary in order to rationalize the considerable long-term investment flows involved in modern technological developments. The development of more effective means of computer-aided forecasting has thus become an important priority.

The French national organization in charge of energy policy (Agence Française pour la Maîtrise de l'Energie), which is involved in the present project, employs a dense and complex information system as well as elaborate forecasting methods in order to determine its objectives and priorities. The information system includes statistical data on energy consumption and the factors which determine its evolution, as well as more general technical information available from internal documents and through the knowledge of experts.

The tools used for forecasting form part of the main body of long-term energy demand simulation models which have been specially developed to analyse energy demand management policies. They are characterized by the considerable extent to which they are disaggregated, by their need for large amounts of socio-economic and technical information and by the extent to which they use scenarios requiring considerable technical expertise. Until now, emphasis has been given to the scientific quality of the instruments used to input and process information and to the quality of the information itself, rather than to the practical conditions under which the models have been used.

In the long-term this has led to three serious problems: a definite sluggishness which has become more and more of a constraint when using models (difficulties in updating, in evaluating the sensitivity of the model, and in entering new scenarios); the absence of any formal and systematic link between the computer data bases and the models; and difficulties in incorporating available technical expertise and advances in knowledge in the field of energy demand management into the scenarios in a formalized manner.

THE AMIA PROJECT

In the field of long-term energy demand, the only program which model-builders have at their disposal is SAMSS (Système d'Aide à la Modélisation en Sciences Sociales) for the numerical resolution of models. They themselves have to provide the program with: the characteristic components of the modelled phenomenon (indices and operands (variables)); a description of the relations (equations) which exist between these components (the description language is close to PASCAL); a description of the known realizations of certain components (initializations); and a description of requests to edit results.

SAMSS (Brunet *et al.*, 1990) has a very rudimentary user interface, it does not memorize the various versions of a model, and only deals with the current scenario (the files are managed manually by the user, which calls for some knowledge of computing). However, the function of numerical resolution of models does satisfy present users, and is quite indispensable in the modelling process.

The main objective of this project is to make up for these deficiencies, and to provide the program with an intelligence containing part of the expertise of the model builders in the field involved. This will make it possible both to improve the effectiveness of the work of the experts themselves, during the construction of the different possible versions of the model, and to diffuse and provide users with expert knowledge in the chosen field (the objective of diffusing expertise or improving teaching content).

FUNCTIONS OF THE PROPOSED SYSTEM

History: handling of the versions, scenarios and corresponding results for a given model

For the model-builder, a version is a meaningful instance of the model used for a given problem. For one version of a model it is possible to have several scenarios, each scenario being associated with its results; the scenarios make it possible to measure the consequences of the various possible initializations of the command variables. (See Fig. 3.1).

The history is used in each phase of the model-builder's work in order to create and store a version of a scenario (by extracting the memorized elements of the model), and to interpret the results (by confronting the differences in the results with the differences in the corresponding scenarios, for example). The constitution of an easily exploitable history is also an essential element in automating the expertise.

Partial resolution: modularity, reuse of partial results

In order to handle very large models, we have introduced modularity into the model data. We can see a model as a hierarchy of variables: a node is an endogenous variable, and the leaves are exogenous ones. A link between X and Y means 'Y is calculated with X'. An aggregating module is characterized by the variable at the root of a given part of the tree and includes all the variables used in its calculus. A hierarchy of aggregating modules can be superimposed on the variables tree, summarizing it. This feature is very useful for the modeller: he can manage his model by small sets of data and he can order a partial resolution of the model.

We have added reuse of results already calculated by a previous resolution. Such a facility guarantees a limited view wherever the module is located in the hierarchy: if results are present and obtained previously with the same scenario, the subtree issuing from this variable can be cut and replaced by the results, reducing dramatically the contents of the module and the resolution time.

From the user-interface point of view, this facility gives the user a possibility of data selection: only relevant entities of the model are present on the screen, and he is not encumbered with the whole model.

User-friendly interface: direct manipulation, browsing through a large amount of data

Three complementary solutions have been implemented in order to help forecasters to grasp huge simulation models. First of all, direct manipulation of hierarchies (of models, modules, variables, etc.) is possible through graphical representation. The user can perform actions directly by pointing at the objects on their visual representation on the screen. Second, alphabetical and cross-reference lists are provided to browse through the model. Third, documentation facilities are provided: as the objects of the models are numerous, their names are often not clear. Then, for each object, a short comment is available in a pop-up menu, within the graphical or list representation. A larger comment can also be displayed in a separate window.

Memorization of the expertise: constitution of several knowledge bases to be used in the different phases of the modelling process

A base to provide assistance in writing out the relations of the model

In this knowledge base, the system possesses rules which allow it to write out typical relations and sequences of relations. For each choice made by the model-builder, the system indicates which relations are most appropriate for resolving the problem, e.g. linear projection, curve, exponential. The sequencing of relations is carried out by moving upstream from the point of entry (the question which the model-builder wishes to resolve). Similarly, the side-effects are also taken into account. The effects which have been pinpointed beforehand (and built upon as the knowledge in the field is improved) are set out in a graph which leads to final results whose degree of precision depends on the model-builder's objective.

A base for assisted error diagnosis

At the present time the relation consistency controls may produce a large number of messages which may contain certain errors. The amount of information may dissuade the user from making use of it. An intelligent filtering of the messages would speed up the production of a reliable version.

A base for consistency rules for the scenarios

A semantic control of the scenario command variables for a given version will prevent the user from making contradictory choices which would lead to inaccurate interpretations of the results. These rules, which are made while the version is being created, represent so many constraints which must be taken into account when the scenarios are being set up.

A base for assistance in the presentation and interpretation of results

The user should be able to visualize the result either on its own or in comparison with the result of another scenario. The result of the resolution of a scenario is a (possibly large) set of numerical values. In this case interpretation is no easy matter, and it can be greatly facilitated by filtering the information (by picking out the most significant results) and by seeking the most adequate representation of the problem. This base will also contain information which will be of assistance when comparing results (e.g., the point at which a difference in results on a variable becomes significant).

All these functions will be carried out using a user interface which integrates current proven techniques (menus, window management, mouse, highly developed editor, etc.) while at the same time taking the portability of the overall system into account.

TECHNICAL OPTIONS

In developing the program, we use object-oriented programming (Cox, 1986). Object-oriented Programming (OOP) advantages have proven to be very useful in artificial intelligence (particularly for the representation and use of knowledge), making it possible to structure the knowledge base; separate knowledge from details of the computing system related to the running of the program; integrate different forms of representation; and mix knowledge and reasoning. They also have important advantages from the software engineering point of view: modifications and extensions are made easier; codes which have already been written can be reused thanks to the fact that properties are passed on (code economies) and to the nature of the idea of classes of objects itself (particularly well adapted to a modular approach); and the final choice of system on which the program is to be run can be taken at the last moment, once the program has been perfected.

In recent years, many object-oriented extensions of conventional AI programming languages such as Lisp or Prolog have been marketed. We have chosen the SMECI-MASAI package based on LeLisp (ILOG, 1990). It offers an expert system generator (SMECI) and an interactive user-interface generator (MASAI). These tools permit us to develop separately the data or the knowledge and their processing capabilities, and the external view of the data for the user interface.

It is important to emphasize the ability of the system to support hierarchical models in every field of application. The numerical resolution software, SAMSS, is completely independent of the applications: for every model, it only needs specific entry files (names of variables, model relations texts, initializations, and requests for results editing). The AMIA user-interface facilities are also independent of the supported models. So are, to some extent, the knowledge bases of the AMIA expert system: those which work on helping to write the relations texts in Pseudo PASCAL (the language of the numerical resolution package); on checking the consistency of model modifications; and on comparing the results of two simulation scenarios. Only the knowledge bases working on the consistency of the scenarios and on the interpretation of results will have to be implemented according to the application field specificities. For overall system architecture see Fig. 3.2.

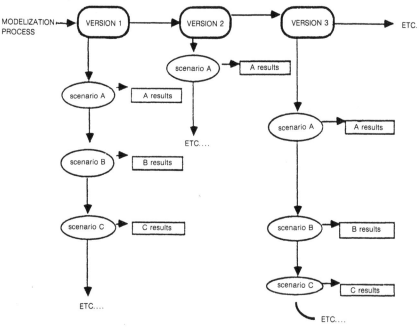

VERSION: model in state concerning the number of variables and the number and nature of the relations
SCENARIOS: for the same version, various instanciations of the command variables of the model

Figure 3.1 Model representations.

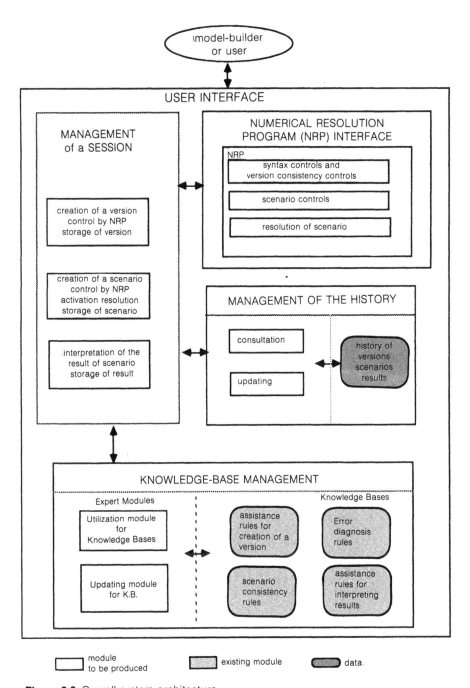

Figure 3.2 Overall system architecture.

AMIA, SAMSS and SMECI-MASAI are implemented on SUN 4 workstations with 8MB RAM and the X-windows environment X11r4.

CONCLUSION

This project is funded by the Development of Statistical Expert Systems (DOSES) programme of the European Communities Statistical Office. The CRISS partners in the project are modellers and model users in the field of long-term energy demand: Agence Française pour la Maîtrise de l'Energie (AFME), Ikerketarao Euskal Institutoa (IKEI), San Sebastian, and Centro de Estudos em Economia de Energia, dos Transportes e do Ambiante (CEETA), Lisbon. The project started in January 1990. The system is used in the field of energy forecasting, but will be able to support applications in other sectors.

REFERENCES

Abdali, A. and Prévôt, P. (1989) Aladin: prototype d'un système expert pour l'analyse des données industrielles, *9th International Avignon Meeting on Expert Systems and their Applications*. Paris: EC2, 269 rue de la Garenne 92000 Nanterre.

Brunet, C., Dumarchez, P., Gonzalez, P. and Robert, S. (1990) *SAMSS: Manuel didactique de référence et de mise en oeuvre*, Grenoble: CRISS.

Cox, B.J. (1986) *Object-oriented Programming: an evolutionary approach*, Reading MA: Addison-Wesley.

Debord, B., Durillon, M.A., Ollivier, M. and Robert, S. (1990) *Spécifications du projet AMIA: besoins des utilisateurs, structure des informations, choix techniques*, Grenoble: CRISS.

Gale, W.A. and Pregibon, D. (1982) *An Expert System for Regression Analysis*, Proceedings of the XIVth Symposium on the Interface, New York: Springer-Verlag.

Gottinger, H.W. (1988) Statistical expert systems, *8th International Avignon Meeting on Expert Systems and their Applications*. Paris: EC2, 269 rue de la Garenne 92000 Nanterre.

Hand, D.J. (1987) The applications of artificial intelligence in statistics, in *Development of Statistical Expert Systems*, Eurostat News Special Edition, Luxembourg, pp. 35–48.

ILOG, S.A. (1990) *SMECI-MASAI: Manuels de réference*, 2 av. Galiéni, 94250 Gentilly, France.

Lapillonne, B. and Chateau, B. (1982) The MEDEE 3 model, in *Energy Demand Analysis*, Report from the French–Swedish Energy Conference, Stockholm, Energy Systems Research Group, University of Stockholm, pp. 124–148.

An architecture for knowledge-based statistical support systems

4

A. Prat, E. Edmonds, J.M. Catot, J. Lores, J. Galmes and P. Fletcher

INTRODUCTION

Open and closed user systems (e.g., statistical packages, optimization packages, numerical algorithm libraries) are widely used in industrial and scientific development environments and represent an enormous body of very complex and valuable knowledge that is becoming increasingly difficult to access. End users to these systems have to cope simultaneously with the intricacies of the software and with the increasing complexity of the application domain problems. For these systems, knowledge-based front-ends (KBFEs) can provide co-operative assistance to end users, enabling them to use the systems successfully while preserving know-how in the libraries and packages and extending their working life.

Front-Ends for Open and Closed User Systems (FOCUS) is an ESPRIT-2 project currently in the third of four budgeted years and is aimed at developing generic tools and techniques for constructing and maintaining KBFEs for open user systems (e.g., libraries, reusable software components) and closed user systems (e.g., free-standing software, packages) for industrial and scientific applications.

OBJECTIVES

The FOCUS project has taken state-of-the-art KBFEs such as GLIMPSE (an advisory KBFE for GLIM), NAXPERT and KASTLE (expert systems for routine selection from the NAG Fortran library), and SISP (a PC-based KBFE for PC packages), along with current models of KBFE architectures, user-interface design and problem-solving environments, as a basis for further development of KBFEs for problems involving numerical computation (e.g., time-series analysis, design of experiments), as defined by the industrial partners.

The kinds of tools that will be built will fall into two categories: run-time tools and development tools. Run-time tools can be either part of the architecture such as the front-end harness and back-end manager or used at the KBFE developer's discretion.

Artificial Intelligence Frontiers in Statistics: AI and statistics III. Edited by D.J. Hand. Published in 1993 by Chapman & Hall, London. ISBN 0 412 40710 8

Development tools such as knowledge-base editors, hypertext tools, output extraction languages, etc., will be developed to define the programmable part of the KBFE.

The KBFEs will be such that the end user will require neither knowledge of the underlying numeric software nor detailed knowledge of the numerical strategies on which this software is based. This knowledge will be encoded within the KBFE. The KBFE itself will therefore act as a kind of intelligent assistant offering strategic and context-sensitive help, while guiding the end user around the complexities of the underlying software.

THE FOCUS ARCHITECTURE

The ability to separate the user interface from its underlying application subsystem has been a subject of investigation for over 20 years. One very important event in the development of separable user-interface architectures was the workshop held in 1983 in Seeheim (Pfaff, 1985). One group at the meeting, consisting of Jan Darksen, Ernest Edmonds, Mark Green, Dan Olsen and Robert Spence, developed an architecture for the user interface (Green, 1985). This architecture has become widely known as the Seeheim model and a good deal of the subsequent debate has used it as the base reference. Its most important feature is that it identifies as number of components of this interface and hence separates out various user-interface issues. As with other aspects of system design, such separation enhances clarity, maintainability and portability. For a full discussion see Edmonds (1991).

Many papers have been written concerning modifications to this model (see, for example, Dance *et al.*, 1990). It is noticeable, however, that most are variations on the Seeheim view, rather than radical departures from it. We may assume, therefore, that the climate of opinion still favours the Seeheim approach as an architecture for the separable user interface. In particular, it can be seen that the user interface is seen to include a substantial part of the software of a system.

The emergence of the separable user interface has occurred partly as an attempt to represent everything that 'shows through to the user' in an identified software module. The degree to which this attempt has been successful remains to be fully assessed, but it is quite clear that the study of separable user interfaces is becoming increasingly important in concentrating our attention on some of the most significant aspects of interactive systems. In particular, the FOCUS project is able to build upon the knowledge gained from the investigations of the last two decades.

Therefore, in the FOCUS project an architecture for KBFEs has been proposed based on an extension to the Seeheim model. In order to maintain the concept of interface separability within this architecture, the front-end harness and the knowledge-based modules concern themselves with the whys and wherefors of what the end user is or should be doing, while the back-end manager takes care of the 'how'.

THE FRONT-END HARNESS

The front end harness (FEH) is a system architecture that will provide a versatile and portable framework which encompasses interaction with the end user (the human–

FRONT-END HARNESS

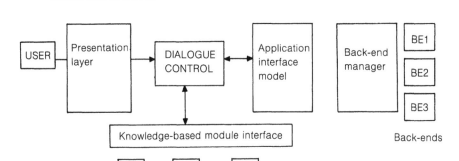

Figure 4.1 The FOCUS architecture.

computer interface), and between the major components of the KBFE system, namely the knowledge-based front-end and the chosen library or package serving as the computational back-end user system.

Essentially, the FEH is made up of a presentation layer, a dialogue control, an application interface model, and a knowledge-based module interface (see Fig. 4.1). The presentation layer is the software component responsible for the management of the surface-level human–computer interface. The dialogue control module is central to the FEH. It manages and co-ordinates communications between the other FEH components, which is achieved through a messaging protocol, and at the same time controls the dialogue with the user. The knowledge-based modules provide the user with co-operative support and guidance, and they communicate with the user through the dialogue control module. The application interface module contains complete representations of the back-end tasks, and allows the dialogue controller to communicate with the back-end manager (BEM).

THE BACK-END MANAGER

The BEM's prime responsibility is that of mapping an application-independent specification of the 'what' into an executable specification of the 'how' with respect to a particular application or package (i.e. mapping from a semantic to a syntactic specification), controlling the execution of the application, and passing the relevant information back to the appropriate KBFE component.

To carry out this role the BEM must be 'programmed' by an expert user with knowledge about the syntax, semantics, functionality and environment of the application software. This is stored in the form of specification objects (tasks and actions), back-end input/output objects, and an environment knowledge base. Figure 4.2 illustrates the structure of the BEM, showing its relationship to the rest of the proposed KBFE architecture.

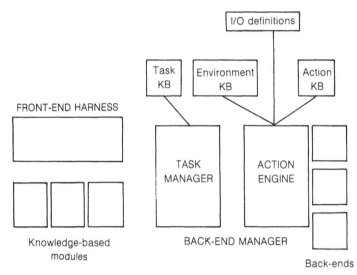

Figure 4.2 The structure of the back-end manager.

The application-independent specification of back-end activity (the 'what') is described in terms of tasks, while the application-specific decomposition of tasks (the 'how') is described in terms of actions (Prat *et al.*, 1990). An action is the control and aggregation of one or more back-end application commands forming a unit perceived by the expert user to be useful for building up tasks. In turn, a task consists of the control and aggregation of one or more actions, for each back-end application whose functionality allows its implementation, forming an application-independent unit perceived by the expert user as able to implement a part of the domain problem-solving strategy.

The reasons for an apparent extra level of abstraction provided by actions are two-fold. First, it allows the clear separation of implementation and domain issues allowing the calling knowledge-based module to request instantiation of a back-end independent task (in the form of a Prolog goal). This means that a KBFE designed to solve a particular domain problem could be rewritten easily to work with a different back-end—the only thing that would change would be the actions and the environment knowledge base; the tasks would remain intact. Second, it gives the package expert user flexibility to group back-end statements into useful units. For someone with 'complete' knowledge of a back-end this may result in many small actions, whereas fewer larger actions may be built by someone with only intermediate knowledge of a back-end.

Basically, the BEM architecture consists of two parts: the task manager and the action engine. The task manager maps from application-independent tasks to application-specific actions utilizing expert user-defined information about the task stored in the task knowledge base in order to drive the action engine.

In order to execute and process actions on a given back-end package the action engine must have facilities to generate back-end commands and process the resulting output. The current methods supported are generating input by a template-inspired approach and processing output by an object-oriented extraction approach. The action engine contains a run-time component called the template engine which interprets off-line developed template programs written in TPL (TemPlate programming Language developed at UPC specifically for the task of generating such input). XTL (eXTraction programming Language) is a UPC-developed object-oriented language for defining the required parts of ASCII output files and an XTL run-time interpreter also lies in the action engine.

The BEM requires a considerable amount of knowledge that, in many ways, is similar to that required by the knowledge-based modules. However, the knowledge available to the BEM must be complete, while that for the modules may be tentative or incomplete. This ensures that the BEM has no need for direct interaction with the user.

SOME PROTOTYPES DEVELOPED WITH THIS ARCHITECTURE

The first prototype toolkit for the architecture been delivered to the KBFE developers and a considerable number of prototypes have been developed using it, among them the following:

1. *SEPSOL.* This is a KBFE aimed at helping experimenters (mainly chemists) to design their experiments. The structure of SEPSOL is based on the methodology used at Solvay to solve experimental design problems.
2. *FAST.* This KBFE can be regarded as the successor to GLIMPSE (a statistical KBFE which was one of the state-of-the-art systems used by the project for initial investigations). A portion of GLIMPSE has been reimplemented and restructured within the FOCUS architecture to produce FAST, a much more modular and user-friendly system. These improvements were a direct consequence of using the FOCUS architecture and will be further developed towards the full functionality of the GLIMPSE system.
3. *IERS.* This is a KBFE for routine selection from the NAG library, improving on NAXPERT and KASTLE.
4. *KAFTS.* This is a KBFE for time-series analysis using the Box–Jenkins strategy. It provides automatically with the best ARIMA model for forecasting. The back-ends used are SCA, GENSTAT and EXCEL (in the PC version).
5. *DOX.* This is a KBFE for experimental design. The system helps engineers in the design and analysis of factorial and fractional designs with factors at two levels using economic and/or technical restrictions. The back-end used is SCA.
6. *REGS.* This is a KBFE for real-estate evaluation using regression techniques. The back-ends used are BMDP2R and BMDP9R.
7. *MERADIS.* This KBFE is aimed at helping the safety engineer of a factory in simulating the dispersion of a toxic gas after an accidental release. The knowledge elicitation is currently in progress and will be followed by the implementation of a prototype using the FOCUS architecture.

The building of these prototypes provided a mechanism for formally evaluating the architecture which was carried out in conjunction with those in FOCUS involved in the development of methodologies for KBFE design. The main objectives of the evaluation were to test the basic design goals of the architecture, principally that of separation and the mechanics of achieving it along with the technical issues of integration.

The results of the evaluation were encouraging but several problem areas were highlighted:

Graphics: little support was provided for graphics, so the next version of the toolkit will provide support for display and manipulation of externally stored graphics in standard formats such as GKS, Postscript, along with the front-end harness providing a basic set of graphical interface objects such as plots, histograms, etc.

Harness programming: the message structure which the front-end harness provided was deemed to be too limited to allow the developers to enhance the harness functionality. As a result, the next version will introduce the concept of abstract interaction objects which may be tailored off-line by the developer without affecting the code of the harness.

Rationalization of BEM notation: Several notational problems were experienced in defining tasks and actions for the back-end manager along with some functional deficiencies in the template definition language and extraction language, which are to be rectified for the next version.

FURTHER RESEARCH

The FOCUS project research program continues in several areas; the most relevant to this chapter are, first, the inclusion of a data management system either to be part of architecture or as a master back-end accessed through the BEM (consisting of an SQL-based raw data management system for managing source data) and a derived data management system for performing a basic set of numerical operations; and second, the possible incorporation of a grammar-based method for generating complex back-end commands, e.g. complex mathematical structures.

CONCLUSIONS

The FOCUS project has made substantial progress so far, with all due deliverables having been produced. The project has entered the system-building phase which aims to provide a thorough test of the architecture and tools, as well as producing exploitable systems. The development of KBFE prototypes provides the focal point for the integration of the various aspects of the FOCUS work programme. Evaluation by both end users and system developers of the tools and techniques emerging from the project is a continuous process. The lessons learnt while developing the systems over the next 24 months will be fed back into the project and used to modify and redevelop tools and architecture. This is the basis on which the project moves forward: an iterative process involving evaluation, refinement of toolkits and architectural components, and development of KBFEs, the majority of them statistical.

ACKNOWLEDGEMENTS

This work has been partially financed by the EEC through ESPRIT II (Project number 2620) and by the CICYT (Spanish government) (TIC 880643).

REFERENCES

Dance, J.R., Ganor, T.E., Hill, R.D., Hudson, S.E., Meads, J., Myers, B.A. and Schulert, A. (1990) The run time structure of UIMS-supported applications, *Computer Graphics*, **21**, 97–101.
Edmonds, E.A. ed. (1991) *The Separable User Interface*, London: Academic Press.
Green, M. (1985) Report on dialog specification tools. In G.E. Pfaff, ed., *User Interface Management Systems*, Springer-Verlag, pp. 9–20.
Pfaff, G. (1985) *User Interface Managements Systems*, Springer-Verlag.
Prat, A., Lores, J., Fletcher, P. and Catot, J.M. (1990) Back end manager: an interface between a knowledge based front end and its application subsystems. *Knowledge-based Systems*, **3**(4).

Enhancing explanation capabilities of statistical expert systems through hypertext

5

P. Hietala

INTRODUCTION

Explanation and help facilities are among the most essential issues in the usability of expert systems. Explanation facilities include the ability of an expert system to provide answers to several types of question, such as questions concerning expert system behaviour ('how' questions) and questions requesting justifications ('why' questions). Facilities for on-line help, on the other hand, are an essential part of any complex software system: expert systems are no exception to this rule. In the area of statistical expert systems the importance of explanation and help facilities has been widely recognized—explanation has been noted as one of the necessary attributes of statistical expert systems (Hand, 1985), and some statistical expert systems adapt their help to the users (Nelder, 1988). Moreover, the characteristics of statistical analysis give rise to new demands for explanation, for example explaining new statistical terms via lexicon (Gale, 1986; Hietala, 1990).

This chapter presents a new approach in order to enhance the explanation and help capabilities of statistical expert systems, namely, the use of hypertext techniques. It is argued that existing hypertext tools provide good facilities for integrating explanation and help under a unified user interface. Many of the advantages obtained by using the hypertext paradigm (e.g., the possibility of combining text and graphics) appear to be exceptionally useful for statistical explanations.

PROBLEMS WITH EXPLANATIONS

Although explanation capabilities are at the very heart of all expert systems, most users as well as expert system builders would agree that these 'machine explanations' are by no means at a level of really useful 'human expert explanations'. Explanations in current systems have several well-known shortcomings, and we mention only a few here (see, for example, Kidd, 1985, for a more complete list): current explanations are too rigid;

Artificial Intelligence Frontiers in Statistics: AI and statistics III. Edited by D.J. Hand. Published in 1993 by Chapman & Hall, London. ISBN 0 412 40710 8

users find dialogues laborious and inefficient; it is difficult to volunteer information in the right order; the reasoning strategies of expert systems are found unnatural; explanations are in the language of experts. To cope with the explanation problem, Chandrasekaran *et al.* (1988) decompose it into three components: the structure of the problem-solving system itself; user modelling; and issues of presentation. Then a generic task methodology is proposed as a way of building expert systems so that the basic explanation constructs are available closer to the conceptual level of the user. The use of 'deeper' model-based domain knowledge has often been suggested as a basis of higher-quality explanations (see, for example, Steels, 1985, for a description of deep models and explanation in 'second-generation expert systems').

The above-mentioned shortcomings are also evident in the domain of statistics. For example, statistical explanation typically involves information from many different sources, so it is difficult for the user to explore it in the right order. One aspect of this problem is the fact that while reading the explanation the user would often like to have access to extra explanatory knowledge (for instance, a link from a 'why' explanation to a lexicon). Moreover, the need for graphics in statistical explanations, justifications and definitions is obvious, while simulations and animations would often be in place to illuminate and explain difficult statistical concepts. However, standard AI techniques used in implementing explanation facilities do not give support to an implementor who wants to include the features outlined above in a statistical expert system.

In the following the use of hypertext techniques is advocated as a unifying method for implementing both the help and explanation subsystems. It appears that this approach would solve most of the above outlined problems present in statistical expert systems. Let us first briefly review the basic properties of hypertext systems.

HYPERTEXT FROM THE VIEWPOINT OF STATISTICAL EXPLANATION

Hypertext has been defined as follows: 'hypertext is an approach to information management in which data is stored in a network of nodes connected by links. Nodes can contain text, graphics, audio, video, as well as source code or the other forms of data' (Smith and Weiss, 1988). The builders of expert systems have utilized hypertext as a powerful tool for knowledge acquisition, knowledge-base maintenance and on-line documentation (see, for example, Hypertext Panel, 1989).

According to Conklin (1987), hypertext has many operational advantages, among them ease of tracing references; ease of creating new references; customized documents; modularity of information; consistency of information: task stacking; and collaboration. These general advantages are obviously beneficial in explanation systems; most of them are especially useful for the builder of a system (e.g., ease of creating new references) but a few are also useful for the user (e.g., ease of tracing references). On the other hand, the well-known disadvantages of hypertext (Conklin, 1987) for the most part directly affect the user of the system: disorientation (getting 'lost in hyperspace') and cognitive overhead. Various approaches have been suggested to overcome these disadvantages; for example, Nielsen (1990) describes navigation facilities (overview diagrams, backtracking) and interaction history mechanisms (timestamps, footprints).

For the purposes of statistical explanation the above-mentioned facility for combining text and graphics seems to be the most important. For example, graphical illustrations or additional graphical examples of statistical concepts can easily be constructed and added using hypertext facilities. Moreover, this additional information can, with very little work, be directed to only those users who need these more concrete explanations. This kind of adaptation to the level of the knowledge of the user can also be obtained in the other types of explanation. Naturally, both small-scale animation and model-based simulations can often be more illustrative than rigid textual chains of expert system deductions. The hypertext tools already available appear to provide a good starting point in implementing the features outlined above. To make our comments more concrete we will next describe our initial efforts in using a hypertext tool in reimplementing the explanation/help subsystem of an existing statistical expert system.

AN APPROACH COMBINING HYPERTEXT AND EXPLANATIONS

In this section we briefly describe a unified explanation/help subsystem within an existing statistical expert system. This subsystem binds together the various types of explanation and user help mentioned above, namely, 'how' and 'why' explanations, lexicon, and facilities for on-line menu help. Drawing on the advantages of hypertext outlined above, we decided to use a hypertext tool to implement our prototype. This brought about a unified user interface over the different parts of the system. The statistical expert system whose existing explanation/help facilities have been re-implemented into a unified subsystem is the latest version of our statistical expert system called ESTES. ESTES runs on the Apple Macintosh^{TM} (for a description of an earlier version of ESTES, see Hietala, 1990) and the hypertext features have been implemented using Apple HyperCard®. The structure of the explanation/help system and the communication between its modules is shown in Fig. 5.1.

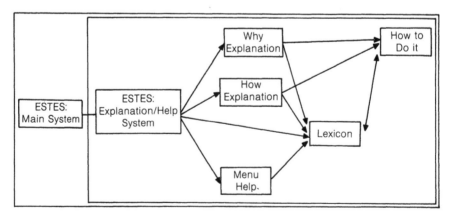

Figure 5.1 The structure of the explanation/help subsystem.

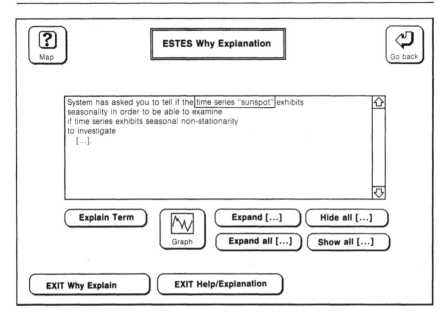

Figure 5.2 'Why explanation' display.

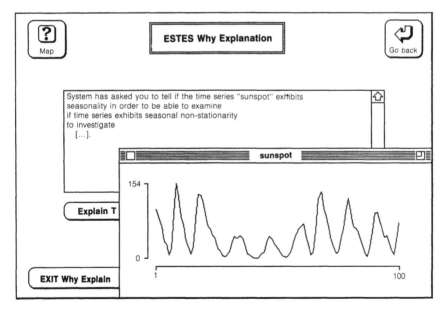

Figure 5.3 'Why explanation' display containing a graphical representation of time series.

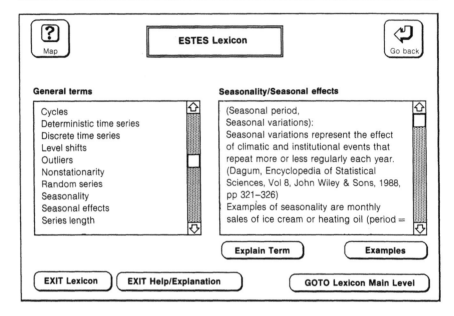

Figure 5.4 Lexicon display.

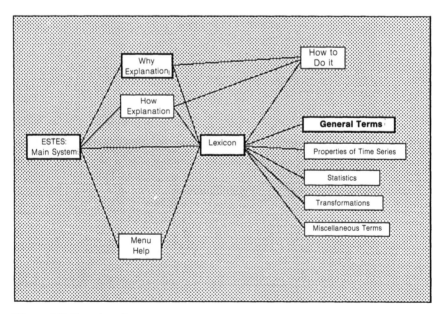

Figure 5.5 Overview diagram.

Figures 5.2–5.5 illustrate the nature of the explanation/help subsystem. In a dialogue preceding the situation in Fig. 5.2 the expert system has asked the user whether the time series 'sunspot' of the user is seasonal. After that the user has in turn asked *why* this information is needed and the expert system has produced a 'why' explanation (i.e. explaining how the system is going to utilize the answer of the user) through its explanation subsystem. A part of this explanation can be seen in Fig. 5.2. A typical display of an explanation includes *buttons* for more information, for graphical illustrations or for exits to the other levels of the system. Moreover, *test fields* can be used for textual explanation. Next in our example situation, the user wants to obtain a graphical representation of his time series, so he first selects the name of his series in the text field and then presses the 'Graph' button, after which the system shows a graphical display of his data in a new window (see Fig. 5.3).

Let us assume that our user wants to know what the term 'seasonality' in the 'why' explanation means. So he selects the word 'seasonality' in the explanation field and asks the system to explain this term (presses the button 'Explain Term'). In Fig. 5.4 our system has produced a description of the meaning of the term 'seasonality' and also allows examples of seasonal time series to be displayed (see the 'Examples' button in Fig. 5.4). Since the user is now in the lexicon part of the system he may also continue by requesting explanations for other terms. However, let us assume that our user wants to know where the currently is located within the system; he presses the 'Map' button. Now an overview diagram of the system (with the name of the current node in bold) is seen on the screen (Fig. 5.5). After studying this overview diagram, our user may select his next step: to go back to exploring terms in the lexicon (by clicking 'Lexicon' in the overview diagram); to return to the 'why' explanation (by clicking 'Why Explanation'); or to return to the expert system's question concerning seasonality of the time series (by clicking 'ESTES: Main System'). These active nodes describing the path of the user are highlighted in the overview diagram. The user can return to any of them just by clicking them. All the other nodes in the overview diagram (e.g, 'Menu Help') are inactive: in this situation the user is not able to click them. They are displayed in order to enable the user to orientate himself.

As we mentioned earlier, our example system utilizes the explanation facilities (i.e. 'how' and 'why' explanations) of an existing statistical expert system (Hietala, 1990). These explanations are composed of standard rule chains (see Fig. 5.2) and possess the well-known shortcomings of 'machine explanations' mentioned in the previous section. Moreover, the new explanation/help subsystem adopts the hypertext principle of the free navigation of the user in the system. However, there are still other possibilities to improve the user friendliness of explanation. For example, our system does not try to guess (via user modelling) the appropriate level of explanation of each user. Further work is needed to modify these explanations so that the system is able more closely to meet the requirements of individual users.

CONCLUDING REMARKS

Let us briefly consider the explanation/help subsystem with respect to the above-mentioned advantages and disadvantages of hypertext. The possibility of combining graphical illustrations with textual explanation was claimed to be one of the most essential advantages. In our system, if the user wants to see graphical information concerning a term in the 'why' (or 'how') explanations, a graphical display of the term in question is produced (see the graphical representation of time series in Fig. 5.3). Moreover, while visiting the lexicon part of the system the user can request more examples of the statistical term being considered (see the 'Examples' button in the lower right-hand part of Fig. 5.4). Now the system dynamically generates (via model-based simulation) a new time series illustrating the statistical term. These example time series are displayed graphically.

The known problems concerning the user's disorientation and cognitive overhead when working with hypertext have also been taken into consideration. We have tried to alleviate the navigation problems by using overview diagrams and backtracking techniques. An overview diagram (see Fig. 5.5) shows the current location in the system, and backtracking (see the 'Go back' button in Fig. 5.4, for example) makes it possible for the user to retrace his steps to any location he chooses.

As mentioned earlier, the adaptation to the level of the user's knowledge is possible through hypertext techniques. We have tried to meet this requirement in the following manner. At the beginning of each session with our system, introductory information is provided to the user. During the session additional information is available if requested (i.e. more examples ('Example' button) or extended explanations ('Expand []' button)). So, additional information is directed only to those who themselves know they need more explanation and who request it. However, in our opinion, this kind of approach does not sufficiently take into account the level of the user's knowledge.

In fact, future work with our system includes empirical as well as theoretical efforts to find out appropriate levels and ways of presenting additional information to the user. These efforts, we hope, will also result in deeper-level explanations. The prototype described in this chapter will hopefully serve as a useful tool in this quest for better explanations in statistical expert systems.

ACKNOWLEDGEMENT

The author would like to thank an anonymous referee for valuable and stimulating comments upon an earlier version of this chapter.

REFERENCES

Chandrasekaran, B., Tanner, M.C. and Josephson, J.R. (1988) Explanation: the role of control strategies and deep models, in *Expert Systems: The User Interface*, J.A. Hendler (ed.), Norwood, NJ: Ablex, 219–247.

Conklin, J. (1987) Hypertext: an introduction and survey. *IEEE Computer*, **20**(9), 17–41.

Gale, W.A. (1986) REX review, in *Artificial Intelligence & Statistics*, W.A. Gale (ed.), Reading, MA: Addison-Wesley, 173–227.

Hand, D.J. (1985) Statistical expert systems: necessary attributes. *Journal of Applied Statistics*, **12**(1), 19–27.

Hietala, P. (1990) ESTES: a statistical expert system for time series analysis. *Annals of Mathematics and Artificial Intelligence*, **2**, 221–235.

Hypertext Panel (1989) Expert systems and hypertext. *Proc. Hypertext '89*, special issue of *ACM SIGCHI Bulletin*, 391–392.

Kidd, A.L. (1985) What do users ask?—Some thoughts on diagnostic advice, in *Expert Systems '85*, M. Merry (ed.), Cambridge: Cambridge University Press, 296–310.

Nelder, J.A. (1988) How should the statistical expert system and its user see each other? in *Proc. 8th Symp. Computational Statistics*, D. Edwards and N.E. Raun (eds), Heidelberg: Physica Verlag, 107–116.

Nielsen, J. (1990) The art of navigation through hypertext. *Communications of the ACM*, **33**, 296–310.

Smith, J.B. and Weiss, S.F. (1988) Hypertext. *Communications of the ACM*, **31**, 816–819.

Steels, L. (1985) Second generation expert systems. *Future Generation Computer Systems*, **1**, 213–221.

Measurement scales as metadata

6

D.J. Hand

INTRODUCTION

Data are the values of measurements taken on properties of objects. However, data themselves also have properties. These properties of data are called **metadata**. Metadata are thus items of information about data. They can be accessed by an analyst (or a computer program) to direct and mediate the course of a statistical analysis.

Metadata have been approached from a number of directions (see, for example, Fessey, 1990; Finney, 1975; Lawson, 1989; Lubinsky, 1989; McCarthy, 1982). Oldford (1987: 2) says: 'There is a good deal of contextual information wrapped up in what is meant by a statistical observation. As this information can often be crucial to the statistical analysis, it should be recorded on the representation of a statistical observation whenever possible.'

Annotations and interpretive comments are metadata, and indeed they are the only form of metadata which is currently used extensively in statistical packages—for example, as variable names and labels. But note that this form of metadata cannot typically be used by the program to influence the course of the analysis.

Metadata may apply at a number of levels:

(a) to variables, for example,
- measurement scale: *religion* is a nominal variable
- in defining derived variables: *body mass index* is related to *weight* and *height* (*bmi* $= w$ (kg)/h^2(m^2))
- bounds on values: age is strictly positive;
(b) to individual cases, for example,
- a zero is structural: no males have had a hysterectomy;
(c) to sets of data, for example,
- the scores in group A are matched with those in group B
- variables x_1 to x_5 are repeated measures, taken at 1, 2, 3, 4 and 5 minutes after treatment

Artificial Intelligence Frontiers in Statistics: AI and statistics III. Edited by D.J. Hand. Published in 1993 by Chapman & Hall, London. ISBN 0 412 40710 8

- by assumption, the height distribution of this population is normal, with the data being a random sample from it.

Note that the distinction between data and metadata can be blurred. For example, the metadata may label cases as class A or as class B—but this could equally be included in the data using an indicator variable.

Note that some data sets may be structured, with correspondingly structured metadata. In a collection of education data we may have data about geographical areas, schools within those areas, years within those schools, classes within those years, and pupils within those classes. The metadata for one level will be different from those for another level and often not derivable from it.

Just as one has to decide, before collecting data, precisely what variables to measure, so it is necessary to decide what metadata are important. This is not always a straightforward task. If the data are to be archived and analysed at a later date, perhaps by someone other than those who collected it and perhaps to answer questions the original users had not considered, it will be difficult to decide a priori what to measure. There is the classic case of the researcher taking data to the statistician, but with the vital piece of information, without which the research question is unanswerable, missing. Similarly, just as the number of potential variables is unlimited, so is the amount of metadata.

Data are manipulated during the course of a statistical analysis, and so will the metadata need to be manipulated. Since metadata come in many forms —textual, numerical and, above all, structured—the manipulation of metadata is best tackled by AI approaches rather than statistical approaches. For example, economic surveys produce aggregated data. Somehow the metadata attached to the sampled microdata have to be condensed properly to apply to the aggregates. A statement that 29% of a population are unemployed has to be accompanied by a metadata definition of 'unemployed', and this may involve a merger of a number of different definitions. Sometimes the problems are not at all straightforward: in comparing the unemployment rates of different countries, or their inflation rates, it is quite likely that subtly different definitions will be used. Stephenson (1990) gives the example of interpreting pension data, in which it is 'necessary to know the rates of benefit, the conditions of eligibility, how the payments are made, their inter-relation with other social security benefits such as invalidity payments, and how well these factors change over time'.

Metadata can be used in a variety of ways, including:

1. Data validation. This is a relatively well-established use in areas such as survey processing. 'Simple edits' examine values of individual variables—for example to see if they are outside certain bounds. 'Logical edits' can see if a set of variable values together imply a contradiction (for example, a five-year-old with two children).
2. Data imputation. Much the same applies here as in (1)—see Fellegi and Holt (1976).
3. To guide an analysis. Metadata can tell us what is possible and what is impermissible.
4. To aid in interpretation.

One possible difficulty is that the best way of representing the metadata inside the computer may depend on the use to which they will be put. It is well established that different knowledge representations are convenient for tackling different problems. This is not an issue which is addressed in this chapter.

We have already mentioned measurement scale as a type of metadata. This chapter is concerned with that type alone, and the next section presents some basic concepts of measurement scales. Although the type of scale is determined by knowledge of the substantive area in question, the implications the type has for what can legitimately be said about the variables and statements using them is a distinct body of knowledge— measurement theory. In principle this theory can be coded into the knowledge base, independently of the substantive knowledge, just as statistical expertise can be independently coded. We can then use this to direct the statistical analysis. The third section provides some illustrations of this direction. The fourth section, however, points out that there are subtle issues arising involving the interaction between the substantive area and the measurement theory. For example, a number of issues associated with measurement scale are closely bound to the precise research question. Without such issues being made adequately clear—and often they are not, even in apparently straightforward situations—statistical expert systems will only be able to play a subsidiary assistant role.

MEASUREMENT THEORY

The best-developed theory of measurement is the **representational theory**. This seeks isomorphisms from the objects being measured into a number system (we shall take this to be the reals) and which preserve a set of relationships between the objects. Transformations between the real representations preserving the relationships are called **admissible** transformations. The set of real representations preserving the relationships between objects is called a **scale**, and its **scale type** is defined by the nature of the admissible transformations. Thus if only the order relationship matters then the admissible transformations are arbitrary monotonic increasing transformations and the scale is said to be of **ordinal** type. If the structure is preserved by affine transformations then the scale type is **interval**. If structure is preserved by similarity transformations then the scale type is **ratio**. Moreover, and this is the important point, a statistical statement is said to be **meaningful** if and only if its truth value is invariant to admissible transformations of the measurement scales involved. If a statement were to be true for some particular numerical representation but false for some other equally legitimate representation—perhaps true if measured in inches and false if measured in centimetres—then it would be **meaningless.**

Note also that a statement relating several variables cannot have a truth value which is invariant if all of the variables involved can undergo independent transformations, even if these are similarity transformations. This fact has been made use of in physics in what is termed *dimensional analysis*. There the various (ratio) measurement scales are related by multiplicative relationships, and so do not transform independently. We can select certain dimensions as fundamental and express all others as combinations of these (see Krantz *et al.*, 1971, Chapter 10; and Luce *et al.*, 1990, Chapter 22). For example, letting Q represent charge, θ temperature, M mass, L length and T time, we have:

density $\sim ML^{-3}$
frequency $\sim T^{-1}$
force $\sim MLT^{-2}$
current $\sim QT^{-1}$
entropy $\sim ML^2\theta^{-1}T^{-2}$

The chosen fundamental dimensions thus form a basis for the space of physical quantities.

The power of the measurement scale as metadata arises because it identifies meaningful statements. The analyst or computer can identify, just from consideration of the type of measurement, what kind of statements it is legitimate to make. The next section provides some illustrations of this.

Before proceeding with those, however, it is worth remarking that in the last two decades a considerable amount of research has been done on representational measurement theory, and some very powerful results have been obtained. The above outline does not even begin to scratch the surface. Among recent results are:

- a classification of scale types which leads to a demonstration that only certain types can exist. This classification is in terms of **homogeneity** (the degree to which any ordered set of points can be mapped to any other similarly ordered set) and **uniqueness** (the number of distinct values which are required to identify uniquely the mapping from the objects onto the reals). What is particularly interesting is that, subject to certain constraints (degree of homogeneity greater than 0 and uniqueness finite), only ratio, interval, and discrete interval scales exist. Thus the early work of Stevens (1946), although not based on the rigorous modern mathematics, proved surprisingly accurate.
- the establishment of axiomatic structures other than extensive measurement for which interval scale measurement is possible, an example being additive conjoint measurement.

This chapter only makes use of basic concepts of measurement, which will be familiar to most statisticians without a background in formal measurement theory. However, it seems likely that the power of the measurement scale as metadata concept will be significantly extended when modern results are taken into account.

MEASUREMENT SCALES AND RELATIONSHIPS: SOME EXAMPLES

Example 6.1

All statisticians will be aware of the problems associated with applying certain numerical operations to ordinal scales. Since observations from ordinal scales can be replaced by other values obtained by arbitrary monotonic increasing transformations, means of values can change arbitrarily. It is easy enough to contrive a situation where the mean of group A, $\bar{x}_A$, is less than the mean of group B, $\bar{x}_B$, for one set of scores, but where $\bar{x}_A > \bar{x}_B$ on another equally valid set. Hence, with ordinal scales the statement

$\bar{x}_A > \bar{x}_B$ is meaningless. The implication is that if ordinal scales are involved then means are an inappropriate statistic.

Example 6.2
As a simple example of dimensional analysis in physics, consider the oscillations of a simple pendulum, and suppose our aim is to find an expression for the period, t. Factors which may influence this are the length, l, of the string, the mass, m, of the bob, the acceleration, g, due to gravity, and the initial arc, s, of the swing. We thus need to find a, b, c, and d in

$$t \propto l^a m^b g^c s^d$$

In terms of the basic dimensions of the previous section, we have

$$T = L^a M^b [LT^{-2}]^c L^d$$

and thus by dimensional homogeneity

$$1 = -2c$$
$$0 = a + c + d$$
$$0 = b$$

yielding $c = -\frac{1}{2}$, $b = 0$, $a + d = \frac{1}{2}$. Thus

$$t = k l^{1/2-d} g^{-(1/2)} s^d$$

$$= k \sqrt{\frac{l}{g}} \left(\frac{s}{l}\right)^d$$

(In fact, since both s and l are of dimension L, their ratio is dimensionless, so that $(s/l)^d$ may be replaced by any finite or infinite series of powers of s/l.) Note that since the ratio s/l is involved, it is the angle of deflection which matters.

Example 6.3
Finney (1977) gives an example of dimensional analysis being used to correct a functional form in statistics. One of the forms suggested (see Harding *et al.*, 1973) for the function of dose that represents expected count in radioimmunoassay can be written

$$\alpha + \beta/[1 + \gamma \ln\{1 + \exp(x - \delta)\}]$$

where x is a dose (e.g., in weight or weight per volume) of dimension D, say. However, exp is a transcendental function, so this cannot be a correct general form since x is a dimensional quality (a change in units of x would change the value of the expected count). To remedy this it is necessary to insert a fifth parameter θ, with dimension D^{-1}:

$$\alpha + \beta/[1 + \gamma \ln\{1 + \exp(\theta x - \delta)\}]$$

Example 6.4

An example of how dimensional analysis can be applied to check an economic equation and to suggest corrected forms if it appears to be wrong is given in De Jong (1967: 27). The suggested equation was

$$\text{stock of money} = \text{prices} \times \text{flow of goods}$$

with dimensions

$$\text{M} \qquad \text{MR}^{-1} \qquad \text{RT}^{-1}$$

where M is the dimension of stock of money, R is the dimension of goods, and T is the dimension of time. Clearly this is not dimensionally homogeneous, and is thus invalid. A factor of dimensionality T^{-1} is missing from the left-hand side (or, de Jong suggests, stock of money will need to be replaced by dM/dt). Economic theory suggests that the former amendment is appropriate—and that a factor 'income velocity of circulation of money', with dimension T^{-1}, is missing from the left-hand side.

Example 6.5

Luce (1959) pointed out that it is sometimes possible to derive the laws relating two scales if one assumes only continuity and knows the scale types. This work has been criticized (Rozeboom, 1962) but has been developed further (Luce, 1962; 1990) and illustrates the sort of power that measurement scales can bring when used as metadata. Suppose that we wish to relate two scales, f and g. The idea is based on the assumption that an admissible transformation of f should lead to an admissible transformation of g. For example, if both f and g are ratio scales then a similarity transformation of f must result in a similarity transformation of g. Thus, if x is an object with values $f(x)$ and $g(x)$ and if $f(x) = \psi(g(x))$ then

$$\psi(\alpha g(x)) = F(\alpha) f(x)$$

for some function F. Given that ψ must satisfy this, it is possible to work out that ψ must have the form $\psi(z) = \alpha z^{\beta} (\alpha > 0)$. Similar results apply where either f or g is interval and the other ratio, and for the case of both interval.

Example 6.6

We have two boxes, x and y, of sticks and we want to know if the mean length of sticks in box x is greater than the mean length of sticks in box y. To explore this we give each child in a primary school one stick from each box to measure, with the aim of calculating the x and y means of these measurements. The rulers that the children use have been constructed by having each child cut a small piece of card and use this as the unit length in marking up graduations to make a ruler. Let x_i be the measurement obtained by child i on her x stick, and y_i her measurement on her y stick. Each child is clearly using a ratio scale so it might be thought legitimate to make statements such as $\bar{x} > \bar{y}$. However, a little thought shows that this is not the case.

We can imagine similarity transforms such that although

$$\frac{1}{n}\sum x_i > \frac{1}{n}\sum y_i$$

it is not true that

$$\frac{1}{n}\sum \alpha_i x_i > \frac{1}{n}\sum \alpha_i y_i$$

The confusion has obviously arisen because we are in effect averaging scores measured on different units.

BEYOND MEASUREMENT THEORY

We have seen how measurement scales can be used as metadata to assist in determining the kinds of statement which can be made about variables and their relationships, and hence assist in fitting statistical models. However, measurement scale, as determined by representational measurement theory, does not determine all the answers.

First, there appear to be other attributes of variables, not determined by the conventional theory, which are made use of in statistical analysis. In part these other attributes seem to have their genesis in the categorical nature of many of the variables subjected to statistical analysis (representational measurement theory is chiefly based on real representations).

For example, Nelder (1990) describes variables as coming in different **modes**, distinguishing continuous counts, continuous ratios, count ratios, and categorical modes, with the last being divided into three subtypes (nominal, ordered on the basis of an underlying continuous scale, and ordered without an underlying scale).

Anderson (1984) suggests that there are two major types of observed ordered categorical variable: those related to a single underlying continuous variable (= grouped continuous); and a variable defined by the judgement of an assessor, who 'processes an indeterminate amount of information before providing his judgement of the grade of the ordered variable'.

McCullagh (1980) points out that there may not always be a continuous latent variable where categories may be thought of as contiguous intervals, but in his models the existence of such a scale is not required for interpretation.

Bartholomew (1987) distinguishes **metrical** from **categorical** variables, and other classifications have also been suggested.

It is interesting to speculate on how much these various distinctions have grown up in response to the forms of data that can arise, and how much as a consequence of the statistical models that are available.

Secondly, there are cases when the measurement issues seem straightforward initially, but prove to be more complex on closer examination, and involve very careful interpretation of the aims of the research. We present some examples to illustrate this.

Example 6.7

In a study of pain relief following wisdom tooth extraction, 71 patients were scored on two measures of pain relief. One was a visual analogue scale, ranging from complete relief on the left to no relief on the right, with the score being the distance in millimetres from the left of the scale. The other was a five-point categorical scale ranging from 0 for no relief to 4 for complete relief. Scores were taken at several times between the time when the patient chose to have an anaesthetic (when the anaesthetic used for the operation began to wear off) and two hours after the operation. For each patient and each scale average pain relief was calculated, yielding $\bar{x}_i$ and $\bar{y}_i$ say, for the ith patient. The correlation between these scores was -0.815: high because they are measuring the same thing and negative because one is scored backwards relative to the other.

The researcher also wanted measures of 'total pain relief' for each patient. This was defined as the area under the pain relief curve and can be calculated as $\bar{x}_i t_i$ and $\bar{y}_i t_i$ where t_i is the time that the ith patient was observed. The correlation between these two 'total' measures was found to be only -0.039, which seemed rather counter-intuitive. We are multiplying two highly (if negatively) correlated variables by the same factors (for each subject) so we might expect to retain a high correlation.

Indeed, we would retain the high correlation if $t_i = t$ for all subjects—it would then remain as -0.815. However, when the t_i are allowed to vary the correlation can be changed arbitrarily—though the most likely situation is that a negative correlation is reduced or destroyed.

Telling a researcher that the 'total' scores are uncorrelated is a rather unsatisfactory resolution to this problem. It is much better to note that the problem can also be regarded as arising because the initial correlation is negative. If the scoring method of one of the variables is reversed, so that the initial correlation is $+0.815$, then the correlation between $\bar{x}_i t_i$ and $\bar{y}_i t_i$ will (normally) also be high. Indeed, in the study in question this correlation was $+0.911$.

The problem here is that there is a relationship between the two measures in question. This relationship, that one is scored backwards relative to the other, could be contained in the metadata. If so, it is a property relating to relationships between variables and not to any individual variable. More to the point is whether it *would* be contained in the metadata. There comes a limit to the amount of information that the metadata might reasonably be expected to contain, and this may well be beyond it. Of course, this example is a particularly simple one: but what about multivariate questions relating several tens or hundreds of variables? Can the metadata really be expected to contain information of this sort?

The problem here seems to be an intrinsically substantive one, beyond the range of measurement theory. This example is discussed in more detail in Thompson *et al.* (1991).

Example 6.8.

Given two alternative treatments A and B, we wish to answer the question 'which should we use on a future patient?', which we interpret as meaning 'which is more likely to give a greater improvement?'. Without loss of generality, we shall assume that a larger score is a better score. We deal with the particular situation in which the illness and treatments are of such a nature that only one treatment can be given to each patient. To address the question we conduct a clinical trial, assigning patients to treatment A or treatment B at random.

What we really want to do is compare two hypothetical responses, z_A and z_B, for a patient randomly sampled from the patient population, and answer the question Is $P(z_A - z_B > 0) > \frac{1}{2}$?'. Alternatively, we would like to ask 'Is $Q(z_A - z_B) > \frac{1}{2}$?', where $Q(x) = P(x > 0)$.

However, from the above, in the clinical trial we do not have matched (z_A, z_B) scores, but independent sets of scores x_A and y_B, so that we are forced to estimate $Q(x_A - y_B)$. In fact, the Wilcoxon two-sample test estimates just this, and compares it with $\frac{1}{2}$. The two questions 'Is $Q(z_A - z_B) > \frac{1}{2}$?' and 'Is $Q(x_A - y_B) > \frac{1}{2}$?' are different because of the independence of the x and y in the latter. And one can be true while the other is false. Numerical examples are given in Hand (1992).

Note that the phenomenon is a structural one, not merely a sampling one—arbitrary degrees of statistical significance in the wrong direction can be achieved by inflating the sample sizes.

The issue here is not one of inadequate use of metadata, nor one of measurement scale; rather it is one of the precise research question. It is not clear how a statistical expert system can be expected to resolve such issues.

Example 6.9

Two researchers wish to compare the education systems of two countries in terms of the efficiency of the average schools in the two countries. They take samples of n schools from each country, measuring for each school the annual running cost and the annual student intake.

The first researcher calculates the cost per student for each school and averages these for each country. The second researcher calculates the number of students per $1000 for the same schools, averaging these values within each country.

To keep things simple, suppose that they both work in units of $1000 and 1 student, and that only two schools are involved in each country.

Researcher R1 finds the efficiencies to be 1 (i.e. $1000 per student) and 4 (i.e. $4000 per student) for the schools in one country (C1) and 2 and 2 for the schools in the other country (C2). The arithmetic means are 2.5 and 2, respectively, from which he concludes that C1 has on average less cost-efficient schools (i.e. more dollars per student).

Researcher R2, however, using the same data, will find efficiencies of 1 (student per $1000) and 0.25 for C1 and 0.5 and 0.5 for C2. His arithmetic means are 0.625 and 0.5, respectively, so that now C2 appears to be less efficient (i.e. less students per dollar, or more dollars per student).

The difference between the two conclusions, which are based on the same data, clearly arises because of the nonlinearity of the relationship between the two scales (being reciprocals). In general, if f is nonlinear, then $f(E(x)) \neq E(f(x))$. However, although this tells us why there is a difference, it does not tell us which is the correct approach.

Sometimes, in problems of this kind, substantive knowledge will lead to a choice. For example, while the English measure a car's fuel efficiency in miles per gallon, the Germans use the metric equivalent of gallons per mile. The appropriate ratio to use will depend on whether one is concerned with how much fuel it takes to cover a certain distance, or how far one can go on a certain amount of fuel. I would argue that people would normally want to know the former. The metadata of measurement scale alone cannot help here (though more complex metadata structures, more intimately linked to the substantive area, may be able to help).

An alternative approach is to say that we are really concerned with 'efficiency' and the two measures we have are merely proxy variables for it. They are (one assumes) monotonically related to it. In this case, given the weak relationship between the variable of interest and its proxies, which preserves only order, if we wish to make a valid (meaningful) statement about efficiency we must use techniques which are invariant to monotonic increasing transformations. And this is so whether or not a particular proxy variable is a ratio variable. The measurement scale metadata could here lead us astray.

(In fact, in example 6.9 there are even more possibilities. The example was deliberately worded to indicate that the concern was with the average of the individual efficiencies of the schools, and not with the overall average of the education within each country. If the latter had been the aim then efficiency would have been estimated using the ratio of the sum of costs to the sum of student numbers (or its reciprocal) and the two researchers would have reached identical conclusions.)

CONCLUSION

There is no doubt that measurement scale is a powerful form of metadata, imposing strong restrictions on the statistical statements which can be made about variables and their relationships, and on the models which can be fitted. However, as the later examples demonstrated, sometimes apparently straightforward problems have, upon closer examination, complications which neither measurement scale metadata nor statistical expertise *per se* can resolve.

It seems that resolution of these problems requires substantive knowledge coupled with a deep examination of the precise objectives of the study. Moreover, it is not clear that the researchers do, as a matter of course, make the deep examinations. Again it seems (see Hand, 1990) that statistical expert systems should be built with the aim of being used by an expert statistician, who can help the researcher undertake these examinations, rather than a statistically naive user.

ACKNOWLEDGEMENTS

I would like to thank Professor Ivo Molenaar and Dr Shelley Channon for comments on an earlier version of this chapter.

REFERENCES

Anderson, J.A. (1984) Regression and ordered categorical variables. *Journal of the Royal Statistical Society, B,* **46**, 1–30.

Bartholomew, D.J. (1987) *Latent Variable Models and Factor Analysis,* London: Griffin.

Cox, C. (1987) Threshold dose-response models in toxicology. *Biometrics,* **43**, 511–523.

De Jong, F.J. (1967) *Dimensional Analysis for Economists,* Amsterdam: North-Holland.

Fellegi, I.P. and Holt, D. (1976) A systematic approach to automatic edit and imputation. *Journal of the American Statistical Association,* **71**, 17–35.

Fessey, M.C., ed. (1990) *Expert Systems and Artificial Intelligence: the need for information about data,* London: Library Association.

Finney, D.J. (1975) Numbers and data. *Biometrics,* **31**, 375–386.

Finney, D.J. (1977) Dimensions of statistics. *Applied Statistics,* **26**, 285–289.

Hand, D.J. (1990) Emergent themes in statistical expert systems, in *Knowledge, Data, and Computer-assisted Decisions,* M. Schader and W. Gaul (eds), Berlin: Springer-Verlag, 279–288.

Hand, D.J. (1992) On comparing two treatments. *The American Statistician,* **46**, 190–192.

Harding, B.R., Thompson, R. and Curtis, A.R. (1973) A new mathematical model for fitting an HPL radioimmunoassay curve. *Journal of Clinical Pathology,* **26**, 273–276.

Krantz, D.H., Luce, R.D., Suppes, P. and Tversky, A. (1971) *Foundations of Measurement, I,* New York: Academic Press.

Lawson, K.W. (1989) A semantic modelling approach to knowledge based statistical software. PhD thesis, Aston University, UK.

Lubinsky, D. (1989) Integrating statistical theory with statistical databases. *Annals of Mathematics and Artificial Intelligence,* **2**, 245–259.

Luce, R.D. (1959) On the possible psychophysical laws. *Psychological Review,* **66**, 81–95.

Luce, R.D. (1982) Comments on Rozeboom's criticisms of 'On the possible psychophysical laws'. *Psychological Review,* **69**, 548–551.

Luce, R.D. (1990) 'On the possible psychophysical laws' revisited: remarks on cross-modal matching. *Psychological Review,* **97**, 66–77.

Luce, R.D., Krantz, D.H., Suppes, P. and Tversky, A. (1990) *Foundations of Measurement, III.* San Diego: Academic Press.

McCarthy, J.L. (1982) Metadata management for large statistical databases. In *Proceedings of the 8th International Conference on Very Large Databases,* September.

McCullagh, P. (1980) Regression models for ordinal data. *Journal of the Royal Statistical Society, B,* **42**, 109–142.

Nelder, J.A. (1990) The knowledge needed to computerise the analysis and interpretation of statistical information, in *Expert Systems and Artificial Intelligence: the need for information about data,* M.C. Fessey (ed.), London: Library Association, 23–27.

Oldford, R.W. (1987) Object oriented software representations for statistical data. University of Waterloo Technical Report, STAT-87-18.

Rozeboom, W.W. (1962) The untenability of Luce's principle. *Psychological Review,* **69**, 542–547.

Stephenson, G. (1990) Access to statistical information, in *Expert Systems and Artificial Intelligence: the need for information about data,* M.C. Fessey (ed.), London: Library Association.

Stevens, S.S. (1946) On the theory of scales of measurement. *Science,* **103**, 677–680.

Thompson, M.J., Hand, D.J. and Everitt, B.S. (1991) Contradictory correlations between derived scales. *Statistics in Medicine,* **10**, 369–372.

PART TWO
Belief networks

On the design of belief networks for knowledge-based systems

7

B. Abramson

INTRODUCTION

Knowledge acquisition, knowledge representation, and **inference** are common terms in the artificial intelligence (AI) literature; they are the central elements of a knowledge-based (or expert) system. AI, however, is not alone in its concern with these issues. Decision analysis (DA) and other statistical decision sciences have also investigated the elicitation of information from human experts and the formal modelling of this information to infer useful recommendations. The approach traditionally taken by AI has been quite different from that favoured by DA. Perhaps the most fundamental distinction between them has been their psychological motivation. Many of the procedures popular in AI are rooted in *descriptive* psychology, or observations about the way that people behave. The prevalent paradigm for expert system design is a case in point; production rules attempt to model the problem-solving strategies employed by experts. In other words, an ideal rule-based system would mimic the behaviour of a human expert. DA, on the other hand, is a sophisticated outgrowth of task analyses of inferences, evaluations and decisions. In so far as it relates to human behaviour, DA prescribes what decision makers *should* do, rather than describing what they do. Decision-analytic task analyses often lead to prescriptions that differ systematically from what people do (see, for example, Kahneman *et al.*, 1982); such effects have been called **cognitive illusions**, as in von Winterfeldt and Edwards (1986).

Formally optimal prescriptions growing out of task analyses, then, should have definite implications for the design of knowledge-based systems. A computer, after all, is a psychological *tabula rasa*; it should not suffer from the cognitive illusions that systematically cause humans—even experts—to err. A few recent expert systems have chosen to accentuate the normative over the descriptive. Most of these systems are based on **belief networks**—graphical embodiments of hierarchical Bayesian analysis. Bayesian statistics differs from classical statistics in its view of probabilities as

Artificial Intelligence Frontiers in Statistics: AI and statistics III. Edited by D.J. Hand. Published in 1993 by Chapman & Hall, London. ISBN 0 412 40710 8

orderly opinions rather than as frequencies of occurrence (see, for example, Edwards *et al.*, 1963). The single most important implication of the Bayesian position to automated system design is that a Bayesian may reasonably ask an expert for an opinion about a rare event; a classical (non-Bayesian) statistician can elicit no such probability in the absence of voluminous data. For this and other less obvious reasons, many of the existing probability-based systems are Bayesian.

Belief networks are a large family of models. Any directed, acyclic, graphical representation of a problem or a domain in which (i) nodes represent individual variables, items, characteristics, or knowledge sources, (ii) arcs demonstrate influence among the nodes, and (iii) functions associated with the arcs indicate the nature of that influence, qualifies as a member of the family. Although belief networks can be traced back at least as far as Wright (1921), the only family member that attracted much attention prior to the 1980s was the decision tree. Decision trees have been studied in conjuction with virtually every decision problem and statistical technique imaginable; they have been proven useful for some purposes, but not quite sophisticated enough for others. Recent work in AI and DA, however, has uncovered a pair of more powerful belief networks, surveyed in Horvitz *et al.* (1988). In DA, **influence diagrams**—deriving their name from the influence indicated by the network's arcs—were introduced in Howard and Matheson (1984) to model complex decision problems. Meanwhile, a group of AI researchers, notably Pearl (1988), have put forward **Bayes nets** as a mathematically precise method for managing uncertain information in an expert system.

The recent wave of relevant literature has concentrated on but a few topics. First, substantial attention has been paid to the comparative epistemological adequacy of probability theory and alternative systems for representing uncertainty (see, for example, Cheeseman, 1988; or Ng and Abramson, 1990). Second, a variety of mathematical studies in Shachter (1986; 1988), Pearl (1988) and elsewhere have investigated the inferential and algorithmic power latent in belief networks. Third, systems like Heckerman *et al.'s*, (1990) PATHFINDER and Abramson and Finizza's (1991) ARCO1, both of which will be discussed throughout this chapter, have shown that belief networks can be implemented in (at least) a few specific domains. Implicit in this literature is an extremely powerful pair of claims:

1. If a belief network accurately captures the domain that it claims to be modelling, then its implications will be powerful and efficient.
2. Useful belief networks can be constructed for at least some interesting domains.

Proof of the first claim can be found in the literature discussing inference and algorithms, proof of the second from the systems already in existence. Taken together, they motivate an investigation of the general representational aspects of belief networks, with an emphasis on the central elements of an expert system: knowledge acquisition, knowledge representation, and inference.

KNOWLEDGE ACQUISITION

This chapter will concentrate on two belief network-based systems: PATHFINDER and ARCO1. These two systems have more in common than simply a mathematical basis;

much of the software used to construct ARCO1 was borrowed from PATHFINDER, and modified as necessary. The shift from PATHFINDER's heavily symbolic setting in medical diagnosis to ARCO1's primarily numeric task of financial forecasting raised many interesting issues that would have gone unnoticed in any single domain. Thus, the appropriate forum for forwarding belief networks as a paradigmatic basis for knowledge-based systems lies neither in PATHFINDER nor in ARCO1, but rather in the ground between them. Most of the issues discussed in this chapter are independent of either system's domain; the interesting points all arise from a consideration of their similarities and differences.

The knowledge acquisition phase of both systems was heavily biased towards set-up time. The most time-consuming components of knowledge acquisition were the convergence of domain experts and system designers to a common language and a specific approach, and the development of software tools that reflected that convergence. As the use of systems based on belief networks matures, development time for software tools will decrease. The education of the system designers and domain experts in each other's languages and approaches, however, will probably always require substantial time and effort.

With the systems' framework set and the design teams comfortable collaborating, actual elicitation could begin. From this point on, neither system required more than about 40 hours of expert time to construct a belief network. The information necessary to model a domain as a belief network can be broken into four categories: variables, values, influence, and dependence. Variables are simply the different objects within the domain that need to be captured by the model; eliciting them should rarely be difficult. Values are the ranges taken on by each variable; some may be binary, others multi-valued or even continuous. The only difficulties inherent in value elicitation arise from the occasional discretization of continuous ranges. Even in the most difficult cases, however, the domain experts of both PATHFINDER and ARCO1 were able to arrive at value sets with which they were comfortable. Influence in a belief network is indicated by an arc; elicitation requires considering the existence (or non-existence) of direct relationships between pairs of variables. Dependence is indicated by assigning specific relationships to sets of arcs. The value taken on by a variable is dependent on the values taken on by all variables that influence it (i.e. all incident arcs). PATHFINDER represented all dependencies as conditional probabilities; ARCO1 extended the concept to include deterministic algebraic dependence as well. Regardless of the types of allowable functions, however, dependence is the most controversial and difficult information to elicit. Fortunately, the technologies of DA provide many tools for eliciting and refining probability estimates. The power of these elicitation techniques — and the consistency with which they have helped decision-makers overcome cognitive illusions and reach good decisions (as described by von Winterfeldt and Edwards, 1986; and Howard and Matheson, 1983)—are among the greatest benefits accrued by modelling a domain as a belief network rather than as a production system.

The basic procedure for modelling a domain as a belief network, then, can be summarized as:

1. Select a specific problem.

2. List relevant variables.
3. Define the range of values taken on by each variable.
4. Indicate influence among the variables.
5. Specify the (functional or probabilistic) dependence of each variable on all variables that influence it.

This procedure should work for a wide variety of domains, although the specifics will certainly vary.

PATHFINDER

PATHFINDER, the result of over five years of ongoing collaboration between researchers at Stanford University and the University of Southern California (USC), helps pathologists diagnose diseases by guiding their interpretations of microscopic features appearing in a section of tissue (for details, see Heckerman *et al.*, 1990). Physicians at community hospitals are rarely experts in all aspects of pathology; they frequently need the help of subspecialists to reach accurate diagnoses. PATHFINDER was designed to capture the knowledge of one subspeciality—haematopathology—so that community hospitals could obtain expert input without calling directly on the subspecialists. All tests run on PATHFINDER suggest that it has met its primary objective: PATHFINDER's diagnoses, given accurately recorded observations, are consistently accurate and on par with those derived by top haematopathologists. (INTELLIPATH, a commercial version of PATHFINDER, comes equipped with a laser-disk library of symptomatic cell features, designed to ensure that the community hospital's observations are identified accurately.)

One of the major breakthroughs in the system's development occurred during the first step: the selection of a specific problem and approach. The decision to use probabilistic modelling and DA elicitation techniques was reached through trial and error. Several representations of the uncertain medical data were attempted; the best performance was obtained with probabilities, as reported in Heckerman (1988). Identifying a specific problem, however, was somewhat trickier. PATHFINDER's domain contains around 60 diseases and 100 symptoms. Although 160 variables and their value ranges could easily be elicited, discussions of influence and dependence are obviously impractical. This difficulty helped refine the problem from 'how can a disease be diagnosed?' to 'how can we differentiate between a pair of similar diseases?'. This refinement, in turn, led to the creation of the **similarity network**, an important addition to DA's collection of modelling tools (Heckerman, 1990). In addition to facilitating the design of PATHFINDER, similarity networks remove many of the theoretical difficulties that earned probabilistic systems a reputation for being unwieldy and

Figure 7.1 PATHFINDER's entire global belief network, minus arcs from the node marked 'DISEASE'. This node, which is the target of all diagnoses, actually points to *every* other node in the network; all symptoms are dependent on the disease. These arcs have been omitted from this diagram, however, to highlight the conditional dependencies among the symptoms.

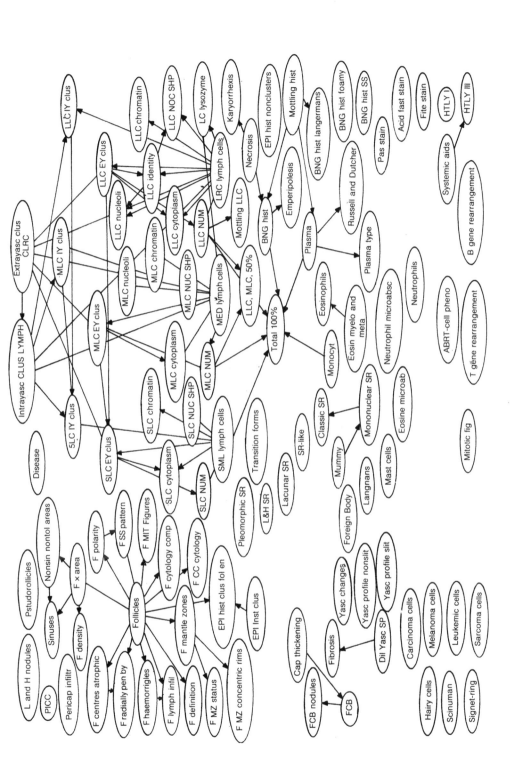

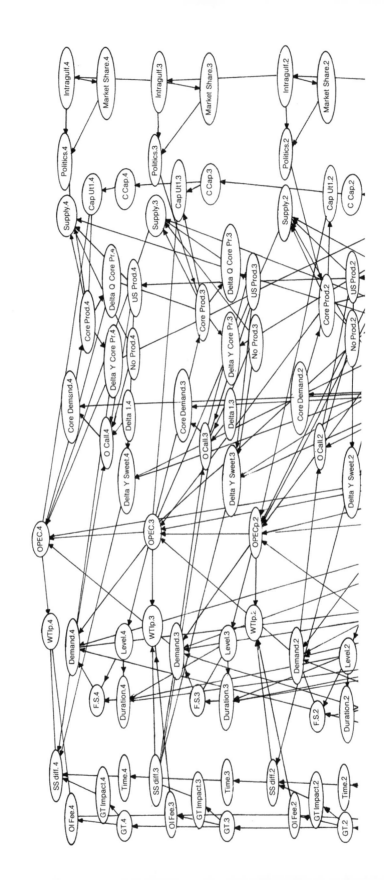

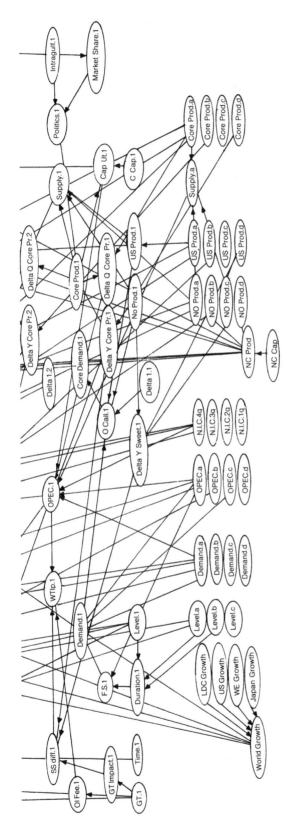

Figure 7.2 ARCO1's belief network model of the 1990 oil market. Monte Carlo simulations of the network yielded price forecasts for each of the four quarters of 1990. Global political and economic considerations have already led to modifications in the model; these changes, however, were relatively minor and easily effected.

impractical. PATHFINDER's similarity network listed all pairs of similar diseases; variables, values, influence, and dependence information was then elicited to differentiate between all similar pairs. These 'local' belief networks were then coalesced into a single, coherent, 'global' belief network that modelled the entire domain. The information necessary to construct the local networks was elicited rapidly, and in a manner that made the domain expert comfortable and confident with his assessments (see Heckerman *et al.*, 1990). PATHFINDER's global belief network is shown in Fig. 7.1.

ARCO1

ARCO1, currently under development at the Atlantic Richfield Company (ARCO) and USC, models the oil market. It is being designed as an aid to the members of ARCO's corporate planning group who forecast the price of crude oil. The first round of the system was completed in mid-1990 and then updated in light of Iraq's invasion of Kuwait and the subsequent world reaction (for details, see Abramson and Finizza, 1991). And a copy of the network underlying ACRO1's original analysis of 1990 is shown in Fig. 7.2. Initial tests on the system were quite positive. The point of primary interest to this chapter, however, is not the system's power, but rather the technique used in its design. The development of ARCO1 began with Abramson's (1988) analysis of ARCO's forecasting models in light of recent developments in AI and DA. The first phase of this analysis focused on (i) identifying the strengths and weaknesses of the existing models, and (ii) decision-analytic problem structuring. Since the current forecasts are based primarily on time series and standard econometric models, many of their general characteristics are already well known; they are strong at projecting continued trends, but much weaker at identifying technical aberrations (e.g., turning points, discontinuities, spikes). Another area of difficulty—and one more specific to the oil industry—is that many of the variables affecting the market are essentially political. Whereas most economic variables are numeric, and thus easily modelled as time series, political considerations tend to be qualitative. As a result, political and economic implications of the market are difficult to combine; political projections are usually used to temper the output of an economic model, and thus rarely given appropriate consideration in the ultimate forecast.

   The oil market contains many interesting questions; market models could be used for corporate resource allocation, individual investment strategies, government policies, and so on. The need for a highly specific problem, however, directed us towards price forecasting—the first and most basic issue addressed by any model of the market, regardless of its ultimate purpose. The selection of a specific approach was somewhat more difficult than the selection of a specific problem. A one-year model of the oil market contains between 100 and 150 nodes, ranging in character from highly precise and public economic data to subjective and occasionally vague interpretations of OPEC's internal politics. Results from the problem structuring phase convinced us the belief networks were appropriate models. In order to capture the domain elegantly, however, the networks had to be extended from the strictly probabilistic character that was appropriate for PATHFINDER to more general algebraic models. Although the mathematics behind this shift was trivial, its implications to knowledge representation and inference are not.

Another important decision was to model the market one quarter at a time, and to worry about the relationship between quarters after they were all modelled. Once again, this decision is more significant than it sounds; it was instrumental in breaking the type of circular thinking that says that demand affects price affects demand. Since cycles are not allowed in belief networks, truly circular reasoning can never be modelled. Division into quarters allowed us to show influence from first-quarter demand to first-quarter price to second-quarter demand, and so on. These decisions all took place during the set-up phase, in which the system designer and domain experts were learning each other's languages and concerns. Once we agreed upon appropriate terminology and representation, the actual elicitation of variables, values, influences, and dependencies proceeded smoothly and quickly.

KNOWLEDGE REPRESENTATION

One of the most significant contributions of expert systems research has been the conceptual division between a system's knowledge base and its inference engine. The construction of a decision-analytic model is expensive and time-consuming. Since relatively few decisions are important enough to warrant the expense, DA techniques have been (more or less) restricted to a few classes of crucial decision in corporate, governmental, military and medical settings. One of the most attractive features of an expert system's knowledge base is its reusability. Knowledge bases model entire domains, and plug in data on a specific problem whenever necessary. This feature allows the design cost to be amortized over the lifetime of the system, thereby greatly extending the range of problems in which decision-analytic modelling is practical. Belief networks, like the rule bases of conventional expert systems (e.g., those designed with commercially available shells), are knowledge bases; as such, they offer a forum in which system designers can simultaneously avail themselves of the modelling prowess of DA and the cost-effectiveness of AI.

In addition to facilitating collaboration between researchers in AI and DA, belief networks also offer an environment in which system designers and domain experts should be able to communicate comfortably. Few restrictions are placed on the types of knowledge that can be captured by a network's nodes; the only requirements are that they all have well-defined inputs and outputs, and that the inputs of one node be defined in terms of the outputs of its predecessors. The format of each node should be guided by the variable that it models. This practice, which is in marked contrast to the shell-based approach of forcing all information into rules, should increase an expert's willingness to participate in the exercise. Thus, a second merit of belief networks is that each variable in a belief network is encoded in its most natural representation. Analyses leading to the construction of these nodes may be directed by the terminology and approaches with which the domain expert is most comfortable.

The variables in PATHFINDER, for example, describe symptoms and diseases. In the global belief network, the diseases are collected into a single multi-valued 'hypothesis' node', in which each of the 63 possible diseases defines a distinct value. Symptoms are each modelled as nodes with a relatively small set of discrete values. Functional relationships among these variables are almost exclusively conditional dependencies. In ARCO1, the variables are somewhat more varied; the mix of numeric and symbolic items in a complex politico-economic domain provides a wider variety of variable types than do most medical domains. Functional relationships in ARCO1 are similarly broad; they include an assortment of algebraic equations as well as conditional dependence.

Although the analyses necessary to construct the PATHFINDER and ARCO1 belief networks obviously had no overlap, the basic software for constructing belief networks transferred easily from PATHFINDER to ARCO1. The only areas that required substantial work were those in which PATHFINDER's software assumed that all data would be symbolic, and did not allow ARCO1's heavily numeric data to be entered in the most efficient manner. (The range of possible values taken on by many economic variables, for example, can be described by a minimum, a maximum and an interval. Since this type of description is rarely meaningful for symbolic variables, the software developed for PATHFINDER contained no provisions for specifying variable ranges in this way.) Another advantage of belief networks, then, is that the form of a belief network, and the software used to construct it, are essentially independent both of the domain itself and of the types of question typically asked in the domain.

INFERENCE AND INFORMATION PROCESSING

The questions that a system is expected to answer must, of course, have a substantial impact on its design. Diagnosis and forecasting, for example, are very different types of questions; medical diagnosis involves observing symptoms and projecting them backwards to determine which disease is already present, while forecasting involves projecting forwards to predict the future value of some variable. This distinction is completely transparent during the construction of the knowledge base—although it will, of necessity, colour the discussion between system designers and domain experts. It is only in the inference engine that these considerations come into play, and it is in the inference engine that differentiation among systems occurs; engines designed to answer different types of question must be driven by different information processing procedures. This ability to shift drivers can be sharply contrasted to the match–select–act cycle that drives nearly all rule-based engines. It also suggests expanding the notion of an inference engine to that a **processing engine**, where any procedure that relates a node's outputs to those of its predecessors is a viable driver for a processing engine. Thus, belief network processing engines may be readily tailored to different types of domains.

Most AI research has focused on symbolic, diagnostic domains, such as medicine. Since the questions of interest to diagnosticians are quite different from those that interest forecasters, engines will not transfer easily from diagnostic to forecasting domains. To appreciate this distinction, consider that in a medical setting, influence will generally be indicated from disease to symptom. Evidence, however, tends to be collected as symptoms are observed. In other words, a doctor might specify the probability with which disease A will cause symptom B to appear, or $P(B|A)$. When a specific case is being considered, however, symptom B will be observed, and the posterior probability of disease A, or $P(A|B)$, will be required. Thus, a diagnostic engine must work *against* the arrows in the belief network. Bayes's rule, which relates $P(A|B)$ to $P(B|A)$, is thus the obvious driver for a diagnostic engine.

In a forecast, on the other hand, influence is generally given in terms of either temporal precedence or ease of observation; evidence tends to be accumulated in the same direction, and the processing engine must work *along* the network's arrows. Thus, forecasting engines (such as that of ARCO1) must differ substantially from diagnostic engines (like PATHFINDER's). These graphical differences reflect a fundamental distinction between diagnoses and forecasts. Consider, for example, a system designed to forecast crude oil prices in the year 2000. As part of the knowledge acquisition phase, prices in 2000 may have been specified in terms of supply and demand in 2000, which, in turn, were dependent on some variables back in 1999, and so on, back to the present. An ideal output for the system would, of course, be a distribution of oil prices in the year 2000, given only the values of the currently observable variables. Unfortunately, this response can rarely be computed for anything but a very simple model. The complexity of multiplying all the relevant distributions is the product of the number of variables, values and arcs in the network. A Monte Carlo analysis of the years between 1990 and 2000, on the other hand, should yield a reasonable approximation in an acceptable length of time. The simulation proceeds in a straightforward manner: Each rooted node (i.e. a node with no incident arcs) in the network is instantiated according to its specified priors, effectively 'removing' the rooted nodes from consideration; all nodes that had depended solely on rooted nodes are now no longer conditioned on anything, and they may also be instantiated. This procedure continues until all variables but the one being forecast have taken on definite values. Although the subject of the forecast could, of course, also be instantiated, the nature of these analyses makes it preferable to leave it undetermined, with the instatiation of its inputs describing it probabilistically. Thus, each run through the network yields a scenario describing the future, along with a distribution describing what the target variable (price in 2000) might look like in that future. Although no single scenario is likely to describe the future accurately, the distributions obtained by combining the results of multiple runs should define a meaningful 'expected future'. Thus, Monte Carlo analyses may replace Bayes's rule as the driving force behind a forecasting engine.

Drivers other than Bayes's rule and Monte Carlo are, of course, possible. In many respects, in fact, these procedures describe opposite ends of a spectrum. In a purely diagnostic system, all information will flow from leaf nodes (those with no outgoing arcs), against the network's arcs, to the rooted nodes. Thus, Bayes's rule is the only driver necessary. In a purely forecasting system, all information will flow from rooted nodes, along the network's arcs, to the leaf nodes; only Monte Carlo is needed. Many settings, however, fall into neither category. When evidence is found in mid-network, mechanisms for simultaneously transmitting information forwards and backwards are necessary; Henrion's (1988) **logic sampling** and Pearl's (1988) **stochastic simulation** are both reasonable candidates. They are not, however, the only ones. The make-up of the processing engine's driver, like that of individual nodes in a network, should be guided by the conventions of the domain in which the system is attempting to function, and the types of question that it is expected to answer.

PRELIMINARY RESULTS

Although discussions of design paradigms, knowledge acquisition, knowledge representation, information, and information processing may be interesting in their own right, they are only truly useful if they lead to good performance. Evaluation of knowledge-based systems, however, is traditionally difficult. What does it means for a system to perform 'well'? Preliminary evaluations have already been conducted on both PATHFINDER (in Heckerman *et al.*, 1990) and ARCO1 (in Abramson and Finizza, 1991). Although these tests remain incomplete, the initial findings are uniformly encouraging.

PATHFINDER was rated along two metrics: expert opinion and (decision-theoretic) information loss. Recall that PATHFINDER outputs its diagnoses as probability distributions across a space of 63 possible lymphomas. Output distributions were studied for 53 cases selected, in sequence, from a large library of cases that had been referred by pathologists at community hospitals to the specialists at USC/County Medical Center. For each distribution, a rating was provided by the expert (Dr Nathwani), who evaluated it on a subjective 1–10 scale; the decision-theoretic information loss metric was calculated by inputting the distribution into a general utility model. Both metrics were calculated for two versions of the system, a simple one that made assumptions of universal independence among symptoms (i.e. that all symptoms are mutually independent, given the disease), and the sophisticated model based on belief networks as outlined above; the studies revealed that the use of the sophisticated model improved the mean expert rating from 7.99 to 8.94, and added an average patient utility of about $6000 per case. The details of these experiments, or of the underlying utility models, are beyond the scope of this chapter. A thorough discussion can be found in Heckerman *et al.* (1990). The key point of interest, however, is that under the conditions studied in these tests, a system based on a sophisticated belief network proved to be much stronger (and more valuable) than one that was simply aware of the diseases and symptoms.

The preliminary tests on ARCO1 were informal, but informative. ARCO1's base case was a model of the 1990 crude oil market. It was run in early 1990, and generated distributional forecasts for the year's four quarters, and for the entire year. Without getting into specific details, the implications of the forecast were that WTI* prices would probably remain in the high teens to low twenties throughout the year (61.5% of the scenarios envisioned prices between $18 and $21). This forecast did not surprise anyone; the range corresponded to where the prices were at the end of 1989, and it was fairly consistent with the forecasts being produced by industry analysts at the time. The

---

*West Texas Instruments, the benchmark US crude, or the grade of oil, most commonly cited when oil prices are quoted.

surprise came in ARCO1's assessment of outlying scenarios. Although $20 WTI represented close to a five-year high, ARCO1 projected almost no possibility of a runaway downside (the lowest scenario envisioned was $14, and only a 0.25 probability was assigned to prices below $17), but a significant chance of a runaway upside (a 0.3 probability of prices of $22 or more, with a high value of $40). The mean values of its forecast distribution (by quarter), were $20.87, $20.62, $21.23, and $21.84. When Iraq invaded Kuwait and the world's consumers reacted with an embargo, we realized that one of the potential upside scenarios had been encountered. The underlying belief network was quickly revised, and the third- and fourth-quarter forecasts were recalculated as $25 and between $29 and $31, respectively. Actual average prices for 1990 (by quarter) were $21.70, $17.76, $26.30, and $31.91. Once again, although these tests can hardly claim to be complete, the initial performance of a belief-network-based system was extremely encouraging. For further details of this analysis, see Abramson and Finizza (1991).

Evaluation of knowledge-based systems remains difficult. By adopting (and adapting) well-accepted decision-analytic techniques and other mathematical/statistical constructs, systems based on belief networks may help advance the field. Incontrovertible studies, however, will take a while to develop. One particularly tantalizing possibility is the comparison of performance of belief-network-based systems with knowledge-based systems designed through competing paradigms (as well as human performance). Unfortunately, experiments of this scope are extremely difficult, expensive and time-consuming to design. One such attempt is currently underway at the National Oceanic and Atmospheric Administration (NOAA), whose Shootout project, first described in Moninger *et al.* (1989), is investigating and comparing the performance of severe weather forecasting systems. It may be several years, however, until this study has been completed. In the meantime, evaluations and comparisons will have to remain less formal.

SUMMARY

This chapter has discussed a new general paradigm for the design of knowledge-based systems. Belief networks are already recognized as precise mathematical models on which many interesting algorithms can be developed. They have also been forwarded as general mechanisms for managing the types of uncertainty with which all AI systems must cope. PATHFINDER and ARCO1 provide individual examples of the usefulness of belief networks as an overall framework for knowledge-based systems. This chapter has abstracted several principles of system design from specific instances of existing systems. Knowledge acquisition, knowledge representation, and information process-ing were all discussed; belief networks were shown to suggest clear approaches to all three. The phrase that best summarizes these approaches might well be 'form follows function': specific languages and tasks are selected to maximize the comfort of domain experts, knowledge sources are modelled as individualized nodes (whose only requirements are that they have well-defined inputs and outputs), and processing engines are designed to fit the specific types of question anticipated within the domain.

All of these approaches emerge from the beliefs that systems should be based on task analyses and designed to solve problems, and that the best way to do this is to think about the problem itself rather than the methods employed by human (expert) problem-solvers. The convergence of concerns among AI and DA researchers is a powerful one; belief network technology should help expand the range, power, and acceptability of knowledge-based systems.

ACKNOWLEDGEMENTS

This research was supported in part by the National Science Foundation under grant IRI-8910173. Discussions of PATHFINDER are based on information provided by two of its designers, Bharat Nathwani and David Heckerman. Discussions of ARCO1 are based on the author's personal experience as system designer; domain expertise was provided by Anthony Finizza, Peter Jaquette, Mikkal Herberg, and Paul Tossetti, of ARCO's Economic and Environmental Analysis group. A large part of the modelling software for both PATHFINDER and ARCO1 was developed by Keung-Chi Ng.

REFERENCES

Abramson, B. (1988) Towards a unified decision technology: areas of common interest to artificial intelligence and business forecasting. Technical Report 88-01/CS, University of Southern California.

Abramson, B. and Finizza, A.J. (1991) Using belief networks to forecast oil prices. *International Journal of Forecasting*, **7**(3), 299–316.

Cheeseman, P. (1988) An inquiry into computer understanding. *Computational Intelligence*, **4**(1), 58–66, 129–142.

Edwards, W., Lindman, H. and Savage, L.J. (1963) Bayesian statistical inference for psychological research, *Psychological Review*, **70**(3), 193–242.

Heckerman, D.E. (1988) An empirical comparison of three inference methods, in *Proceedings of the Fourth Workshop on Uncertainty in Artificial Intelligence*, 158–169.

Heckerman, D.E. (1990) Probabilistic similarity networks. PhD thesis, Stanford University.

Heckerman, D.E., Horvitz, E.J. and Nathwani, B.N. (1990) Toward normative expert systems: the Pathfinder project. Technical Report KSL-90-08, Stanford University.

Henrion, M. (1988) Propagating uncertainty in Bayesian networks by probabilistic logic sampling, in *Uncertainty in Artificial Intelligence 2*, L. Kanal and J. Lemmer (eds), Amsterdam: North-Holland, 149–163.

Horvitz, E.J., Breese, J.S. and Henrion, M. (1988) Decision theory in expert systems and artificial intelligence. *International Journal of Approximate Reasoning*, **2**, 247–302.

Howard, R.A. and Matheson, J.E., eds (1983) *Readings on the Principles and Applications of Decision Analysis*, Menlo Park, CA: Strategic Decisions Group.

Howard, R.A. and Matheson, J.E. (1984) Influence diagrams, in *Readings on the Principles and Applications of Decision Analysis, Vol. II*, R.A. Howard and J.E. Matheson (eds), Menlo Park, CA: Strategic Decisions Group, 721–762.

Kahneman, D., Slovic, P. and Tversky, A., eds (1982) *Judgement Under Uncertainty: Heuristics and Biases*, Cambridge: Cambridge University Press.

Moninger, W.R., Flueck, J.A., Lusk, C. and Roberts, W.F. (1989) SHOOTOUT-89: a comparative evaluation of knowledge-based systems that forecast severe weather, in *Proceedings of the Fifth Workshop on Uncertainty in Artificial Intelligence*, 265–271.

Ng, K.-C. and Abramson, B. (1990) Uncertainty management in expert systems. *IEEE Expert*, **5**(2), 29–48.

Pearl, J. (1988) *Probabilistic Reasoning in Intelligent Systems*, Palo Alto, CA: Morgan Kaufmann.

Shachter, R.D. (1986) Evaluating influence diagrams. *Operations Research*, **34**, 871–882.

Shachter, R.D. (1988) Probabilistic inference and influence diagrams. *Operations Research*, **36**, 589–604.

von Winterfeldt, D. and Edwards, W. (1986) *Decision Analysis and Behavioral Research*, Cambridge: Cambridge University Press.

Wright, S. (1921) Correlation and causation. *Journal of Agricultural Research*, **20**, 557–585.

Lack-of-information-based control in graphical belief systems

8

R.G. Almond

THE BELIEF FUNCTION MODEL

Dempster (1968) suggested a method of inference using upper and lower probabilities, which was extended and named **belief functions** by Shafer (1976). These models provide many simple expressions of familiar concepts from statistics, logic and natural language (see Almond, 1991). They are based on complementary set functions $\mathrm{BEL}(A)$ and $\mathrm{PL}(A)$, the **belief** and **plausibility**, respectively, which provide lower and upper bounds on the probability that the outcome will lie in the set A. When $\mathrm{BEL}(A) = \mathrm{PL}(A)$, then the belief function behaves like an ordinary probability distribution and is said to be **Bayesian**. Corresponding to the upper and lower probabilities, upper and lower expectations, $\underline{E}(U)$ and $\bar{E}(U)$, can be defined for an unknown variable U.

Actually, not all upper and lower probability functions are belief functions. To move from upper- and lower-bound probabilities to belief functions requires additional regularity constraints; in particular, we must explicitly assume the following constraint:

$$\mathrm{BEL}\left(\bigcup_{i=1}^{k} A_i\right) \geqslant \sum_{i=1}^{k} \mathrm{BEL}(A_i) - \sum_{i=1}^{k} \sum_{j=i+1}^{k} . \,\mathrm{BEL}(A_i \cup A_j)$$

$$+ \cdots + (-1)^{n+1}\mathrm{BEL}\left(\bigcap_{i=1}^{k} A_i\right) \qquad \forall k$$

This condition, known as *k*-**monotonicity**, allows the possibility of an alternate representation of belief functions, the **mass function**. The mass function plays a similar role in the theory of belief functions to that of the probability function ordinary probability theory. However, where the probability function maps a single outcome into the interval $[0, 1]$, the mass function maps an arbitrary set of outcomes (for the sake of simplicity we will only consider finite frames of discernment (outcome spaces)) onto the interval $[0, 1]$. The belief and plausibility are then defined in terms of this mass function as follows:

Artificial Intelligence Frontiers in Statistics: AI and statistics III. Edited by D.J. Hand. Published in 1993 by Chapman & Hall, London. ISBN 0 412 40710 8

Definition 8.1

Let Θ be a frame and let $m(\cdot)$ be mass function over that frame. Let $B \subseteq \Theta$. Then the **belief function** corresponding to m is defined as follows:

$$\text{BEL}(B) = \sum_{A \subseteq B} m(A) \qquad (8.1)$$

Definition 8.2

Let Θ be a frame and let $m(\cdot)$ be mass function over that frame. Let $B \subseteq \Theta$. Then the **plausibility function** corresponding to m is defined as follows:

$$\text{PL}(B) = \sum_{A \cap B \neq \varnothing} m(A) \qquad (8.2)$$

The subsets of the outcome space which have non-zero mass are called **focal elements**. Notice that when the focal elements are all singleton sets corresponding to individual outcomes, the mass function looks very much like a probability function. In fact, it is exactly when this occurs that a belief function is Bayesian.

Consider the following very simple decision problem. The Bahia Mar Hotel has a policy that if the chance of rain is greater than 50% they will move indoors banquets scheduled for outdoors. (This example is based on a conversation overheard between Bill DuMouchel and the hotel sales manager.) Suppose that the weatherman appears on television and says that there is a 40–60% chance of rain tonight. What should the hotel do? If the weatherman had said 'The chance of rain tonight is 60–70%', they would move the banquet indoors; if he had said 'There is less than 10% chance of rain tonight', they would stay outdoors. The lack of decisiveness in the implied model: BEL (rain tonight) = 0.4, PL(rain tonight) = 0.6, is a fundamental problem with belief functions. On the other hand, if the hotel staff were fully Bayesian, they would construct some arbitrary model for the probability of rain tonight (such as a beta over the interval [0.2, 0.4]) and make their decision based on the expected chance of rain tonight. This method has the drawback of requiring a more elaborate model and basing a decision on that, often arbitrary, model.

More generally, consider a simple belief function decision problem. Imagine that there are two alternatives: a deterministic alternative, with fixed utility 0, and a random alternative, with unknown utility U. If $\underline{E}(U) > 0$, then the random alternative is clearly better; if $\overline{E}(U) < 0$, then the fixed alternative is clearly better. In either case, we are able to make a strong decision. If $\underline{E}(U) < 0 < \overline{E}(U)$ then it is unclear which alternative is better and the model is too weak. If the model were Bayesian, then a decision could be made on the basis of the expected utility $E(U)$ and there always exists a strong (Bayes) decision.

On the other hand, a strong decision made with a weak belief function model has certain advantages. While the Bayesian model identifies the best decision under one specific prior, the Belief function model identifies the best decision under a whole class of priors (those falling within the belief and plausibility bounds, subject to the k-monotonicity restraint). Thus, if a strong decision can be made with a weak belief function, it might represent the joint decision of a class of decision-makers (e.g., the members of a committee).

VAGUENESS AND DECISION DISTANCE

The preceding section suggests that a belief function model is weaker than a Bayesian model, and that this weakness might be exploited. To measure that weakness, define the function $V(\mathrm{BEL}_U) = \bar{E}(U) - \underline{E}(U)$, the **vagueness** of the belief function BEL_U. A Bayesian belief function will always have vagueness 0, and a vacuous belief function will have vagueness as large as possible for the problem domain.

Although vagueness as defined in the preceding section is well defined for belief functions over utilities (or other random variables), it is not well defined for arbitrary belief functions. Knowledge of the strength of quite arbitrary belief functions is necessary for criticism (and control, as we will see later) of complex models. Therefore, we must produce some measure of the strength of an arbitrary belief function.

Let us consider a belief function BEL, defined over a frame Θ, and let θ be any particular outcome. Consider the decision about whether or not θ occurs. Let the utility of θ occurring be 1 and the utility of $\Theta - \{\theta\}$ be zero. Then we have that the θ-vagueness of a particular belief function is:

$$V_\theta(\mathrm{BEL}) = \mathrm{PL}(\{\theta\}) - \mathrm{BEL}(\{\theta\}) \tag{8.3}$$

We can therefore define the θ-vagueness for primitive event in the frame Θ.

Now consider a set of weights $R = \{r_\theta : \theta \in \Theta\}$ (which may or may not sum to 1). We can use these weights to produce a weighted average of the θ-vaguenesses. We will call such a weighted average the **R-weighted vagueness** or **R-vagueness** and define it as:

$$V_R(\mathrm{BEL}) = \sum_{\theta \in \Theta} r_\theta V_\theta(\mathrm{BEL}) = \sum_{\theta \in \Theta} r_\theta[\mathrm{PL}(\{\theta\}) - \mathrm{BEL}(\{\theta\})] \tag{8.4}$$

Equivalently, this quantity can be defined in terms of the mass function:

$$V_R(\mathrm{BEL}) = \sum_{\substack{A \subset \Theta \\ |A| > 1}} m(A) \sum_{\theta \in A} r_\theta \tag{8.5}$$

Of course, the previous equations depend on the existence of the weights. Some choices of weights may be more useful than others. One obvious choice of weight is to use a uniform distribution, that is $r_\theta = 1/|\Theta|$. This would define a uniformly weighted vagueness. A far more useful choice may depend on the utility of the various outcomes and how far the belief and plausibility are from the decision point.

Let us consider a very simple decision problem. There are two decisions, d_1 and d_2, and two random outcomes θ_1 and θ_2. The utility, $u(d, \theta)$, has four values: $u(d_1, \theta_1) = u_{11}$, $u(d_2, \theta_1) = u_{21}$, $u(d_1, \theta_2) = u_{12}$, and $u(d_2, \theta_2) = u_{22}$. Let $p = P(\theta_1)$ be the probability that θ_1 occurs, and express the expected utility associated with decision d_i by $U(d_i, p) = pu_{i1} + (1-p)u_{i2}$. There are two cases: either one of the decisions

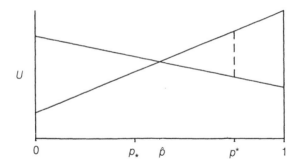

Figure 8.1 Plot of utility functions $U(d_1, p)$, $U(d_2, p)$.

has uniformly maximum utility over all values of p (in which case we select that decision), or else there exist a $\hat{p}$ such that for $p \leqslant \hat{p}$ one decision is better, and for $p > \hat{p}$ the other is better. We will assume that there does not exist a uniformly maximum utility, and that d_1 is the optimal decision for $p \leqslant \hat{p}$. These utility functions are shown in Fig. 8.1.

Now suppose our information about p is represented by a Bayesian probability distribution. The decision problem then reduces to finding $E(p)$; if $E(p) \leqslant \hat{p}$ then choose d_1, and if $E(p) > \hat{p}$ then choose d_2. Now suppose our information about p is given by a belief function. Let $E_*(p) = p_*$ and $E^*(p) = p^*$. We assume that we are in the weak decision case, that is $p_* < \hat{p} < p^*$. Suppose we make the decision which is conservative with respect to θ_1, that is, we assume that $p = p_*$, and the actual expected value of p is p^*, and compute the loss incurred by assuming the correct decision was d_2 instead of d_1. That loss—the **conservatism loss** with respect to θ_1—is:

$$r_{\theta_1} = U(d_2, p^*) - U(d_1, p^*) \tag{8.6}$$

This is shown as the dotted line in Fig. 8.1.

The conservatism loss as defined in Equation (8.6) forms a set of weights that can be used to form a weighted vagueness (Equation (8.4)). The conservatism-loss-weighted vagueness has some interesting properties. If a strong decision exists ($p_* < p^* < p$ or $p < p_* < p^*$) then the R-vagueness will be zero. More generally, the conservatism-loss-weighted vagueness will be a measure of how weak a particular belief function is with respect to a particular decision problem. Therefore, we define the **decision distance** of a belief function to be the r-vagueness (Equation (8.4)) using the conservatism loss (Equation (8.6)) as weights.

The extension of these results to decisions with more than two critical outcomes is not straightforward and will be left for future work. As the results should be extensions of similar Bayesian results, Braga-Illa (1964) might prove a good starting place. Instead, consider a decision problem over a single binary outcome where that binary outcome is a single variable in a complex multivariate model represented by a graphical model. The rest of this chapter focuses on how the conservatism loss weights and decision distance propagate through the graphical model.

GRAPHICAL MODELS

Complex belief functions are usually expressed as graphical belief functions. Let $\mathscr{A}$ be a collection of variables and let $\mathscr{C}$ be a collection of sets of those variables. For every $C \in \mathscr{C}$ let BEL_C be a belief function describing the relationships among the variables in C. Let $\mathscr{B} = \{\mathrm{BEL}_C | C \in \mathscr{C}\}$. The triple $\mathscr{G} = \langle \mathscr{A}, \mathscr{B}, \mathscr{C} \rangle$ describes a graphical belief model with the corresponding graphical belief function:

$$\mathrm{BEL}_{\mathscr{G}} = \bigoplus_{C \in \mathscr{C}} \mathrm{BEL}_C \tag{8.7}$$

where $\oplus$ is the combination operator (Kong, 1986). The pair $\langle \mathscr{A}, \mathscr{C} \rangle$ forms a hypergraph describing the relationship among the variables.

For computational purposes (Dempster and Kong, 1988; Almond, 1989) it is necessary to eliminate all of the cycle from this graphical structure. This is done by reforming the model into a tree model, where the nodes of the tree are collections of variables and edges connect nodes with maximal overlap. The set of nodes $\mathscr{N}$ of the tree model contains all of the original edges of the graphical model ($\mathscr{C}$) plus some additional nodes which represent filled-in cycles in the hypergraphical structure. The set of belief functions is extended by defining, for all $N \in \mathscr{N}$, $\mathrm{BEL}_N = \mathrm{BEL}_C$ if $N = C$ for some $C \in \mathscr{C}$, otherwise BEL_N is vacuous. In this formation:

$$\mathrm{BEL}_{\mathscr{G}} = \bigoplus_{N \in \mathscr{N}} \mathrm{BEL}_N \tag{8.8}$$

However, local computation strategies (like the fusion and propagation algorithm) can be used to determine margins of the graphical distribution. In particular, such strategies can be used to find the utility of a decision.

If it is not possible to make a strong decision using a given graphical belief model, this suggests that the model is too vague and should be strengthened. This could be done in a graphical belief model by refining any of the components of the model. Tracing the vagueness in utility (the decision distance) back through the tree of cliques should suggest portions of the model whose refinement will most greatly influence the weakness of the model. For example, consider the portion of the tree shown in Fig. 8.2. The vagueness in T can be broken down into contributions from A, contributions from B and contributions from the A, B, T relationship.

Suppose that in addition to its normal message telling about its local value, the node T (possibly containing several variables) sends a message R_T describing the conserv-

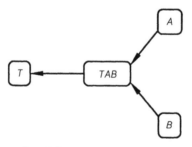

Figure 8.2 Influence from A and B.

ative loss r_{t_i} for each outcome $t_i \in T$. This is passed as an array from the node T to the node TAB. Meanwhile, the nodes A and B have passed messages $\text{BEL}_{A \Rightarrow TAB}$ and $\text{BEL}_{B \Rightarrow TAB}$ to the node TAB which contains local information BEL_{TAB}. The fusion and propagation algorithm combines these two messages with the local information to make a message to send on to the node T as follows:

$$\text{BEL}_{TAB \Rightarrow T} = (\text{BEL}_{TAB} \oplus \text{BEL}_{B \Rightarrow TAB} \oplus \text{BEL}_{A \Rightarrow TAB})_{\downarrow T} \qquad (8.9)$$

where $\downarrow$ is the projection operator. The R_T-vagueness of $\text{BEL}_{TAB \Rightarrow T}$ is the contribution of the information from A, B and TAB to the vagueness about T; call that vagueness v_T. We now must decide how to assign the decision distance up among the two contributors to vagueness about T, A and B.

Let the possible values for node A be $\{a_1, \ldots, a_n\}$, and define for each a_i a belief function BEL_{a_i}, the deterministic belief function asserting that a_i is true. Let $\text{BEL}_{a_i \uparrow TAB}$ be the minimal extension of that belief function onto the frame TAB. Replacing $\text{BEL}_{A \Rightarrow TAB}$ with $\text{BEL}_{a_i \uparrow TAB}$ in Equation (8.9) is equivalent to conditioning on a_i; we therefore define:

$$\text{BEL}_{TAB \Rightarrow T | a_i} = (\text{BEL}_{TAB} \oplus \text{BEL}_{B \Rightarrow TAB} \oplus \text{BEL}_{a_i \uparrow TAB})_{\downarrow T} \qquad (8.10)$$

Intuitively, the R_T-vagueness of this belief function should be smaller than v_T, thus

$$r_{a_i} = v_T - V_{R_T}(\text{BEL}_{TAB \Rightarrow T | a_i}) \qquad (8.11)$$

Equations (8.10) and (8.11) define a set of weights $R_A = \{r_{a_i} | a_i \in A\}$. We can then examine the R_A-weighted vagueness of $\text{BEL}_{A \Rightarrow TAB}$; this is the decision distance for A.

While it is true that r_{a_i} could be less than zero, this would occur only in situations in which knowing that a given event in A occurs would increase the uncertainty in T. For example, if we already knew that A was a_1 then the situation in which A was a_2 might be worse. When the decision distance for A is zero or negative, further specification of A may increase the vagueness with respect to the decision problem and is hence undesirable.

Equations (8.10) and (8.11), with the labels A and B exchanged, define weights R_B and the decision distance for B. The generalization of more complicated trees with more source nodes is straightforward. In the standard fusion and propagation algorithm $\text{BEL}_{B \Rightarrow TAB}$ in Equation (8.9) would be replaced with a sum over all the incoming messages except the one from T and the one from A. The analogous replacement is made in Equation (8.10) to produce the weights R_A, and the weights for the other nodes are produced similarly.

The weights R_A and R_B are passed to the nodes A and B, respectively. If the node A is a leaf of the tree, then we can calculate the decision distance associated with that node. If the node A is the root of a subtree, then we are in an analogous situation to the one pictured in Figure 8.2 and we can continue passing conservative loss weights up the tree. Finally, we choose the leaf with the maximum decision distance to instantiate or otherwise refine our beliefs about.

Note that we have only calculated the vagueness attributable to the univariate leaf nodes, and not to the multivariate nodes representing complex relationships. Here we find that vagueness is not so easily or usefully defined. Usually nodes that represent relationships are given belief functions which are deliberately vague about the margins.

Furthermore, some of those relationships are given vacuous belief functions because they represent filled-in cycles in the original model; such belief functions will always be vague. Although the vagueness of this belief function is important for model criticism, it is not necessary for control of expert systems, and so we leave it for future work.

CONTROL USING VAGUENESS

Consider an expert system with an embedded graphical belief model. We can think of one of the variables as the **decision variable**. The goal of the expert system is to reduce the vagueness of the decision variable to the point where a strong decision can be made. (In practice, the decision variable may be an artificial construct involving the ability to make one of several decisions). Certain variables in the graphical model are observable quantities, possibly background variables or test results. The expert system will try to make a decision on the basis of those observable variables. As the observable variables are represented by leaf nodes in the tree model, we call them **observable leaves**.

Rule-based expert systems have two types of evidence propagation: **forward chaining** and **backward chaining**. Forward chaining takes a piece of evidence and derives as many consequences as possible from those pieces of evidence. In the graphical model-based expert system, forward propagation is carried out by the fusion and propagation algorithm, which propagates the consequences of an observation to the rest of the model. Note that the fusion and propagation model can absorb the impact of several pieces of information at the same time. This can be used to initialize the expert system, by absorbing the effects of all pieces of evidence known at the outset. For example, in a medical diagnosis system, the doctor will have a list of symptoms that the patient complains of and a list of background variables.

Backward chaining is a process whereby the expert system tries to find a variable whose instantiation would enable it to reach a goal (or subgoal). The expert system works backwards through the rule base until it discovers a variable which can be observed, and then it queries the user for that node's value. For example, in a medical diagnostic system, it may query the doctor about a particular symptom or bit of history, or recommend a test. An expert system based on a graphical belief model would select the observable leaf with the smallest decision distance and query the user about its value. Updating information about this node would, in theory, best enable the expert system to make a strong decision.

Note that the entire tree does not need to be searched to find the maximum decision distance leaf. Instead a branch-and-bound search could be used. At the start label all nodes of the tree unexpanded. First label the decision node (the leaf node containing the margin of interest) as expanded and calculate the decision distance weights, propagating the weights to its neighbour in the tree model. If the node N^* receives weights from N^+, which is closer to the root of the tree, call the value $v_{N+} = V_{RN+}$ ($BEL_{N^* \to N+}$) the decision distance associated with the node N^*. Now at each step, check all unexpanded nodes that have decision distance weights, and select the one that has the largest decision distance. If it is a leaf node, stop and report that node as the next to be instantiated (i.e. query the user about its value). If it is not a leaf node, expand it by marking it as expanded and propagating conservative loss weights towards all of its neighbours further away from the root.

COST AND OTHER EXTENSIONS

The procedure as described above will work very well for engineering or scientific model-building and refinement. Tracing back the vagueness in the system will suggest a component belief function or a collection of components which are the biggest contributors to the weakness of the model. This, then, suggests that studying the systems those components represent—either gathering more data, or more carefully gathering expert opinion—will increase the decision-making power of the system.

For a production expert system, other factors, such as costs, should be taken into consideration. For example, imagine a medical diagnosis system. The doctor enters some background information on the patient and the symptoms. The machine then decides if, on the basis of the information given, a strong decision (diagnosis or treatment plan) can be made. If not, the computer traces back the vagueness in the model to one of the leaf nodes (a bit of background information or a test result). Using that information, the computer suggests a test that will provide the most information. Of course, medical tests are unequal in cost, and an inexpensive test which yields a little information may be preferred to an expensive test which yields a lot of information. In particular, checking patient history is usually much less expensive than a biopsy. The expert-system control structure would continue suggesting choosing them by weighing the cost–benefit structure of each test until a strong decision can be made.

Vagueness is just one possible measure of utility (or quasi-utility). New results by Shenoy (1990) show how utilities and uncertainties can be mixed in the same graphical model. This will open up the possibility of other rigorous control strategies based on information and on lack of information.

REFERENCES

Almond, R.G. (1989) Fusion and propagation in graphical belief models: an implementation and an example. PhD thesis; Department of Statistics Technical Report S-130, Harvard University.

Almond, R.G. (1991) Building blocks for graphical belief models. *Journal of Applied Statistics*, **18**, 63–76.

Braga-Illa, A. (1964) A simple approach to the Bayes choice criterion: the method of critical probabilities. *Journal of the American Statistical Association*, **59**, 1227–1230.

Dempster, A.P. (1968) A generalization of Bayesian inference (with discussion). *Journal of the Royal Statistical Society*, B, **30**, 205–247.

Dempster, A.P. and Kong, A. (1988) Uncertain evidence and artificial analysis, *Journal of Statistical Planning and Inference*, **20**, 355–368.

Kong, A. (1986) Multivariate belief functions and graphical models. PhD thesis; Department of Statistics, Technical Report S-107, Harvard University.

Shafer, G. (1976) *A Mathematical Theory of Evidence*, Princeton, NJ: Princeton University Press.

Shenoy, P.P. (1990) Valuation-based systems for Bayesian decision analysis. Working Paper no. 220, School of Business, University of Kansas, Lawrence, KS. (Derivative paper presented in Chapter 11, this volume.)

Adaptive importance sampling for Bayesian networks applied to filtering problems

9

A.R. Runnalls

INTRODUCTION

Integrated aircraft navigation systems are concerned with combining data from various navigational sensors to arrive at a refined estimate of the position, velocity and other kinematic parameters of one or more air vehicles, along with error bounds for these estimates. The most frequent problem is one of filtering, i.e. estimating the state of a system at a given time based on observational data up to that time, although problems of prediction and smoothing also arise. Consequently, since the 1970s the field has become one of the classic areas of application of Kalman filters and kindred linear smoothing techniques.

However, in many cases the performance of these filters has been disappointing, and many more recent applications are poorly adapted to the framework presupposed for the Kalman filter. Consequently, a major concern of the current research is with exploring novel ways of handling these problems, exploiting the possibilities of parallel computing, and thus freeing the data fusion engineer to utilize a wider range of error models for the sensors, and also allowing the sensor data to be combined with—or interpreted in the light of—information of a less immediately statistical character, such as might emerge from a knowledge-based system for mission management. A broader Bayesian framework is an attractive means for achieving this, and in particular the problems lend themselves to representation in the form of a Bayesian network.

THE PROBLEM

We consider a system $\mathcal{X} = \{X_1, \ldots, X_n\}$ of random variables (nodes) whose dependencies are expressed by means of a directed acyclic graph, thus constituting what is known as an influence diagram or **Bayesian network**. With each node X_i in the system is associated a set, possibly empty, of **parent** nodes; we shall use the notation $\mathcal{P}_i$ to

Artificial Intelligence Frontiers in Statistics: AI and statistics III. Edited by D.J. Hand. Published in 1993 by Chapman & Hall, London. ISBN 0 412 40710 8

denote the set of parents of X_i. We shall assume in this chapter that the X_i are each real-valued random variables, and that for each i we are given $f_i(x_i|\mathcal{P}_i)$, the probability density function for node X_i conditional on the values of the parents of X_i. For a Bayesian network, the joint density of the node values is given by

$$f(x_1,\ldots,x_n) = \prod_{i=1}^{n} f_i(x_i|\mathcal{P}_i) \tag{9.1}$$

In the problems with which we shall be dealing, certain of the nodes in the network will represent measurements to hand, and have known values. In its most general terms, the task then is to determine the expected values of one or more (integrable) functions $T(x_1,\ldots,x_n)$ of the node values, conditional upon these measured values. In other words, if $\mathcal{Z}$ is the set of measured nodes, the task is to determine the expectation

$$E(T|\mathcal{Z}) = \frac{\int_{-\infty}^{\infty}\cdots\int_{-\infty}^{\infty} Tf\Pi_{x_i\notin\mathcal{Z}}\,dx_i}{\int_{-\infty}^{\infty}\cdots\int_{-\infty}^{\infty} f\Pi_{x_i\notin\mathcal{Z}}\,dx_i} = \frac{\mathbf{N}}{\mathbf{D}} \tag{9.2}$$

Typically, the function T with which one is concerned will not depend on the value of every node in the network, but will depend rather on a set—often quite small—of nodes of interest, a point we shall exploit in a later section. (In filtering problems, for example, T will generally depend only on the most recent values of some of the state variables.)

T is a known function, and f is known in virtue of Equation (9.1), so the problem therefore boils down to evaluating the two integrals $\mathbf{N}$ and $\mathbf{D}$. In the problems with which we are concerned, these integrals are not amenable to direct analytical solution. This chapter will explore a technique for determining these integrals by Monte Carlo integration using adaptive importance sampling, using as the importance sampling function a multivariate Gaussian distribution. The importance sampling function is itself implemented as a Bayesian network which is topologically derived from the original network, and thus exploits the conditional independence relationships of the original network. Not only does this radically reduce the number of parameters of the importance sampling function, it also yields advantages in processing observational data sequentially.

AN EXAMPLE: 'CYCLOPS'

We shall illustrate the technique by means of a simple—albeit rather artificial—example illustrated in Fig. 9.1. In the figure, the solid line represents the motion of an object in the plane. Each component of the object's position is described by a non-stationary second-order Gaussian autoregressive process, the two processes being independent. Initially the x-component has mean -1.0 and standard deviation 0.29, while the y-component has mean 0 and standard deviation 0.58. After nine steps (the final time shown), the x-component has mean 1.0 and standard deviation 0.5, while the y-component has mean 0 and standard deviation 1.0. Figure 9.2 shows 25 sample tracks drawn from this prior distribution.

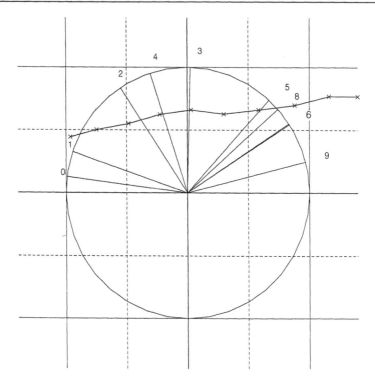

Figure 9.1 Cyclops example, showing the true path of the object and the measured bearings.

At each step in the object's motion (shown by the crosses), its bearing is measured with respect to the centre of the circle—as if by a Cyclopean eye—with the measurements being subject to angular errors uniformly distributed between ± 0.4 radians. The numbered radii in Fig. 9.1 show the actual bearing measurements obtained; it will be noted that several of the measurements, including those at times 3, 4 and 5, are seriously in error.

The problem is now to estimate the path of the object given the bearing measurements and the prior distribution for the motion process. The example is of interest because of the scantiness of the information available, and the non-linear and non-Gaussian nature of the bearing measurements.

The first stage in tackling the problem is to represent it as a Bayesian network; this is shown (for the simplified case of five time steps) in Fig. 9.3.

MAIN CONCEPTS

We shall use the notation $\mathbf{x}$ to denote a vector whose components are the x_i such that $X_i \notin \mathscr{X}$ (i.e. the non-measurements), and the notation $\check{f} = \check{f}(\mathbf{x})$ to denote the function which results from fixing to their known values the arguments x_i of f

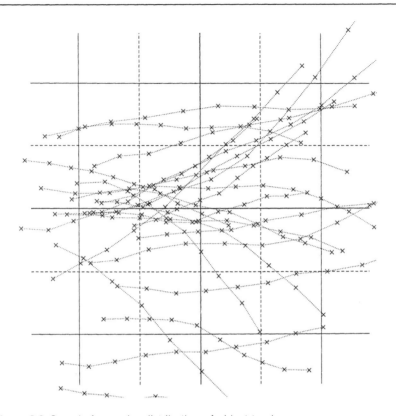

Figure 9.2 Sample from prior distribution of object tracks.

such that $X_i \in \mathcal{X}$. Evidently, $\check{f}/\mathbf{D}$ will be the joint probability density function for the unmeasured nodes, conditional on the measured nodes.

Let us consider first the evaluation of the denominator term $\mathbf{D}$; we shall consider the evaluation of the numerator term in a later section. To evaluate $\mathbf{D}$ by Monte Carlo integration with importance sampling, we choose an **importance sampling function** over the variables of integration, i.e. a pdf $g = g(\mathbf{x})$. We then draw from g a certain number N of pseudorandom samples $\mathbf{x}_k$ for $k = 1, \ldots, N$. $\mathbf{D}$ is then estimated as

$$\hat{\mathbf{D}} = \frac{1}{N} \sum_{k=1}^{N} \frac{\check{f}(\mathbf{x}_k)}{g(\mathbf{x}_k)} \qquad (9.3)$$

g should be chosen to keep the variance of $\check{f}(\mathbf{x}_k)/g(\mathbf{x}_k)$ as small as is practical; this entails, roughly speaking, that g should resemble $\check{f}$, up to a constant of proportionality. Clearly the choice of g depends on $\check{f}$, as well as on considerations of tractability. In some ways, an ideal choice for g would be $\check{f}/\mathbf{D}$: this would in fact reduce the Monte Carlo sampling variance to zero. Unfortunately we do not know $\mathbf{D}$ (cf. Hammersley and Handscomb, 1964: 58); moreover the functional form of $\check{f}$ may not lend itself

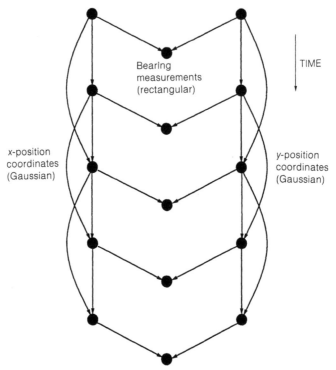

x-position
coordinates
(Gaussian)

Bearing
measurements
(rectangular)

TIME

y-position
coordinates
(Gaussian)

Figure 9.3 Representation of the Cyclops problem as a Bayesian network.

to the generation of pseudorandom variates. Nevertheless, the properties of $\check{f}/\mathbf{D}$ can be a guide in choosing a suitable g.

We take for our g a multivariate Gaussian distribution. In order to limit the number of parameters of g, however, we impose on it the same conditional independence relationships as are implied for $\check{f}$ by its network representation. This means that g can itself be represented as a Bayesian network, constructed by the following steps:

1. From the Bayesian network characterizing f (the **target** network), we construct another network, having identical topology, but in which each node represents a Gaussian approximation to the conditional distribution represented by the corresponding node in the target network. It is desirable that this Gaussian approximation err somewhat on the conservative side.

2. This new network is then topologically transformed into another network representing the same joint distribution, but in which:

   • any parent of a measurement node is itself a measurement;
   • any parent of a node of interest is either itself of interest or a measurement.

3. The distributions of the non-measurement nodes in the resulting network are conditionalized on the measurement values, and the measurement nodes deleted from the network.

The resulting network is known as the **auxiliary network** or **auxnet**, and defines the importance sampling function g.

Monte Carlo integration now proceeds by using the auxiliary network to generate sets of pseudorandom values—one value from each node of the network—which are jointly distributed according to g. These values are then injected into corresponding nodes of the target network, and the joint density $\check{f}$ evaluated using Equation (9.1). This is illustrated for the Cyclops problem in Fig. 9.4. The resulting ratios $\check{f}(\mathbf{x})/g(\mathbf{x})$ are then used in Equation (9.3). However, it is also possible to improve the efficiency of the importance sampling if, initially at least:

4. In the light of the observed ratios $\check{f}(\mathbf{x})/g(\mathbf{x})$ (for the pseudorandom $\mathbf{x}$ drawn from g), the parameters of the auxiliary network are perturbed so as to diminish the variance of this ratio. This perturbation is achieved by a method similar to the training of a neural network by back-propagation (cf. Rumelhart *et al.*, 1986).

AUXILIARY NETWORK

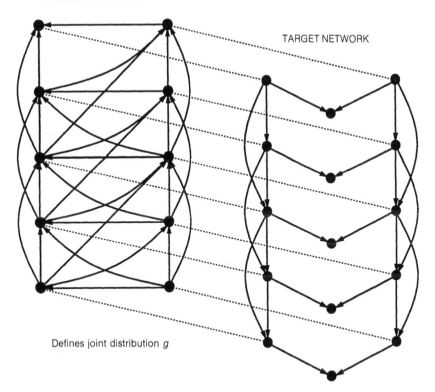

TARGET NETWORK

Defines joint distribution g

Defines joint distribution f

Figure 9.4 Use of the auxiliary network as an importance sampling function.

In the following sections we shall motivate these steps, describe them in somewhat greater detail, and illustrate them in the case of the Cyclops problem.

CONSTRUCTING THE AUXILIARY NETWORK: STEP 1

The objective here is to set up a Gaussian influence diagram, in the sense of Shachter and Kenley (1989). This means that each node must, conditional on the values of its parent nodes, have a Gaussian distribution with its mean a linear combination of the parent values plus a fixed offset, and a fixed[1] variance. Notice that each node in the constructed network requires only *local* information, i.e. information directly accessible to the corresponding node in the target network, namely, its own conditional distribution. This lends itself well to implementation using an object-oriented programming approach, and offers opportunities for parallelizing the computation.

The first step in constructing the auxiliary network for the Cyclops problem is shown in Fig. 9.5(a). The network topology is identical to that of the target network. The nodes of the target network representing the components of the object's position were already Gaussian, and they map onto identical nodes in the auxiliary network, i.e. nodes that have an identical distribution conditional on their respective parents' values. The nodes of the original network representing bearing measurements, which exhibit rectangularly distributed errors, cannot be carried across unchanged, however: instead, we have converted each one into a (univariate) Gaussian distribution whose mean is the component of the object's position (at the corresponding time) at right angles to the measured bearing, and whose standard deviation is proportional to the sine of the semi-angle of the error distribution. These nodes homologous to the bearing measurements are themselves regarded as being measured, each one with the value zero.[2] Thus, whereas the original bearing measurements give rise to likelihood functions whose contours are V-shaped pairs of lines symmetrically straddling the line of the measured bearing, their homologues in the auxiliary network give rise to likelihood functions whose contours are pairs of parallel lines symmetrically straddling the line of the measured bearing.

CONSTRUCTING THE AUXILIARY NETWORK: STEPS 2 AND 3

The purpose of these steps is to convert the result of Step 1 into a first-stab approximation of the distribution $\breve{f}/D$ of the non-measurement nodes conditional on the measurements, and at the same time—by placing the nodes of interest at the head of the ancestral tree—to set up the auxiliary network so that it is suited to computing $\mathbf{N}$ as well as $\mathbf{D}$ (see below). The topological transformations required can be achieved, for example, by the arc reversal procedure of Shachter and Kenley (1989), in which it is shown that if there is an arc from node A to node B, and no other directed path from A to B, then the network can be replaced by an equivalent network in which this arc is reversed, but in which A and B now share all their remaining parents.

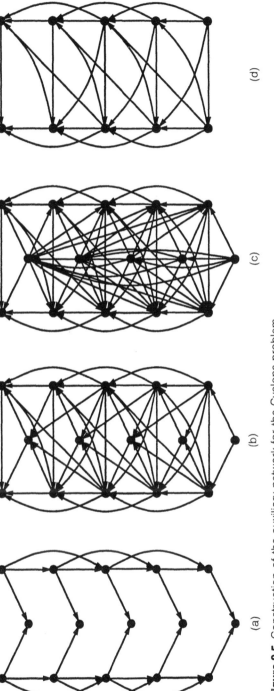

Figure 9.5 Construction of the auxiliary network for the Cyclops problem.

(a)

(b)

(c)

(d)

In Fig. 9.5 we show these transformations for the Cyclops problem in three stages. In Fig. 9.5(b) we have transformed the network in Fig. 9.5(a) into an equivalent network in which all the original arcs have been reversed. Then in Fig. 9.5(c) we further transform the network so that in no case is a measured node a child of a non-measured node. This completes Step 2. From this stage it is then straightforward to conditionalize the measurements out, yielding the network shown in Fig. 9.5(d). There are several reasons for reversing all of the arcs of the original network:

- In filtering applications, it is (by definition) the most recent values of some of the state variables that are the nodes of interest, so that it is desirable to place the corresponding nodes at the head of the network for the reasons brought out in our discussion of the estimation of **N** below.
- In future work, we envisage carrying out the computations described here in a sequential manner. As time proceeds and new measurements arrive, the target and auxiliary networks will be appropriately augmented. Keeping the auxiliary network in 'backward order' greatly reduces the number of arc reversals and other manipulations needed at each time slice to conditionalize the new measurements out of the auxiliary network.
- For the problems we have investigated, the backward arrangement of the auxiliary network also appears to be computationally economical for the training process described in the next section.

TRAINING THE AUXILIARY NETWORK: STEP 4

The variance of the estimator $\hat{\mathbf{D}}$ of Equation (9.3) is V/N, where

$$
V = \int_{-\infty}^{\infty} \cdots \int_{-\infty}^{\infty} \left(\frac{\breve{f}(\mathbf{x})}{g(\mathbf{x})} - \mathbf{D} \right)^2 g(\mathbf{x}) \, d\mathbf{x} \tag{9.4}
$$

The objective of Step 4 is to perturb the parameters θ_i of the auxiliary network, i.e. of g, as the integration proceeds so as progressively to diminish V. (These parameters comprise the node variances, the link weights, and the node mean offsets.) This is achieved by a steepest descent technique based on the derivatives $\partial V / \partial \theta_i$.

Now these derivatives $\partial V / \partial \theta_i$ will normally involve analytically intractable integrals just as much as **D** itself. However, if $\mathbf{x}$ is a random sample from g, it is readily established that, for any value of the constant k,

$$
\left(\widehat{\frac{\partial V}{\partial \theta_i}} \right) = - \left\{ \left(\frac{\breve{f}(\mathbf{x})}{g(\mathbf{x})} \right)^2 - k^2 \right\} \frac{1}{g(\mathbf{x})} \frac{\partial g}{\partial \theta_i} \bigg|_{\mathbf{x}}
$$

is an unbiased estimator of $\partial V / \partial \theta_i$. As k is varied, the variance of this estimator attains a local maximum for $k = 0$; however, the value of k that minimizes the variance depends on f. However, an attractive possibility is to choose as k some estimate $\tilde{\mathbf{D}}$ of **D**.[3]

We shall not work through the details here, but the upshot is that following each drawing of a pseudorandom value $\mathbf{x}$ from g and calculation of the ratio $\breve{f}(\mathbf{x})/g(\mathbf{x})$, each parameter of g will be perturbed by amount proportional to

$$\eta\left\{\left(\frac{\check{f}(\mathbf{x})}{\tilde{\mathbf{D}}g(\mathbf{x})}\right)^2 - 1\right\}$$

where η is a small constant (cf. Rumelhart $et\ al.$, 1986). The constant of proportionality for each parameter can be determined by simple computations using data which are highly local to the node concerned; the method thus lends itself to parallel implementation.

The progress of this training process for the Cyclops problem is shown over 5000 iterations in Fig. 9.6, which relates to the node of the auxiliary network corresponding to the y-component of the object's position at time 0. It shows as the central line the marginal mean of the Gaussian distribution defined by the node, with the marginal standard deviation shown by the lines on either side.[4] The effect of the training process may be examined by considering the variance under g of $\check{f}/g$, i.e. the single-shot variance V of Equation (9.4). When estimated over 10 000 samples before training of the network, this variance was 4.88×10^{-3}, the corresponding estimate after training being 1.05×10^{-4}—a reduction by a factor of 46. Incidentally, the mean value of $\check{f}/g$ from the sample after training was 4.02×10^{-3}, so we may take as our estimate of the denominator term $\hat{\mathbf{D}} = 4.02 \times 10^{-3} \pm 5\%$.[5]

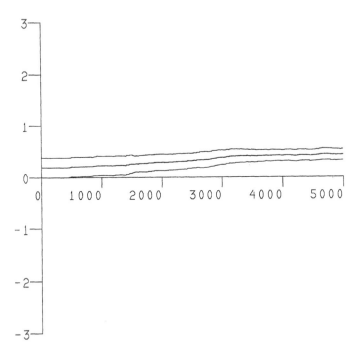

Figure 9.6 Evolution under the training process of the marginal distribution of node y_0.

The result of the training process is an importance sampling function which is well adapted to providing accurate estimates of functionals of the posterior distribution depending on the selected nodes of interest. Moreover, as a result of the training, the importance sampling function *in itself* provides a valuable summary of the characteristics of the posterior distribution, and its parameters can thus serve as a basis for informal reasoning about the system.

ESTIMATING N

One estimator for the numerator in Equation (9.2) is

$$\hat{N}_1 = \frac{1}{N'} \sum_{k=1}^{N'} \frac{T(\mathbf{x}_k) \breve{f}(\mathbf{x}_k)}{g(\mathbf{x}_k)}$$

where the $\mathbf{x}_k$, for $k = 1, \ldots, N'$, are drawn from g as before. This is the approach favoured in Kloek and Van Dijk (1978: 6); it has the merit of being straightforward to implement, and it is possible also to save some computing by using the same pseudorandom samples $\mathbf{x}_k$ for $\hat{N}_1$ and $\hat{D}$.[6] However, the variance of this estimator will generally be greater—often much greater—than the variance of the analogous estimator $\hat{D}$ of Equation (9.3). This is because g has been specifically chosen to resemble $\breve{f}$, with the objective of keeping the variance of $\hat{D}$ down; it has not, however, been chosen to resemble $T\breve{f}$.

An alternative approach would be to choose another importance sampling function, g_T, specifically for use with this integral. Instead of choosing g_T *ab initio*, however, it is worth considering whether we can derive a suitable function from g and T. Suppose that the function T is non-negative,[7] and that the integral of Tg over the appropriate state space (excluding $X_i \in \mathcal{L}$) exists and has a non-zero value K. Then Tg/K is a pdf. Suppose that K is known and that it is feasible to draw pseudorandom samples from the distribution Tg/K. Now

$$\mathbf{N} = \int_{\mathbb{R}^{n'}} T\breve{f}\,\mathrm{d}\mathbf{x} = K \int_{\mathbb{R}^{n'}} \frac{\breve{f}}{g} \frac{Tg}{K}\,\mathrm{d}\mathbf{x}$$

so an unbiased estimator of **N** is

$$\hat{N}_2 = \frac{K}{N'} \sum_{k=1}^{N'} \frac{\breve{f}(\mathbf{x}_k)}{g(\mathbf{x}_k)} \tag{9.5}$$

where the $\mathbf{x}_k$ are pseudorandom samples drawn not now from g but from Tg/K. Now the resemblance between $\breve{f}$ and g may well once again yield dividends in keeping the variance of $\hat{N}_2$ down.[8]

SAMPLING FROM *Tg*

Equation (9.5) required us to draw pseudorandom samples from the distribution with pdf Tg/K. In this section we consider how to choose the topology of the auxiliary network that defines g so as to facilitate this.

The important point is that, as we remarked earlier, the function(s) T of interest will typically not depend on all of the non-measurement variables in the Bayesian network, but on only relatively few 'nodes of interest'. Suppose (without loss of generality) that of the n' non-measurement nodes, T depends only on $X_1, \ldots, X_m$ and that, following Step 2 as described above, the auxiliary network has been organized so that in no case is one of the nodes $X_{m+1}, \ldots, X_{n'}$ a parent of one of the nodes $X_1, \ldots, X_m$. We have two tasks: first to determine the integral K (preferably by analytic means), and second to draw pseudorandom samples from Tg/K. Now we can readily express g in the form

$$g(\mathbf{x}) = g(x_1, \ldots, x_m) \times \prod_{i=m+1}^{n'} g(x_i | \mathcal{P}_i^*)$$

where $\mathcal{P}_i^*$ denotes the set of parents of node X_i within the **auxiliary** network. ($\mathcal{P}_i$ referred to the target network.) Now evidently, in determining the integral K, the product term on the right-hand side of this equation will integrate out to unity, leaving us to determine merely the integral of $Tg(x_1, \ldots, x_m)$, where the second term is the marginal density (under g) of $x_1, \ldots, x_m$. Further, having determined K, if we can then produce a pseudorandom sample $(x_1, \ldots, x_m)$ from the joint density $Tg(x_1, \ldots, x_m)/K$, we can then readily augment this into a complete sample from Tg/K by drawing x_i from $g_i(x_i | \mathcal{P}_i^*)$ for $i = m+1, \ldots, n'$, in each case conditioning on the values already drawn.

This, then, is why the algorithm organizes the auxiliary network so that the nodes of interest are at the head of the ancestral tree.

Let us now illustrate the results of this section and the preceding one in connection with the Cyclops problem. Suppose that it is desired to determine the probability that, at the time of the final bearing measurement, the object is still within the unit circle. In other words, we wish to determine the expectation of T, where

$$T(\mathbf{x}) = \begin{cases} 1 & \text{if } x_9^2 + y_9^2 < 1 \\ 0 & \text{otherwise} \end{cases}$$

To apply the method, we need (i) to determine K, and (ii) to be able to draw samples from Tg. Let us take (ii) first. Referring to Fig. 9.5(d), we can see that in the auxiliary network, for the ten-time-step case, the node corresponding to y_9 will have no parents, while the node corresponding to x_9 will have only the y_9 node as a parent. We can therefore sample from Tg as follows:

1. Sample y_9 from its (unconditional) distribution as defined by the auxiliary network. Repeat this step until $|y_9| < 1$.
2. Sample x_9 from its distribution conditional on y_9 as defined by the auxiliary network. If $x_9^2 + y_9^2 > 1$, go back to Step 1.
3. Otherwise sample the remaining nodes in the auxiliary network from their conditional distributions, working in descending ancestral order.
4. The values thus generated are injected into the corresponding nodes of the target network, and the ratio $\check{f}/g$ calculated.

We continue in this fashion until we have the desired number of samples from Tg. Turning now to task (i), the estimation of K, we may recall that K is simply that probability that a sample from g will fall within the unit circle. This may readily be estimated by counting the number of samples we needed to reject in Steps 1 and 2 of the above procedure in order to obtain the desired number of samples within the unit circle, and applying a standard estimator for the parameter of a negative binomial distribution.

In the event, the mean of $\check{f}/g$ over 10 000 samples obtained as just described was 3.96×10^{-3}, with a sample standard deviation of 1.02×10^{-2}. 119 977 samples were rejected in Steps 1 and 2 to obtain the 10 000 good ones, so K may be estimated as $0.077 \pm 7\%$. Hence, from Equation (9.5), we obtain as our estimate of the numerator $\mathbf{N}$ the value $3.05 \times 10^{-4} \pm 12\%$. Combining this with our earlier estimate of the denominator, we arrive at the estimate $0.076 \pm 17\%$ for the probability that at time 9 the object still lies within the unit circle.

SAMPLING FROM THE POSTERIOR DISTRIBUTION

Finally, we show how the auxiliary network can be used to generate pseudorandom samples from the posterior distribution $\check{f}/\mathbf{D}$, using the method of Metropolis *et al.* (1953). It involves the construction of a Markov sequence, each of whose terms is a vector comprising values for each of the non-measured nodes of the target network (or equivalently, for each of the nodes of the auxiliary network). The construction of the sequence involves the use of a quantity that we shall call the **datum ratio**, initially set to be 0. Successive terms of the sequence are generated as follows:

1. Draw a set of sample values from g (one for each node), and compute the corresponding ratio $\check{f}/g$. We shall refer to this ratio as the **new ratio**.
2. If the new ratio is at least as great as the datum ratio, we accept the set of sample values as the new term of the Markov sequence, and set the datum ratio to equal the new ratio.
3. Otherwise, generate a random number uniformly distributed between 0 and 1. If the new ratio is at least as great as the datum ratio multiplied by this random number, we again accept the set of sample values from g as the new term of the Markov sequence, and set the datum ratio equal to the new ratio.
4. Otherwise, the new term of the sequence is set equal to the previous term, and the datum ratio is left unchanged.

Now it can readily be shown (see, for example, Hastings, 1970) that provided g assigns positive probability to every interval which has positive probability under $\check{f}$, the unique stationary distribution of this Markov chain is the desired posterior distribution $\check{f}/\mathbf{D}$. Hence a sample from the posterior distribution can be obtained by selecting terms from the Markov sequence, with the selected terms being sufficiently widely spaced as to achieve a good coverage of the sample space.[9]

In determining the efficiency of the above procedure, Hastings (1970) recommends considering the **rejection rate**, i.e. the frequency with which Step 4 of the generation

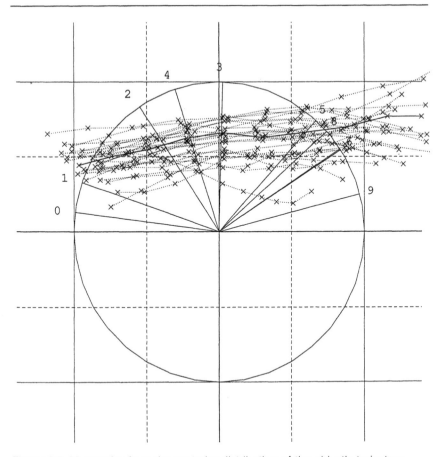

Figure 9.7 25 samples from the posterior distribution of the object's trajectory.

process is invoked. We recommend examining instead the *dwell* period, i.e. the number of successive terms in the Markov chain which are identical in virtue of Step 4 applying, and choosing the interval at which samples are selected from the Markov sequence so as to exceed the longest dwell. (The dwell can often be surprisingly high in relation to the rejection rate.)

Figure 9.7 shows 25 sample tracks obtained by this procedure for the Cyclops problem. 3750 terms of the Markov sequence were computed, with every 150th term being taken as a posterior sample. The rejection rate was 79%, and the longest dwell encountered was 88. This figure can usefully be compared with Fig. 9.2 showing the prior distribution. Notice in particular how well the true bearings at time Steps 3, 4 and 5 have been established—presumably a consequence of the rectangular error distribution and the fact that the measurements at these times have large errors of alternating sense. The direction of the object's travel is also well established.

CONCLUSIONS AND FURTHER WORK

A method has been described by which properties of the posterior distribution may be established for Bayesian problems represented in network (influence diagram) form by adaptive Monte Carlo integration. The method involves the construction of an auxiliary network, topologically related to the original network, which is used as an importance sampling function. As sampling proceeds, the parameters of the auxiliary network are perturbed by a method akin to the training a back-propagation neural network, so that the topography of the density function defined by the auxiliary network increasingly resembles that of the target posterior distribution, thus reducing the variance of the Monte Carlo integration process. The auxiliary network, thus trained, may also be used as the basis for simulating the posterior distribution by the method of Metropolis *et al.* (1953). These techniques have been illustrated for a simple tracking problem, and shown to yield promising results.

Besides applying this technique to a variety of more realistic problems, there are two specific avenues of research that we propose to pursue:

- As described in this chapter, each node in the auxiliary network is characterized by a (univariate) Gaussian distribution. Experience suggests that it would be desirable to use a distribution with heavier tails. The logistic distribution looks tractable in this regard. We also intend to consider the use of mixture distributions, so as to cope better with multimodal posteriors.
- In this chapter, the construction and training of the auxiliary network were each unitary operations, carried out in turn only after the target network was completely defined. We have been concerned throughout, however, that the technique should be modifiable so that these activities can be carried out in stages, as new measurement data come in. At each stage, first the target network will be expanded to include the new measurement data, then the auxiliary network will be augmented correspondingly, and finally the auxiliary network will be subjected to a further bout of training.

ACKNOWLEDGEMENTS

This chapter describes work carried out under a research contract sponsored by the Royal Aerospace Establishment, Farnborough, UK, entitled 'The Use of Artificial Intelligence Techniques in Aircraft Navigation'.

NOTES

1. Describing these values as 'fixed' here means simply that they are independent of the parent values. The values are subject to alteration in the training process of Step 4.
2. The choice of 0 here arises from the particular method of approximating the bearing nodes; it is not a general feature of the auxiliary network concept.
3. This estimate need not necessarily be obtained from Equation (9.3), though this, of course, would be an attractive choice.

4. Notice that, to begin with, the curve is absolutely flat: this is because the sampling process took some time to locate any point where $\check{f}$ was non-zero. This underlines the importance of obtaining a fairly good initial approximation to the posterior distribution.
5. Tolerances are given in this chapter as 2σ levels.
6. If T tends to be high where $\check{f}$ is high, this will also result in more accurate estimation. Conversely, if T tends to be high where $\check{f}$ is low, sharing the $\mathbf{x}_k$ will increase the variance of the estimator.
7. If this is not the case, similar conclusions can be drawn by treating the positive- and negative-going parts of T separately.
8. Not in all cases however. g will have been chosen to resemble $\check{f}$ under the weighting g, not under the weighting Tg/K. If, for example, Tg/K places all its weight on regions of the state space receiving little emphasis under g, then the variance of $\hat{N}_2$ may be considerable. In that case it will be necessary to choose an importance sampling function g_T by another method—perhaps an iterative method taking Tg as its point of departure.
9. With sufficiently wide spacing of the selected terms, the statistical dependency between successive samples can be rendered negligible. It is an advantage of our approach, in which a complete new sample is drawn from g for each term of the Markov sequence, that one does not run into the problems of 'almost reducible' Markov sequences which can arise in techniques such as the Gibbs sampler which rely on piecemeal modification of previous samples. See, for example, Pearl (1988: 222–223).

REFERENCES

Hammersley, J.M. and Handscomb, D.C. (1964) *Monte Carlo Methods*, London: Methuen.

Hastings, W.K. (1970) Monte Carlo sampling methods using Markov chains and their applications. *Biometrika*, **57**(1), 97–109.

Kloek, T. and Van Dijk, H.K. (1978) Bayesian estimates of equation system parameters: an application of integration by Monte Carlo. *Econometrica*, **46**, 1–19.

Metropolis, N., Rosenbluth, A.W., Rosenbluth, M.N., Teller, A.H. and Teller, E. (1953) Equations of state calculations by fast computing machines. *Journal of Chemical Physics*, **21**, 1087–1092.

Pearl, J. (1988) *Probabilistic Reasoning in Intelligent Systems*, San Mateo, CA: Morgan Kaufmann.

Rumelhart, D.C., Hinton, G.E. and Williams, R.J. (1986) Learning internal representations by error propagation, in *Parallel Distributed Processing, Vol. 1*, David E. Rumelhart and James L. McClelland (eds), Cambridge, MA: MIT Press, 318–362.

Shachter, Ross D. and Kenley C.R. (1989) Gaussian influence diagrams. *Management Science*, **35**, 527–550.

Intelligent arc addition, belief propagation and utilization of parallel processors by probabilistic inference engines

10

A. Ranjbar and M. McLeish

INTRODUCTION

Causal probabilistic networks (CPNs) have been widely recognized as suitable knowledge representation tools for several reasons: constructing CPNs is a practical task, their topology reflects independence relationships (independence semantics of knowledge), and they can store dependency measures. These networks, also known as **causal nets** and **Bayesian networks**, are generally directed acyclic graphs (DAGs) with nodes representing probabilistic (propositional) variables and arcs representing direct dependencies among the variables they connect.

Currently no sound algorithm has yet been proposed for propagation of belief through multiply connected causal nets (networks within which there might be more than one path from one variable to another) that does not make use of a secondary structure. Pearl's (1988) construction of the **join-tree** (also referred to as the **clique-tree**) of a CPN and using it as an inference engine (belief absorption and propagation medium) of an expert system, originally suggested in Lauritzen and Spiegelhalter (1988), is now one common approach to the problem. The join-tree of a DAG is a tree structure, whose nodes are cliques of the triangulated DAG. A non-triangulated graph is made chordal by adding chords which eliminate cycles of size greater than 3 within the graph.

Depending on the triangulation method employed—maximum cardinality search (Tarjan and Yannakakis, 1984); lexicographic search (Rose *et al.*, 1976); extended elimination (Fujisawa and Orino, 1974); recursive thinning (Kjaerulff, 1990), etc.— different triangulations (and hence, different join-trees) could result. Nevertheless, regardless of the triangulation method undertaken, the resulting join-tree always has

Artificial Intelligence Frontiers in Statistics: AI and statistics III. Edited by D.J. Hand. Published in 1993 by Chapman & Hall, London. ISBN 0 412 40710 8

the running intersection property. The task of probabilistic inference using belief networks is shown to be an NP-hard problem (Cooper, 1987). Different triangulations have been inspected and compared for their yield of total state space, their behaviour, and their possible superiority (Kjaerulff, 1990).

The material presented here are the results gained through an effort to improve the response time of an expert system that undertakes CPNs as the knowledge representation scheme and which employs the clique-tree of the triangulated CPN as the inference medium. Most of the techniques and ideas presented herein are independent of each other, i.e. one can use one or more of the suggested techniques if desired. First, an algorithm, which produces a (partial) topological ordering of the vertices of a DAG, called CPN-oriented topological ordering (COTO), is introduced. Second, a new algorithm for addition of chords to DAGs, the intelligent arc addition algorithm (IAAA), is introduced. This algorithm makes use of the results of the work done by Rose *et al.* (1976), Fujisawa and Orino (1974) and Kjaerulff (1990), and avoids adding redundant chords to a graph being triangulated. Third, a methodology called intelligent belief propagation (IBP) is illustrated; IBP is designed to allow faster response regarding one or more variables of the CPN, after absorption of some item(s) of evidence, by performing a limited propagation (as opposed to a global propagation) through the clique-tree. Fourth, utilization of multiple processors is discussed. This discussion reviews how independent tasks can be prepared and performed using multiple processors, if the dependencies among the tasks are recognized. Finally, the possibility of using other promising techniques and the merits of such implementations are presented.

NEW METHODS FOR ARC ADDITION AND BELIEF PROPAGATION

CPN-ORIENTED TOPOLOGICAL ORDERING

COTO is a variation of the well-known topological sorting algorithm (see Fig. 10.1). If the nodes of a directed acyclic graph were ordered according to the topological sorting algorithm, the rank (order) of each node would be larger than the rank of its parent nodes, and smaller than its descendant nodes. For this reason, topological sorting is often used for scheduling tasks within a project.

Topological sorting algorithm
Let us assume that a directed acyclic graph $G = (V, E)$ has $n > 0$ vertices. While there still are vertices with no outgoing edges that are not yet ranked (ordered) do: (1) choose one vertex with no outgoing edge and mark it n; (2) delete the node that has just been ordered (ranked) along with all its incoming edges (and $n := n - 1$). This process stops after all the vertices are ordered (ranked).

CPN-oriented topological ordering algorithm
Since we have altered the classical topological ordering algorithm slightly, and used it to obtain a partial ordering of the vertices of CPNs, the phrase 'CPN-oriented' has been

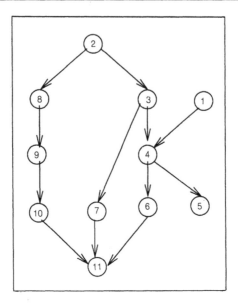

Figure 10.1 An ordering produced by the classical topological ordering algorithm (in 11 steps).

adopted to be added to the name 'topological ordering'. The COTO algorithm starts with the leaf nodes (i.e. the nodes with no outgoing arcs) and labels them $V, V - 1, V - 2, \ldots, V - i + 1$ (when there are i leaf nodes). Note that the order chosen among the leaf nodes is arbitrary. After labelling the leaf nodes, these vertices, along with their corresponding incoming arcs, are eliminated. Note that the word 'eliminated' here only means 'deleted' and should not be confused with terminology used in graph triangulation. It is a trivial task to prove that after each elimination step, among the remaining nodes there will be at least one with no outgoing edge (i.e. although the graph might become disconnected after an elimination step, each piece remains a DAG). The COTO algorithm stops after all the vertices of the graph have been labelled and eliminated.

The only difference between the COTO and the classical topological sorting algorithm is that while COTO labels and eliminates the leaf nodes in one step, the classical topological sorting labels and eliminates the leaf nodes one at a time. Therefore, the orderings produced by these two algorithms will NOT always be compatible. Figure 10.2 depicts the DAG of Fig. 10.1 with the ordering produced by the COTO algorithm. It must be noted that the classical topological ordering algorithm and COTO *may* produce the same ordering by coincidence, but in general they *do not* (compare Fig. 10.1 and 10.2(e)). At the first glance one might mistakenly think that COTO is merely a reverse breadth-first-search ordering. Interestingly enough, COTO may be viewed as a *bottom-up breadth-first search*. In order to keep the topological algorithm in $O(|V| + |E|)$, a search for the leaf nodes at each step must be avoided (Reingold *et al.*, 1977); this is also true for the COTO algorithm. Therefore, at each step

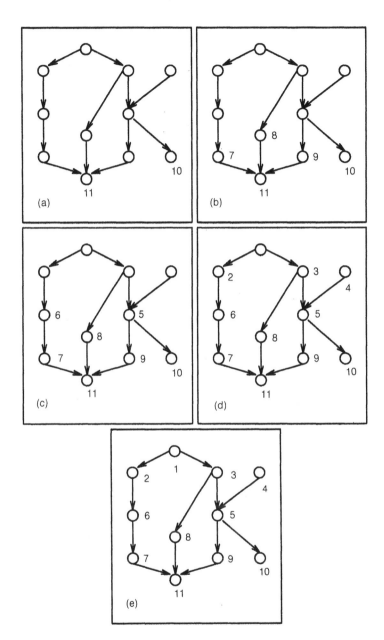

Figure 10.2 The step-by-step process of ordering the vertices of a DAG, using the COTO algorithm: (a) step 1; (b) step (2); (c) step 3; (d) step 4; (e) step 5 (complete).

of the labelling and elimination process, a list of the current leaf nodes and their parent nodes is necessary. This list can be gathered by performing a breadth-first search on the DAG before starting the COTO algorithm. The special property of COTO is that the ordering it produces for the nodes of a DAG reflects the topology of that DAG more closely than does any other ordering algorithm (e.g., classical topological or maximum cardinality). The nodes that are labelled in the same COTO step all have the same depth (considering depth as the number of arcs on the simple path between a node and its most distant 'leaf grandchild'). This fact reveals the reason for us to claim that COTO reflects the topology of the DAG closely.

INTELLIGENT ARC ADDITION ALGORITHM

This algorithm in its pure form is in the same category as the **minimum fill-in** algorithm. The ordering scheme used by this algorithm (for subsequent vertex elimination process) is COTO. It is hence obvious that IAAA is only applicable to DAGs. Furthermore, as opposed to other triangulation methods, instead of backward elimination, IAAA performs the vertex elimination process in the order produced by COTO. The idea behind this algorithm has sprung from Rose *et al.*'s (1976) definition of **redundant chords**. In other words, at each vertex elimination step, a fill-edge is added to the graph after it is confirmed that this edge will *not* be a redundant chord of a 4-cycle. Please refer to Kjaerulff (1990) for a thorough explanation of redundant arcs and removing them. Although this algorithm is not as fast as other triangulation algorithms such as maximum cardinality or lexicographic search, it is still worth using. The reason is that triangulation is only performed once (in the context of CPNs) and, although expensive, IAAA will save the possible future complexities (which repeat over and over) that 'bad' triangulations may cause. Moreover, since the ordering produced by COTO is very predictable, so is the triangulation produced by IAAA. The following is the formal pseudocode for IAAA:

Intelligent arc addition algorithm
- order the vertices of the DAG according to the COTO algorithm
- for vertex i ($i = 1 \ldots V$) do
  - $j := 0$ (counter for chords added in this iteration)
  - if vertex i has multiple children then
    - inspect children pairwise (say X, Y)
    - if X and Y are NOT joined then
      - if there is a chain (arc directions ignored) from X to Y that does not include vertex i then
        - if ((there exist unattached vertices V and W) OR (NO vertices V and W)) such that $\{V, W\} \subseteq \text{adj}\{X\} \cap \text{adj}\{Y\}$ then,
          - $j := j + 1$
          - add a direct arc (t_j) between X and Y with the direction of arc from the vertex with smaller label towards the vertex with larger label.

- while $j > 1$ do
  - delete the chords that intersect with t_j and are redundant
  - $j := j - 1$

Figure 10.3 shows a DAG and its corresponding minimal fill triangulation produced by IAAA.

If the direct children and the descendants of each vertex are found during COTO and stored at each node, no searching (large) will be required at the times when such information is required. Therefore, the average case complexity of IAAA can be expressed as follows: $O(|v| + |e|) + O(|f|^*|D|^*|c|)$, where f is the average number of nodes with more than one child (fan-out greater than 1); D is the complexity of determining if there is a chain between two nodes that does not include one of their specific common parents; and c is a variable that can be an estimation of the complexity of the task of finding pairs of unattached vertices that belong to the intersection of the adjacency sets of two vertices.

This algorithm is in the category of minimum fill-in algorithms. Checking if there is a path between two children (let's say X, Y) of a node (let's say Z) is basically determining the conditional independence of X and Y given Z or the conditional dependence of X and Y given some other node W. Conceptually, evaluating $I(X, Z, Y)$ in a DAG means examining if the vertex Z D-separates X and Y (Pearl, 1988). The main advantage of this algorithm, however, is that it avoids adding redundant chords to the graph that is being triangulated. Note that, as opposed to the other triangulation methods, IAAA inserts directed chords into non-chordal DAGs. The constant f is a crude estimate of the average fan-out of DAG's nodes. Obviously, the fan-out (number of children) of a node changes as a result of addition of directed chords. However, the authors do acknowledge that for any DAG that needs a large number of chords, the triangulation task will be more complex. Hence, f must be estimated carefully. The worst cases for this

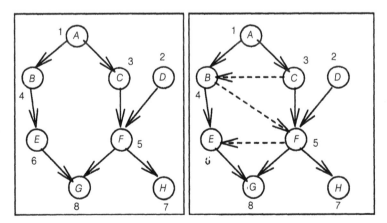

Figure 10.3 (a) A DAG with ordering produced by COTO. (b) A triangulation produced by IAAA.

algorithm occur when the average number of children per vertex (expressed by the estimated constant f) is large and/or when the number of fill-in chords required to make the graph chordal is very large. In the first case, all of the children of each vertex are checked pairwise for connectivity and dependence. In the second case the algorithm will be entered within the deepest inner loop, since for many pairs of vertices it is the case that they are unattached (and conditionally dependent) children of a vertex (i.e. there are many large non-chordal cycles within the DAG). On the other hand, the best case for this algorithm occurs when each vertex of the DAG has only one parent and one child (i.e. the DAG is a long chain). Therefore, we can conjecture that the performance of this algorithm for sparse CPNs, which usually comprise the knowledge representation structure of typical expert systems, will be reasonable.

One difference between IAAA and the method proposed by Lauritzen and Spiegelhalter (1988) is in the addition of moral arcs. A moral arc added to a pair of vertices is *not* always necessary for triangulating a graph. However, in this context, triangulation is solely performed so that a clique-tree can be derived and used as the inference medium. A coherent clique-tree is one that preserves the independence relationships portrayed by the original DAG. Currently there is no theorem proving that the clique-tree obtained from only triangulating the DAG (and not necessarily moralizing it) *does* preserve all the independence relations. Therefore, before such a theorem is available, moralizing the DAG is necessary for the purposes of this context. It is important to note that if one of the other existing triangulation algorithms is used, in order to eliminate the possible redundant arcs from the triangulated DAG, one must introduce yet another heuristic. One such heuristic has already been developed by Kjaerulff (1990). Finally, another advantage of IAAA is that one can add a step to this algorithm and hence be able to choose among the possible alternative chords and achieve better clique size or clique weight.

INTELLIGENT BELIEF PROPAGATION

The IBP algorithm has been developed in order to improve the response time of the expert systems which have adopted clique-trees (or junction-trees) as the structures within which probabilistic inference is performed. The HUGIN expert system shell (Andersen *et al.*, 1989) creates a junction-tree from the original CPN and performs belief absorption, belief propagation, and consistency control by means of making calls to collect-evidence and distribute-evidence procedures (Jensen *et al.*, 1988; Olesen *et al.*, 1989). However, before a query can be answered by this system, immediately after some evidence has been entered, a global update of the belief universes (i.e. the nodes of the junction-tree) is necessary.

Here, a method named intelligent belief propagation (IBP) is presented that will not always update the clique-tree globally. This occurs when evidence has been entered regarding a number of variables and a query has been made regarding the marginal probability (or probabilities) of one or more specific nodes (variables). Obviously, global propagation of belief cannot be avoided altogether. For example, depending on the technique employed at the implementation level, making queries regarding the joint probabilities of a number of variables or performing a case study may need a globally consistent system.

Let us assume that U is the set of variables for which evidence is available, and let V be the set of variables, marginal probabilities of which must be re-evaluated. In other words, there are a number of variables whose marginal probabilities have been changed (elements of U), and there are a number of variables (elements of V) whose new marginal probabilities must be provided by the system. The most simple case occurs when U and V each have one element only (i.e. single evidence available, and there is a query about one variable's new marginal probability). The quickest way one can find the new marginal probability of the element of V is through performing the following:

1. Find the shortest path from the element of U to the element of V from the triangulated CPN.[1]
2. Consider S to be the set of all those vertices along the path found in Step 1.
3. From the clique-tree, find a path of cliques that include all elements of S.
4. From the cliques of Step 3, find one that includes the element of U.
5. Update the clique found in Step 4 and calibrate its neighbours from the path of Step 3.
6. Continue the process of calibrating the neighbours until all the cliques of Step 3 are updated.
7. Marginalize the clique that includes the element of V over its other elements.
8. The result of the calculation in Step 7 is the new marginal probability of the element of V.

Clearly, a locally up-to-date and consistent subtree of the clique-tree can be used to answer queries regarding all of the variables that belong to the cliques of this subtree. Hence, in order to provide a fast response regarding a number of variables, it is sufficient to update appropriately a subtree of the clique-tree that includes all those variables that have been updated and those whose new marginal probabilities have been queried.

We will now discuss the case when multiple items of evidence are available and the new marginal probabilities of a number of variables are queried. This is the general case and involves solving the following problems: first, finding the smallest subtree (the subtree with the smallest number of nodes) of the triangulated CPN that covers all the vertices corresponding to the evidence variables (elements of U) and those which are being queried (elements of V); and second, finding the smallest subtree of the clique-tree that includes a set of cliques which cover the set of vertices of the subtree found in the previous step. The solution to the first problem is given below. The second problem can be easily solved by finding a clique that includes one of the evidence vertices and performing collect-evidence and distribute-evidence by only calling those neighbours that include the variables found on the CPN.

The complexity added by IBP to the conventional belief propagation schemes in the clique-tree is merely dependent on the search algorithms used in Steps 1, 3 and 4 of the IBP algorithm above. However, since these searches are not exhaustive, even for large CPNs and clique-trees, no complexity explosions should be expected.

The algorithm for the solution of the first problem
- let $x_1 \ldots x_n$ be the elements of $U \cup V$
- $i := 1$; $S := \{x_i\}$; $z := x_i$;
- repeat until $i = n$
  - $i := i + 1$
  - find the shortest path from z to x_i from the triangulated CPN
  - $S := S \cup$ Set of elements on the path from z to x_i
  - $z :=$ the vertex in S with smallest label

An illustrative example
The clique-tree depicted in Fig. 10.4(b) corresponds to the CPN depicted in Fig. 10.4(a). The dotted line in Fig. 10.4(b) means that in order to provide the posterior probability of node G after absorption of evidence regarding node E, only cliques (E, B) and (B, F, G) need to be updated.

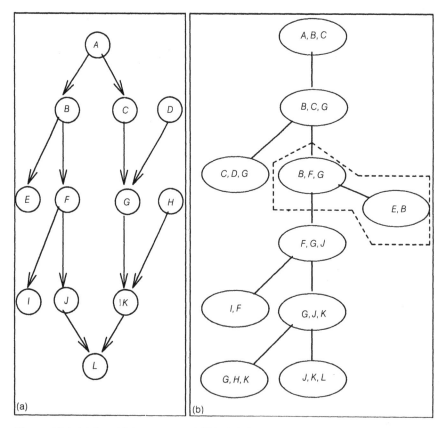

(a)

(b)

Figure 10.4 (a) A multiply-connected CPN. (b) IBP example: the belief regarding E has been revised, and the belief regarding G has been queried (subsequently).

UTILIZATION OF PARALLEL PROCESSORS

The parallel implementation of belief updating/propagation in the clique-tree corresponding to a CPN makes use of multiple processors (CPUs) working independently and in cooperation with each other. Each clique has a message queue through which it receives its 'calls'. The algorithm described herein dedicates one processor (called the 'dispatcher') to oversee the message queues of all of the cliques. The dispatcher also maintains an 'idle-list' of the other processors (called the 'servants'). The dispatcher's duty, therefore, is to find a clique that has a non-empty queue and call the first servant on the idle-list (or wait for one to become available). The servant processor just called is given the job of dequeuing the message queue of the given clique, performing the appropriate calculations (calibration and marginalization) and message generations (propagation messages) and, finally, sending a completion message to the dispatcher (i.e. going to the idle-list). While a servant is performing the above-mentioned tasks, we say that the servant 'simulates' the clique it has been assigned to. After all of the message queues become empty, the clique-tree (and naturally the underlying CPN) is consistently updated and the human user can then make a query, or update again. The formal algorithm for the just explained scenario is presented below. It must be noted that the IBP technique can be easily fused into this algorithm for even faster response time.

An algorithm for belief propagation using parallel processors
*Servant = Clique Simulator
- Wait for call from **Dispatcher**
- Identify the clique to simulate
- Read message [identify the calling clique]
- Perform required task [calibrate?]
- Send message(s) to neighbour(s)
  [Except the one from which a message has just been received]
- Send completion message to **Dispatcher**

*Dispatcher = Process on a separate machine
- Wait for user request
- Identify some clique(s) that contain the node just received update information about [i.e. new evidence]
- Produce a message [with dummy source: 'user'] and put it on the message queue of the node(s) chosen [in last step]
- While there is (are) non-empty message queue(s) do
  - If idle-list is empty, wait;
  - Update idle-list and .pick the first idle **Servant**
  - Dequeue the first message from the chosen queue
  - Give queue-id [clique to be simulated] and the dequeued message [calling clique's id] to the chosen **Servant**

OTHER PROMISING TECHNIQUES

Within the framework of the constraints imposed by the present methodologies and schemes, some shortcuts have been found which can improve performance in certain situations. This section will describe these techniques.

Allowing multiple parents in clique-trees

Consider the clique-tree shown in Fig. 10.5. As it can be seen, α and β are the separators for four cliques (running intersection property preserved). Let us assume that the belief regarding d changes and the new belief in a is queried. Even using the IBP scheme, it is obvious that at least four cliques must be manipulated before the new belief in a can be provided for the user.

   Hence, it is only natural to propose that in clique-trees, two cliques that are not adjacent should be connected in the scenarios similar to that depicted in Fig. 10.5. For example, the clique (α, β, d) in Fig. 10.5, in addition to (α, β, c), should also have (α, β, b) and (α, β, a) as its parents also. We must emphasize that while this technique is very simple, it does *not* violate any rules, nor can it cause any inconsistencies. Furthermore, it must be remembered that one can only take advantage of this technique if IBP is the belief propagation scheme employed.

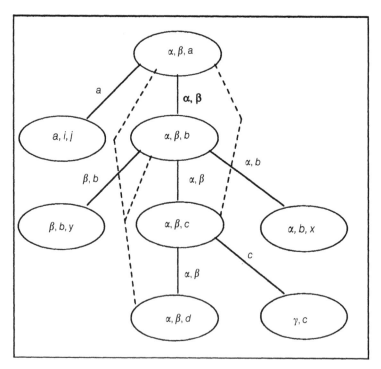

Figure 10.5 Allowing multiple connection within clique-trees.

Usage of redundant cliques

It is clear that graph triangulation can vary depending on the ordering of the vertices. Figure 10.6 shows two different triangulations for a DAG. In the clique-tree located on the left-hand side of Fig. 10.6, if evidence is entered regarding the node d, a query regarding the node c can be answered without a need for any belief propagation; however, if the other clique-tree is used, a propagation from the clique (B, D, E) to the clique (B, C, E) is necessary. In addition to the above two, another clique structure, which can be perceived as a hybrid of the two normal cliques, is shown in Fig. 10.6. The cliques within the dotted box all have two vertices in common pairwise. This architecture rectifies the problem that may occur if any one of the normal clique-trees is used alone. It is important to notice that at each instance, it is required (for consistency purposes) that at least two of the four cliques within the dotted box of Fig. 10.6 be up to date. This technique is not too complex for implementational purposes, yet it can further improve the response time of inference engines.

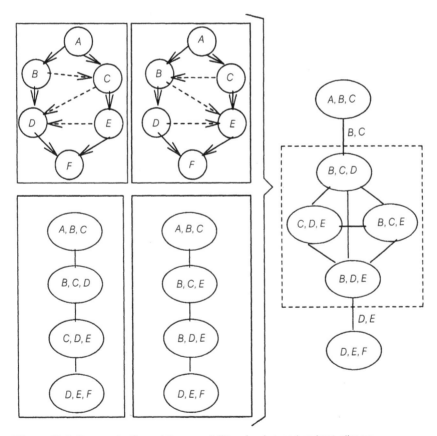

Figure 10.6 Demonstration of the possibility of using redundant cliques.

CONCLUSIONS

A partial ordering algorithm (COTO) has been introduced, which is particularly well suited for directed acyclic graphs and runs in $O(|V| + |E|)$. The new triangulation algorithm (IAAA) presented in this chapter uses the ordering produced by COTO and at each step avoids (and deletes) the redundant chords of 4-cycles. Whether the triangulation produced by IAAA is sufficient (without moralization) for building a clique-tree that preserves the independence relations embedded within the original CPN is a subject to be investigated by the author. Another topic that was given special attention in this chapter was improving the response time of probabilistic expert systems. IBP and utilization of parallel processors, along with some other techniques, were illustrated and shown to be promising methods for achieving better system response time.

NOTE

1. By 'triangulated CPN' we mean the original CPN that has been triangulated using IAAA.

REFERENCES

Andersen, S.K., Olesen, S.K., Jensen, F.V. and Jensen, F. (1989) HUGIN—a shell for building Bayesian belief universes for expert systems, *Proc. 12th International Conference on AI (IJCAI-89)*, 1080–1085.

Cooper, G.F. (1987) Probabilistic inference using belief networks is NP-hard. Technical Report KSL-87-27, Medical Computer Science Group, Stanford University.

Fujisawa, T. and Orino, H. (1974) An efficient algorithm for finding a minimal triangulation of a graph, *IEEE International Symposium on Circuits and Systems*, 171–175.

Jensen, F.V., Olesen, K.G. and Andersen, S.K. (1988) An algebra of Bayesian belief universes for knowledge based systems. Research Report R-88-25, Institute of Electronic Systems, Aalborg University, Denmark.

Kjaerulff, U. (1990) Triangulation of graphs—algorithms giving small total space. Research Report R-90-09, Institute for Electronic Systems, Aalborg University, Denmark.

Lauritzen, S.L. and Spiegelhalter, D.J. (1988) Local computations with probabilities on graphical structures and their application to expert systems, *Journal of the Royal Statistical Society, B*, **50**(2), 157–224.

Olesen, K.G., Kjaerulff, U., Jensen, F., Jensen, F.V., Falck, B., Andreassen, S. and Andersen, S.K. (1989) A MUNIN network for the median nerve—a case study on loops, *Applied Artificial Intelligence*, Special Issue: *Towards Causal AI Models in Practice*.

Pearl, J. (1988) *Probabilistic Reasoning in Intelligent Systems: networks of plausible inference*, San Mateo, CA: Morgan Kaufmann.

Reingold, E.M., Nievergelt, J. and Deo, N. (1977) *Combinatorial Algorithms*, Englewood Cliffs, NJ: Prentice Hall.

Rose, D.J., Tarjan, R.E. and Leuker, G.S. (1976) Algorithmic aspects of vertex elimination on graphs, *SIAM Journal of Computing*, **5**, 266–283.

Tarjan, R.E. and Yannakakis, M. (1984) Simple linear-time algorithms to test chordality of graphs, test acyclicity of hypergraphs, and selectively reduce hypergraphs, *SIAM Journal of Computing*, **13**(3).

A new method for representing and solving Bayesian decision problems

11

P.P. Shenoy

INTRODUCTION

The main goal of this chapter is to describe a new method for representing and solving Bayesian decision problems. A new representation of a decision problem called a **valuation-based system** is described. A graphical depiction of a valuation-based system is called a **valuation network**. Valuation networks are similar in some respects to influence diagrams. Like influence diagrams, valuation networks are a compact representation emphasizing qualitative features of symmetric decision problems. Also, like influence diagrams, valuation networks allow representation of symmetric decision problems without any preprocessing. But there are some differences. Whereas influence diagrams emphasize conditional independence among random variables, valuation networks emphasize factorizations of joint probability distributions. Also, the representation method of influence diagrams allows only conditional probabilities. While conditional probabilities are readily available in pure causal models, they are not always readily available in other graphical models (see, for example, Darroch *et al.*, 1980; Wermuth and Lauritzen, 1983; Edwards and Kreiner, 1983; Kiiveri *et al.*, 1984; Whittaker, 1990). The representation method of valuation-based systems is more general and allows direct representation of all probability models.

We also describe a new computational method for solving decision problems called a **fusion algorithm**. The fusion algorithm is a hybrid of local computational methods for computation of marginals of joint probability distributions and local computational methods for discrete optimization. Local computational methods for computation of marginals of joint probability distributions have been proposed by, for example, Pearl (1988), Lauritzen and Spiegelhalter (1988), Shafer and Shenoy (1988; 1990), and Jensen *et al.* (1990). Local computational methods for discrete optimization are also called **non-serial dynamic programming** (Bertele and Brioschi, 1972). Viewed abstractly using the framework of valuation-based systems, these two local computational methods are actually similar. Shenoy and Shafer (1990) and Shenoy (1991b) show that the same three axioms justify the use of local computation in both these cases.

Artificial Intelligence Frontiers in Statistics: AI and statistics III. Edited by D.J. Hand. Published in 1993 by Chapman & Hall, London. ISBN 0 412 40710 8

Valuation-based systems are described in Shenoy (1989; 1991c). In valuation-based systems, we represent knowledge by functions called **valuations.** We draw inferences from such systems using two operations called **combination** and **marginalization.** Drawing inferences can be described simply as marginalizing all variables out of the joint valuation. The joint valuation is the result of combining all valuations. The framework of valuation-based systems is powerful enough to include also the Dempster–Shafer theory of belief functions (Shenoy and Shafer, 1986; 1990), Spohn's theory of epistemic beliefs (Shenoy, 1991a; 1991c), possibility theory (Zadeh, 1979; Dubois and Prade, 1990), propositional logic (Shenoy, 1990), and constraint satisfaction problems (Shenoy and Shafer, 1988).

Our method for representing and solving decision problems has many similarities to influence diagram methodology (Howard and Matheson, 1984; Olmsted, 1983; Shachter, 1986; Ezawa, 1986; Tatman, 1986). But there are also many differences both in representation and solution. A detailed comparison of our method with decision tree and influence diagram methods is given in Shenoy (1991d).

The fusion algorithm described in this chapter applies to representations that contain only one utility valuation. The assumption of one utility valuation is similar to the assumption of one value node in influence diagrams (Shachter, 1986). If the utility valuation decomposes multiplicatively into several smaller utility valuations, then the fusion algorithm described in this chapter applies also to this case. But if the utility valuation decomposes additively into several smaller utility valuations, then the fusion algorithm does not apply directly to this case. Before we can apply the fusion algorithm described in this chapter, we have to combine the valuations to obtain one joint utility valuation. Thus the fusion algorithm described in this chapter is unable to take computational advantage of an additive decomposition of the utility valuation. Shenoy (1992) describes a modified fusion algorithm that is capable of taking advantage of an additive decomposition of the utility valuation. The modification involves some divisions.

We describe our new method using a diabetes diagnosis problem. The next section gives a statement of this problem and shows a decision tree representation and solution. The third section describes a valuation-based representation of a decision problem. The fourth describes the process of solving valuation-based systems. The fifth describes a fusion algorithm for solving valuation-based systems using local computation. Finally, the sixth section contains proofs of all results.

A DIABETES DIAGNOSIS PROBLEM

A medical intern is trying to decide a policy for treating patients suspected of suffering from diabetes. The intern first observes whether a patient exhibits two symptoms of diabetes—blue toe and glucose in the urine. After she observes the presence or absence of these symptoms, she then either prescribes a treatment for diabetes or does not.

Table 11.1 shows the intern's utility function. For the population of patients served by the intern, the prior probability of diabetes is 10%. Furthermore, for patients known to suffer from diabetes, 1.4% exhibit blue toe, and 90% exhibit glucose in the urine.

Table 11.1 The intern's utility function for all state–act pairs

		Act	
	Intern's utilities (v)	treat for diabetes (t)	not treat ($\sim t$)
State	has diabetes (d)	10	0
	no diabetes ($\sim d$)	5	10

On the other hand, for patients known not to suffer from diabetes, 0.6% exhibit blue toe, and 1% exhibit glucose in the urine.

Consider three random variables D, B, and G representing diabetes, blue toe, and glucose in the urine, respectively. Each variable has two possible values. $D = d$ will represent the proposition *patient has diabetes*, and $D = \sim d$ will represent the proposition *patient does not have diabetes*; similarly for B and G. Assume that variables B and G are conditionally independent given D.

Figure 11.1 shows the preprocessing of probabilities in the decision tree. Figure 11.2 shows a decision tree representation and solution of this problem. Thus an optimal strategy for the medical intern is to treat the patient for diabetes if and only if there is glucose in the patient's urine. The expected utility of this strategy is 9.86.

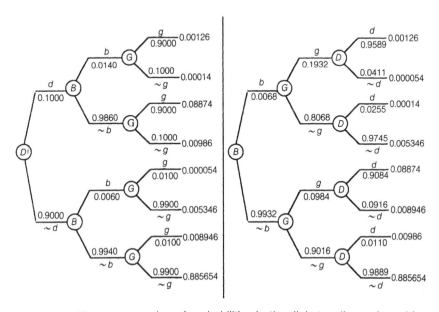

Figure 11.1 The preprocessing of probabilities in the diabetes diagnosis problem.

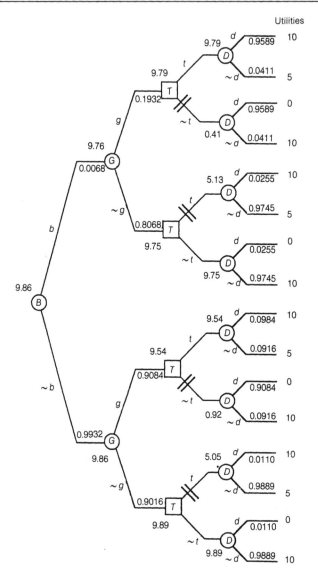

Figure 11.2 A decision tree representation and solution of the diabetes diagnosis problem.

A count of the computations involved in the solution of this problem (including the preprocessing of probabilities) reveals that we do 17 additions, 38 multiplications, 12 divisions, and 4 comparisons, for a total of 71 operations. If we solve this problem using Shachter's (1986) arc-reversal method, we do exactly the same operations (and hence the same number of operations) as the decision tree method. As we will see in a later section, we will solve this problem using our method with only 43 operations.

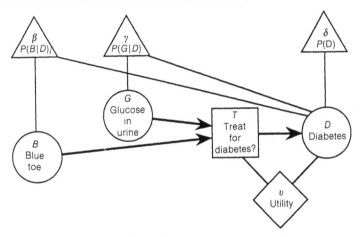

Figure 11.3 A valuation network for diabetes diagnosis problem.

VALUATION-BASED SYSTEM REPRESENTATION

In this section, we describe a valuation-based system (VBS) representation of a decision problem. A VBS representation consists of decision variables, random variables, frames, utility valuations, potentials, and precedence constraints. A graphical depiction of a VBS is called a **valuation network.** Figure 11.3 shows a valuation network for the diabetes diagnosis problem.

VARIABLES, FRAMES AND CONFIGURATIONS

A decision node is represented as a variable. The possible values of a decision variable represent the acts available at that point. We use the symbol $\mathscr{W}_D$ for the set of possible values of decision variable D. We assume that the decision-maker has to pick one and only one of the elements of $\mathscr{W}_D$ as a decision. We call $\mathscr{W}_D$ the **frame** for D. Decision variables are represented in valuation networks by rectangular nodes. In the *diabetes diagnosis* problem, there is one decision node T. The frame for T has two elements: Treat the patient for diabetes (t), and not treat ($\sim t$).

If R is a random variable, we will use the symbol $\mathscr{W}_R$ to denote its possible values. We assume that one and only one of the elements of $\mathscr{W}_R$ can be the true value of R. We can $\mathscr{W}_R$ the frame for R. Random variables are represented in valuation networks by circular nodes. In the diabetes diagnosis problem, there are three random variables: blue toe (B), glucose in the urine (G) and diabetes (D). Each variable has a frame consisting of two elements.

Let $\mathscr{X}_D$ denote the set of all decision variables, let $\mathscr{X}_R$ denote the set of all random variables, and let $\mathscr{X} = \mathscr{X}_D \cup \mathscr{X}_R$ denote the set of all variables. We will often deal with non-empty subsets of variables in $\mathscr{X}$. Given a non-empty subset h of $\mathscr{X}$, let $\mathscr{W}_h$ denote the Cartesian product of $\mathscr{W}_X$ for X in h, i.e. $\mathscr{W}_h = \times \{\mathscr{W}_X | X \in h\}$. We can think of the set $\mathscr{W}_h$ as the set of possible values of the joint variable h. Accordingly, we call $\mathscr{W}_h$

the frame for h. Also, we will refer to elements of $\mathcal{W}_h$ as **configurations** of h. We will use lower-case, bold-faced letters such as $\mathbf{x}, \mathbf{y}$, etc. to denote configurations. Also, if $\mathbf{x}$ is a configuration of g, $\mathbf{y}$ is a configuration of h, and $g \cap h = \varnothing$, then $(\mathbf{x}, \mathbf{y})$ will denote a configuration of $g \cup h$.

It will be convenient to extend this terminology to the case where the set of variables h is empty. We will adopt the convention that the frame for the empty set $\varnothing$ consists of a single configuration, and we will use the symbol $\blacklozenge$ to name that configuration; $\mathcal{W}_\varnothing = \{\blacklozenge\}$. To be consistent with our notation above, we will adopt the convention that if $\mathbf{x}$ is a configuration for g, then $(\mathbf{x}, \blacklozenge) = \mathbf{x}$.

VALUATIONS

Suppose $h \subseteq \mathcal{X}$. A **utility valuation** υ for h is a function from $\mathcal{W}_h$ to $\mathbb{R}$, where $\mathbb{R}$ denotes the set of real numbers. The values of utility valuations are utilities. If $h = d \cup r$ where $d \subseteq \mathcal{X}_D$ and $r \subseteq \mathcal{X}_R$, $\mathbf{x} \in \mathcal{W}_d$, and $\mathbf{y} \in \mathcal{W}_r$, then $\upsilon(\mathbf{x}, \mathbf{y})$ denotes the utility to the decision-maker if the decision-maker chooses configuration $\mathbf{x}$ and the true configuration of r is $\mathbf{y}$. If υ is a utility valuation for h and $X \in h$, then we will say that υ **bears on** X.

In a valuation network, a utility valuation is represented by a diamond-shaped node. To permit the identification of all valuations that bear on a variable, we will draw undirected edges between the utility valuation node and all the variable nodes it bears on. In the diabetes diagnosis problem, there is one utility valuation υ as shown in Fig. 11.3. Table 11.1 shows the values of this utility valuation.

Suppose $h \subseteq \mathcal{X}$. A **probability valuation** (or, simply, a **potential**) ρ for h is a function from $\mathcal{W}_h$ to the unit interval $[0, 1]$. The values of potentials are probabilities. In a valuation network, a potential is represented by a triangular node. Again, to identify the variables related by a potential, we draw undirected edges between the potential node and all the variable nodes it bears on. In the diabetes diagnosis problem, there are three potentials β, γ, and δ, as shown in Figure 11.3. Table 11.2 shows the details of these potentials. Note that β is a potential for $\{B, D\}$, γ is a potential for $\{G, D\}$, and δ is a potential for $\{D\}$.

PRECEDENCE CONSTRAINTS

Besides acts, states, probabilities, and utilities, an important ingredient of problems in decision analysis is the chronology, or structure, of information constraints. Some

Table 11.2 The potentials δ, β and γ

				B			G	
D	δ	β	b	$\sim b$	γ	g	$\sim g$
d	0.10		d	0.014	0.986		d	0.90	0.10
		D				D		
$\sim d$	0.90		$\sim d$	0.006	0.994		$\sim d$	0.01	0.99

decisions have to be made before the observation of some uncertain states, and some decisions can be postponed until after some states are observed. In the diabetes diagnosis problem, for example, the medical intern does not know whether the patient has diabetes or not. And the decision whether to treat the patient for diabetes or not may be postponed until after the observation of the blue toe and glucose in urine.

If a decision-maker expects to be informed of the true value of random variable R before making a decision D, then we represent this situation by the binary relation $R \rightarrow D$ (read as R **precedes** D). On the other hand, if a random variable R is only revealed after a decision D is made or perhaps never revealed, then we represent this situation by the binary relation $D \rightarrow R$. In the diabetes diagnosis problem, we have the precedence constraints $B \rightarrow T, G \rightarrow T, T \rightarrow D$. The decision whether to treat the patient for diabetes or not (T) is only made after observing blue toe (B) and glucose in the urine (G). And, diabetes (D) is not known at the time the decision whether to treat the patient for diabetes (T) has to be made.

Suppose $>$ is a binary relation on $\mathcal{X}$ such that it is the transitive closure of $\rightarrow$, i.e. $X > Y$ if either $X \rightarrow Y$, or there exists a $Z \in \mathcal{X}$ such that $X > Z$ and $Z > Y$. First, we will assume that $>$ is a partial order of $\mathcal{X}$ (otherwise the decision problem is ill defined and not solvable). Second, we will require that this partial order $>$ is such that for any $D \in \mathcal{X}_D$ and any $R \in \mathcal{X}_R$, either $D > R$ or $R > D$. We will refer to this second condition as the **perfect recall condition**. The reason for the perfect recall condition is as follows. Given the meaning of the precedence relation $\rightarrow$, for any decision variable D and any random variable R, either R is known when decision D has to be made, or not. This translates to either $R > D$ or $D > R$.

Next, we will define two operations called combination and marginalization. We use these operations to solve the valuation-based system representation. First we start with some notation.

PROJECTION OF CONFIGURATIONS

Projection of configurations simply means dropping extra coordinates; if (w, x, y, z) is a configuration of $\{W, X, Y, Z\}$, for example, then the projection of (w, x, y, z) onto $\{W, X\}$ is simply (w, x), which is a configuration of $\{W, X\}$.

If g and h are sets of variables, $h \subseteq g$, and $\mathbf{x}$ is a configuration of g, then we will let $\mathbf{x}^{\downarrow h}$ denote the projection of $\mathbf{x}$ onto h. The projection $\mathbf{x}^{\downarrow h}$ is always a configuration of h. If $h = g$ and $\mathbf{x}$ is a configuration of g, then $\mathbf{x}^{\downarrow h} = \mathbf{x}$. If $h = \varnothing$, then $\mathbf{x}^{\downarrow h} = \blacklozenge$.

COMBINATION

The definition of **combination** will depend on the type of valuations being combined. Suppose h and g are subsets of $\mathcal{X}$, suppose υ_i is a utility valuation for h, and suppose ρ_j is a potential for g. Then the combination of υ_i and ρ_j, denoted by $\upsilon_i \otimes \rho_j$, is a utility valuation for $h \cup g$ obtained by pointwise multiplication of υ_i and ρ_j, i.e. $(\upsilon_i \otimes \rho_j)(\mathbf{x}) = \upsilon_i(\mathbf{x}^{\downarrow h}) \rho_j(\mathbf{x}^{\downarrow g})$ for all $\mathbf{x} \in \mathcal{W}_{h \cup g}$. See Table 11.3 for an example.

Table 11.3 The computation of the combinations $\beta \oplus \gamma \oplus \delta$ and $\upsilon \oplus \beta \oplus \gamma \oplus \delta$

$\mathcal{W}_{\{B,G,T,D\}}$				υ	β	γ	δ	$\beta \oplus \gamma \oplus \delta$	$\upsilon \oplus \beta \oplus \gamma \oplus \delta$ $= \tau$
b	g	t	d	10	0.014	0.90	0.10	0.00126	0.0126
b	g	t	$\sim d$	5	0.006	0.01	0.90	0.000054	0.00027
b	g	$\sim t$	d	0	0.014	0.90	0.10	0.00126	0
b	g	$\sim t$	$\sim d$	10	0.006	0.01	0.90	0.000054	0.00054
b	$\sim g$	t	d	10	0.014	0.10	0.10	0.00014	0.0014
b	$\sim g$	t	$\sim d$	5	0.006	0.99	0.90	0.005346	0.02673
b	$\sim g$	$\sim t$	d	0	0.014	0.10	0.10	0.00014	0
b	$\sim g$	$\sim t$	$\sim d$	10	0.006	0.99	0.90	0.005346	0.05346
$\sim b$	g	t	d	10	0.986	0.90	0.10	0.08874	0.8874
$\sim b$	g	t	$\sim d$	5	0.994	0.01	0.90	0.008946	0.04473
$\sim b$	g	$\sim t$	d	0	0.986	0.90	0.10	0.08874	0
$\sim b$	g	$\sim t$	$\sim d$	10	0.994	0.01	0.90	0.008946	0.08646
$\sim b$	$\sim g$	t	d	10	0.986	0.10	0.10	0.00986	0.0986
$\sim b$	$\sim g$	t	$\sim d$	5	0.994	0.99	0.90	0.885654	4.42827
$\sim b$	$\sim g$	$\sim t$	d	0	0.986	0.10	0.10	0.00986	0
$\sim b$	$\sim g$	$\sim t$	$\sim d$	10	0.994	0.99	0.90	0.885654	8.86554

Suppose h and g are subsets of $\mathcal{X}$, suppose ρ_i is a potential for h, and suppose ρ_j is a potential for g. Then the combination of ρ_i and ρ_j, denoted by $\rho_i \otimes \rho_j$, is a potential for $h \cup g$ obtained by pointwise multiplication of ρ_i and ρ_j, i.e. $(\rho_i \otimes \rho_j)(\mathbf{x}) = \rho_i(\mathbf{x}^{\downarrow h})\rho_j(\mathbf{x}^{\downarrow g})$ for all $\mathbf{x} \in \mathcal{W}_{h \cup g}$. See Table 11.3 for an example.

Note that combination is commutative and associative. Thus, if $\{\alpha_1, \ldots, \alpha_k\}$ is a set of valuations, we will write $\otimes \{\alpha_1, \ldots, \alpha_k\}$ or $\alpha_1 \otimes \cdots \otimes \alpha_k$ to mean the combination of valuations in $\{\alpha_1, \ldots, \alpha_k\}$ in some sequence.

MARGINALIZATION

Suppose h is a subset of variables, and suppose α is a valuation for h. Marginalization is an operation where we reduce valuation α to a valuation $\alpha^{\downarrow(h-\{X\})}$ for $h - \{X\}$. $\alpha^{\downarrow(h-\{X\})}$ is called the **marginal** of α for $h - \{X\}$. Unlike combination, the definition of marginalization does not depend on the nature of α. But the definition of marginalization does depend on whether X is a decision or a random variable.

If R is a random variable, $\alpha^{\downarrow(h-\{R\})}$ is obtained by summing α over the frame for R, i.e. $\alpha^{\downarrow(h-\{R\})}(\mathbf{c}) = \Sigma\{\alpha(\mathbf{c}, \mathbf{r}) | \mathbf{r} \in \mathcal{W}_R\}$ for all $\mathbf{c} \in \mathcal{W}_{h-\{R\}}$. Here, α could be either a utility valuation or a potential. See Table 11.4 for an example.

Table 11.4 The computation of $\tau^{\downarrow\{B,G,T\}}$, $\tau^{\downarrow\{B,G\}}$, Ψ_T, $\tau^{\downarrow\{B\}}$, and $\tau^{\downarrow\varnothing}(\blacklozenge)$. τ denotes the joint valuation $\upsilon\otimes\beta\otimes\gamma\otimes\delta$

$\mathscr{W}_{\{B,G,T,D\}}$				τ	$\tau^{\downarrow\{B,G,T\}}$	$\tau^{\downarrow\{B,G\}}$	Ψ_T	$\tau^{\downarrow\{B\}}$	$\tau^{\downarrow\varnothing}(\blacklozenge)$
b	g	t	d	0.0126	0.01287	0.01287	t	0.06633	9.864
b	g	t	~d	0.00027					
b	g	~t	d	0	0.00054				
b	g	~t	~d	0.00054					
b	~g	t	d	0.0014	0.02813	0.05346	~t		
b	~g	t	~d	0.02673					
b	~g	~t	d	0	0.05346				
b	~g	~t	~d	0.05346					
~b	g	t	d	0.8874	0.93213	0.93213	t	9.79767	
~b	g	t	~d	0.04473					
~b	g	~t	d	0	0.08646				
~b	g	~t	~d	0.08646					
~b	~g	t	d	0.0986	4.52687	8.86554	~t		
~b	~g	t	~d	4.42827					
~b	~g	~t	d	0	8.86554				
~b	~g	~t	~d	8.86554					

If D is a decision variable, $\alpha^{\downarrow(h-\{D\})}$ is obtained by maximizing α over the frame for D, i.e. $\alpha^{\downarrow(h-\{D\})}(\mathbf{c}) = \max\{\alpha(\mathbf{c},\mathbf{d})\,|\,\mathbf{d}\in\mathscr{W}_D\}$ for all $\mathbf{c}\in\mathscr{W}_{h-\{D\}}$. Here, α must be a utility valuation. See Table 11.4 for an example.

We now state three lemmas regarding the marginalization operation. Lemma 1 states that in marginalizing two decision variables out of a valuation, the order in which the variables are eliminated does not affect the result. Lemma 2 states a similar result for marginalizing two random variables out of a valuation. Lemma 3 states that in marginalizing a decision variable and a random variable out of a valuation, the order in which the two variables are eliminated may make a difference.

Lemma 11.1

Suppose h is a subset of $\mathscr{X}$ containing decision variables D_1 and D_2, and suppose α is a utility valuation for h. Then

$$(\alpha^{\downarrow(h-\{D_1\})})^{\downarrow(h-\{D_1,D_2\})}(\mathbf{c}) = (\alpha^{\downarrow(h-\{D_2\})})^{\downarrow(h-\{D_1,D_2\})}(\mathbf{c})$$

for all $\mathbf{c}\in\mathscr{W}_{h-\{D_1,D_2\}}$.

Lemma 11.2

Suppose h is a subset of $\mathscr{X}$ containing random variables R_1 and R_2, and suppose

α is a valuation for h. Then

$$(\alpha^{\downarrow(h-\{R_1\})})^{\downarrow(h-\{R_1,R_2\})}(\mathbf{c}) = (\alpha^{\downarrow(h-\{R_2\})})^{\downarrow(h-\{R_1,R_2\})}(\mathbf{c})$$

for all $\mathbf{c}\in\mathcal{W}_{h-\{R_1,R_2\}}$.

Lemma 11.3

Suppose h is a subset of $\mathcal{X}$ containing decision variable D and random variable R, and suppose α is a utility valuation for h. Then

$$(\alpha^{\downarrow(h-\{D\})})^{\downarrow(h-\{R,D\})}(\mathbf{c}) \geqslant (\alpha^{\downarrow(h-\{R\})})^{\downarrow(h-\{R,D\})}(\mathbf{c})$$

for all $\mathbf{c}\in\mathcal{W}_{h-\{R,D\}}$.

It is clear from Lemma 11.3, that in marginalizing more than one variable, the order of elimination of the variables may make a difference. As we shall see shortly, we need to marginalize all variables out of the joint valuation. What sequence should we use? This is where the precedence constraints come into play. We will define marginalization such that variable Y is marginalized before X whenever $X > Y$.

Suppose h and g are non-empty subsets of $\mathcal{X}$ such that g is a proper subset of h, suppose α is a valuation for h, and suppose $>$ is a partial order on $\mathcal{X}$ satisfying the perfect recall condition. The **marginal** of α for g **with respect to the partial order** $>$, denoted by $\alpha^{\downarrow g}$, is a valuation for g defined as follows:

$$\alpha^{\downarrow g} = (((\alpha^{\downarrow(h-\{X_1\})})^{\downarrow(h-\{X_1,X_2\})})\ldots)^{\downarrow(h-\{X_1,X_2,\ldots,X_k\})} \tag{11.1}$$

where $h-g = \{X_1,\ldots,X_k\}$ and $X_1 X_2 \ldots X_k$ is a sequence of variables in $h-g$ such that with respect to the partial order $>$, X_1 is a minimal element of $h-g$, X_2 is a minimal element of $h-g-\{X_1\}$, etc.

The marginalization sequence $X_1 X_2 \ldots X_k$ may not be unique since $>$ is only a partial order. But, since $>$ satisfies the perfect recall condition, it is clear from Lemmas 1 and 2 that the definition of $\alpha^{\downarrow g}$ in Equation (11.1) is well defined.

STRATEGY

The main objective in solving a decision problem is to compute an optimal strategy. What constitutes a strategy? Intuitively, a strategy is a choice of an act for each decision variable D as a function of configurations of random variables R such that $R > D$. Let $\Pr(D) = \{R\in\mathcal{X}_R | R > D\}$. We shall refer to $\Pr(D)$ as the **predecessors** of D. Thus a **strategy** σ is a collection of functions $\{\xi_D\}_{D\in\mathcal{X}_D}$ where $\xi_D\colon \mathcal{W}_{\Pr(D)} \to \mathcal{W}_D$.

SOLUTION FOR A VARIABLE

Computing an optimal strategy is a matter of bookkeeping. Each time we marginalize a decision variable out of a utility valuation using maximization, we store a table of optimal values of the decision variable where the maxima are achieved. We can think of this table as a function. We will call this function a **solution** for the decision variable. Suppose h is a subset of variables such that decision variable $D\in h$, and suppose v is a utility valuation for h. A function $\Psi_D: \mathcal{W}_{h-\{D\}} \to \mathcal{W}_D$ is called a solution for D (with respect to v) if $v^{\downarrow(h-\{D\})}(\mathbf{c}) = v(\mathbf{c}, \Psi_D(\mathbf{c}))$ for all $\mathbf{c} \in \mathcal{W}_{h-\{D\}}$. See Table 11.4 for an example.

SOLVING A VALUATION-BASED SYSTEM

Suppose $\Delta = \{\mathcal{X}_D, \mathcal{X}_R, \{\mathcal{W}_X\}_{X \in \mathcal{X}}, \{v_1\}, \{\rho_1, \ldots, \rho_n\}, \to\}$ is a VBS representation of a decision problem consisting of one utility valuation and n potentials. What do the potentials represent? And how do we solve Δ? We will answer these two related questions in terms of a canonical decision problem.

CANONICAL DECISION PROBLEM

A **canonical decision problem** Δ_C consists of a single decision variable D with a finite frame $\mathcal{W}_D$, a single random variable R with a finite frame $\mathcal{W}_R$, a single utility valuation v for $\{D, R\}$, a single potential ρ for $\{R, D\}$ such that

$$\Sigma\{\rho(\mathbf{d}, \mathbf{r}) | \mathbf{r} \in \mathcal{W}_R\} = 1 \qquad \text{for all } \mathbf{d} \in \mathcal{W}_D \qquad (11.2)$$

and a precedence relation $\to$ defined by $D \to R$. Figure 11.4 shows a valuation network and a decision tree representation of the canonical decision problem.

The meaning of the canonical decision problem is as follows. The elements of $\mathcal{W}_D$ are acts, and the elements of $\mathcal{W}_R$ are states of nature. The potential ρ is a family of probability distributions for R, one for each act $\mathbf{d} \in \mathcal{W}_D$, i.e. $\Sigma\{\rho(\mathbf{d}, \mathbf{r}) | \mathbf{r} \in \mathcal{W}_R\} = 1$ for all $\mathbf{d} \in \mathcal{W}_D$. In other words, the probability distribution of random variable R is conditioned on the act $\mathbf{d}$ chosen by the decision-maker. The probability $\rho(\mathbf{d}, \mathbf{r})$ can be interpreted as the conditional probability of $R = \mathbf{r}$ given that $D = \mathbf{d}$.

The utility valuation v is a utility function—if the decision-maker chooses act $\mathbf{d}$ and the state of nature $\mathbf{r}$ prevails, then the utility to the decision-maker is $v(\mathbf{d}, \mathbf{r})$.

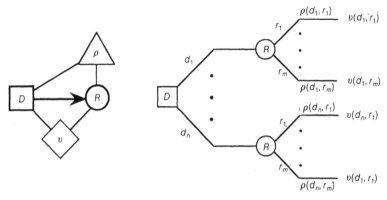

Figure 11.4 A valuation network and a decision tree representation of the canonical decision problem.

The precedence relation $\rightarrow$ states that the true state of nature is revealed to the decision-maker only after the decision-maker has chosen an act.

Solving a canonical decision problem using the criterion of maximizing expected utility is easy. The expected utility associated with act $\mathbf{d}$ is $\Sigma\{(v \oplus \rho)(\mathbf{d}, \mathbf{r}) \,|\, \mathbf{r} \in \mathcal{W}_R\} = (v \oplus \rho)^{\downarrow\{D\}}(\mathbf{d})$. The maximum expected utility (associated with an optimal act, say $\mathbf{d}^*$) is $\max \{(v \oplus \rho)^{\downarrow\{D\}}(\mathbf{d}) \,|\, \mathbf{d} \in \mathcal{W}_D\} = ((v \oplus \rho)^{\downarrow\{D\}})^{\downarrow\varnothing}(\blacklozenge) = (v \oplus \rho)^{\downarrow\varnothing}(\blacklozenge)$. Finally, act $\mathbf{d}^*$ is optimal if and only if $(v \oplus \rho)^{\downarrow\{D\}}(\mathbf{d}^*) = (v \oplus \rho)^{\downarrow\varnothing}(\blacklozenge)$.

Consider the decision problem $\Delta = \{\mathcal{X}_D, \mathcal{X}_R, \{\mathcal{W}_X\}_{X \in \mathcal{X}}, \{v_1\}, \{\rho_1, \ldots, \rho_n\}, \rightarrow\}$. We will explain the meaning of Δ by reducing it to an equivalent canonical decision problem $\Delta_C = \{\{D\}, \{R\}, \{\mathcal{W}_D, \mathcal{W}_R\}, \{v\}, \{\rho\}, \rightarrow\}$. To define Δ_C, we need to define $\mathcal{W}_D, \mathcal{W}_R, v$ and ρ. Define $\mathcal{W}_D$ such that, for each distinct strategy σ of Δ, there is a corresponding act $\mathbf{d}_\sigma$ in $\mathcal{W}_D$. Define $\mathcal{W}_R$ such that for each distinct configuration $\mathbf{y}$ of $\mathcal{X}_R$ in Δ, there is a corresponding configuration $\mathbf{r_y}$ in $\mathcal{W}_R$.

Before we define utility valuation v for $\{D, R\}$, we need some notation. Suppose $\sigma = \{\xi_D\}_{D \in \mathcal{X}_D}$ is a strategy, and suppose $\mathbf{y}$ is a configuration of $\mathcal{X}_R$. Then together σ and $\mathbf{y}$ determine a unique configuration of $\mathcal{X}_D$. We will let $\mathbf{a}_{\sigma,\mathbf{y}}$ denote this unique configuration of $\mathcal{X}_D$. By definition, $\mathbf{a}_{\sigma,\mathbf{y}}^{\downarrow\{D\}} = \xi_D(\mathbf{y}^{\downarrow \Pr(D)})$ for all $D \in \mathcal{X}_D$.

Consider the utility valuation v_1 in Δ. Assume that the domain of this valuation includes all of $\mathcal{X}_D$. Typically the domain of this valuation will include also some (or all) random variables. Let p denote the subset of random variables included in the domain of the joint utility valuation, i.e. $p \subseteq \mathcal{X}_R$ such that v_1 is a utility valuation for $\mathcal{X}_D \cup p$. Define a utility valuation v for $\{D, R\}$ such that $v(\mathbf{d}_\sigma, \mathbf{r_y}) = v_1(\mathbf{a}_{\sigma,\mathbf{y}}, \mathbf{y}^{\downarrow p})$ for all strategies σ of Δ, and all configurations $\mathbf{y} \in \mathcal{W}_{\mathcal{X}_R}$. Remember that $\mathbf{a}_{\sigma,\mathbf{y}}$ is the unique configuration of $\mathcal{X}_D$ determined by σ and $\mathbf{y}$.

Consider the joint potential $\rho_1 \otimes \ldots \otimes \rho_n$. Assume that this potential includes all random variables in its domain. Let q denote the subset of decision variables included in the domain of the joint potential, i.e. $q \subseteq \mathcal{X}_D$ such that $\rho_1 \otimes \ldots \otimes \rho_n$ is a potential for $q \cup \mathcal{X}_R$. Note that q could be empty. Define potential ρ for $\{D, R\}$, such that $\rho(\mathbf{d}_\sigma, \mathbf{r_y}) = (\rho_1 \otimes \ldots \otimes \rho_n)(\mathbf{a}_{\sigma,\mathbf{y}}^{\downarrow q}, \mathbf{y})$ for all strategies σ and all configurations $\mathbf{y} \in \mathcal{W}_{\mathcal{X}_R}$. Δ_C, as defined above, will be a canonical decision problem only if ρ satisfies Condition (11.2). This motivates the following definition. Δ is a **well-defined VBS representation of a decision problem** if and only if $\Sigma\{(\rho_1 \otimes \ldots \otimes \rho_n)(\mathbf{x}, \mathbf{y}) \,|\, \mathbf{y} \in \mathcal{W}_{\mathcal{X}_R}\} = 1$ for every $\mathbf{x} = \mathcal{W}_q$.

In summary, the potentials $\{\rho_1, \ldots, \rho_n\}$ represents the factors of a family of probability distributions. It is easy to verify that the VBS representation of the diabetes diagnosis problem is well defined since $\delta \otimes \beta \otimes \gamma$ is a joint probability distribution for $\{D, B, G\}$ (see Table 11.3).

THE DECISION PROBLEM

Suppose $\Delta = \{\mathcal{X}_D, \mathcal{X}_R, \{\mathcal{W}_X\}_{X \in \mathcal{X}}, \{v_1\}, \{\rho_1, \ldots, \rho_n\}, \rightarrow\}$ is a well-defined decision problem. Let $\Delta_C = \{\{D\}, \{R\}, \{\mathcal{W}_D, \mathcal{W}_R\}, \{v\}, \{\rho\}, \rightarrow\}$ represents an equivalent canonical decision problem. In the canonical decision problem Δ_C, the two computations that are of interest are, first, the computation of the maximum expected value

$(v \otimes \rho)^{\downarrow \varnothing}(\blacklozenge)$, and second, the computation of an optimal act $\mathbf{d}_{\sigma^*}$ such that $(v \otimes \rho)^{\downarrow \{D\}}(\mathbf{d}_{\sigma^*}) = (v \otimes \rho)^{\downarrow \varnothing}(\blacklozenge)$. Since we know the mapping between Δ and Δ_C, we can now formally define the questions posed in a decision problem Δ. There are two computations of interest.

First, we would like to compute the maximum expected utility. The maximum expected utility is given by $(\otimes \{v_1, \rho_1, \ldots, \rho_n\})^{\downarrow \varnothing}(\blacklozenge)$. Second, we would like to compute an optimal strategy σ^* that gives us the maximum expected value $(\otimes \{v_1, \rho_1, \ldots, \rho_n\})^{\downarrow \varnothing}(\blacklozenge)$. A strategy σ^* of Δ is **optimal** if $(v \otimes \rho)^{\downarrow \{D\}}(\mathbf{d}_{\sigma^*}) = (\otimes \{v_1, \rho_1, \ldots, \rho_n\})^{\downarrow \varnothing}(\blacklozenge)$, where v, ρ, and D refer to the equivalent canonical decision problem Δ_C.

In the diabetes diagnosis problem, we have four valuations v, β, γ, and δ. Also, from the precedence constraints, we have $B > T, G > T, T > D$. Thus we need to compute either

$$((((v \otimes \beta \otimes \gamma \otimes \delta)^{\downarrow \{B, G, T\}})^{\downarrow \{B, G\}})^{\downarrow \{B\}})^{\downarrow \varnothing} \quad \text{or} \quad ((((v \otimes \beta \otimes \gamma \otimes \delta)^{\downarrow \{B, G, T\}})^{\downarrow \{B, G\}})^{\downarrow \{G\}})^{\downarrow \varnothing}$$

In either case, we get the same answers. Tables 11.3 and 11.4 display the former computations. As seen from Table 11.4, the optimal expected utility is 9.864. Also, from Ψ_T, the solution for T (shown in Table 11.4), the optimal act is to treat the patient for diabetes if and only if the patient exhibits glucose in the urine.

Note that no divisions were done in the solution process, only additions and multiplications. But both decision tree and influence diagram methodologies involve unnecessary divisions and unnecessary multiplications to compensate for the unnecessary divisions. It is this feature of valuation-based systems that makes it more efficient than decision trees and influence diagrams. In solving the diabetes diagnosis problem using our method, we do only 11 additions, 28 multiplications and 4 comparisons, for a total of 43 operations. This is a savings of 40% over the decision tree and influence diagram methodologies, which required a total of 71 operations.

A FUSION ALGORITHM FOR SOLVING VALUATION-BASED SYSTEMS USING LOCAL COMPUTATION

In this section, we will describe a fusion algorithm for solving a VBS using local computation. The solution for the diabetes diagnosis problem shown in Tables 11.3 and 11.4 involves combination on the space $\mathcal{W}_{\mathcal{X}}$. While this is possible for small problems, it is computationally not tractable for problems with many variables. Given the structure of the diabetes diagnosis problem, it is not possible to avoid the combination operation on the space of all four variables, B, G, T and D. But in some problems it may be possible to avoid such global computations.

The basic idea of the method is successively to delete all variables from the VBS. The sequence in which variables are deleted must respect the precedence constraints in the sense that if $X > Y$, then Y must be deleted before X. Since $>$ is only a partial order, a problem may allow several deletion sequences. Any allowable deletion sequence may be used. All allowable deletion sequences will lead to the same answers.

But different deletion sequences may involve different computational costs. We will comment on good deletion sequences at the end of this section.

When we delete a variable, we have to do a 'fusion' operation on the valuations. Consider a set of k valuations $\alpha_1, ..., \alpha_k$. Suppose α_i is a valuation for h_i. Let $\text{Fus}_X\{\alpha_1, ..., \alpha_k\}$ denote the collection of valuations after fusing the valuations in the set $\{\alpha_1, ..., \alpha_k\}$ with respect to variable X. Then

$$\text{Fus}_X\{\alpha_1, ..., \alpha_k\} = \{\alpha^{\downarrow\{(h - \{X\})\}}\} \cup \{\alpha_i | X \notin h_i\}$$

where $\alpha = \otimes\{\alpha_i | X \in h_i\}$, and $h = \cup(h_i | X \in h_i)$. After fusion, the set of valuations is changed as follows. All valuations that bear on X are combined, and the resulting valuation is marginalized such that X is eliminated from its domain. The valuations that do not bear on X remain unchanged.

We are ready to state the main theorem.

Theorem 11.1

Suppose $\Delta = \{\mathcal{X}_D, \mathcal{X}_R, \{\mathcal{W}_X\}_{X \in \mathcal{X}}, \{v_1\}, \{\rho_1, ..., \rho_n\}, \rightarrow\}$ is a well-defined decision problem. Suppose $X_1, X_2...X_k$ is a sequence of variables in $\mathcal{X} = \mathcal{X}_D \cup \mathcal{X}_R$ such that, with respect to the partial order $>$, X_1 is a minimal element of $\mathcal{X}$, X_2 is a minimal element of $\mathcal{X} - \{X_1\}$, etc. Then $\{(\otimes\{v_1, \rho_1, ..., \rho_n\})^{\downarrow\varnothing}\} = \text{Fus}_{X_k}\{...\text{Fus}_{X_2}\{\text{Fus}_{X_1}\{v_1, \rho_1, ..., \rho_n\}\}\}$.

To illustrate Theorem 11.1, consider a VBS as shown in Fig. 11.5 for a generic medical diagnosis problem. In this VBS, there are three random variables, D, P, and S, and one decision variable, T. D represents a disease, P represents a pathological state caused by the disease, and S represents a symptom caused by the pathological state. We assume that S and D are conditionally independent given P. The potential δ is the prior probability of D, the potential π is the conditional probability of P given D, and the potential σ is the conditional probability of S given P. A medical intern first observes the symptom S and then either treats the patient for the disease

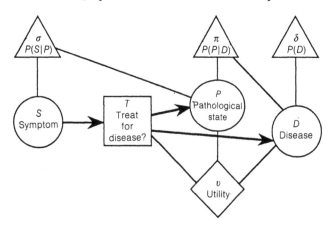

Figure 11.5 A valuation network for the medical diagnosis problem.

and pathological state or not. The utility valuation υ bears on the intern's action T, the pathological state P, and the disease variable D (see Shenoy, 1991d, for a more detailed description of this problem).

Figure 11.6 shows the results of the fusion algorithm for this problem. The deletion sequence used is $DPTS$. The valuation network labelled 0 in Fig. 11.6 is the same as the one in Fig. 11.5. The valuation network labelled 1 is the result after deletion of D and the resulting fusion. The combination involved in the fusion operation only involves variables D, P and T. The valuation network labelled 2 is the result after deletion of P. The combination operation involved in the corresponding fusion operation involves only three variables, P, T and S. The valuation network labelled 3 is the result after deletion of T. There is no combination involved here, only marginalization on the frame of $\{S, T\}$. The valuation network labelled 4 is the result after deletion of S. Again, there is no combination involved here, only marginalization on the frame of $\{S\}$. The maximum expected utility value is given by

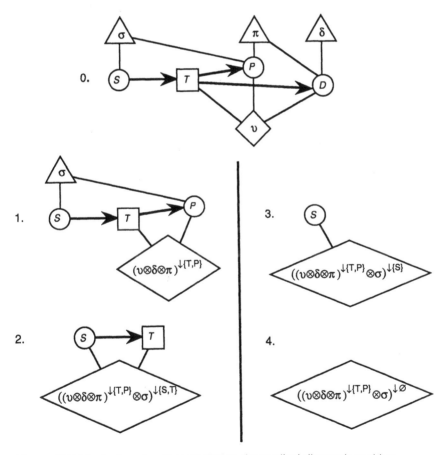

Figure 11.6 The fusion algorithm applied to the medical diagnosis problem.

$((v \otimes \delta \otimes \pi)^{\downarrow\{T,P\}} \otimes \sigma)^{\downarrow\varnothing}$ (◆). An optimal strategy is given by the solution for T with respect to $((v \otimes \delta \otimes \pi)^{\downarrow\{T,P\}} \otimes \sigma)^{\downarrow\{S,T\}}$ computed during fusion with respect to T. Note that in this problem, the fusion algorithm avoids computation on the frame of all four variables.

In solving the medical diagnosis problem using our method, we do a total of 31 operations (Shenoy, 1991d). On the other hand, for this problem, the decision tree solution method requires 59 operations (Shenoy, 1991d). Thus, for this problem, our method results in a savings of 47% over the decision tree methodology. If we use the influence diagram methodology for this problem, we do 49 operations (Shenoy, 1991d). Thus, for this problem, our method results in a savings of 20% over the influence diagram methodology.

The fusion method described in this section applies when there is one utility valuation in the VBS. This method will also apply unchanged in problems where the joint utility valuation factors multiplicatively into several utility valuations. In this case, we can define combination of utility valuations as pointwise multiplication, i.e. if v_i is a utility valuation for h_i and v_j is a utility valuation for h_j, then $v_i \otimes v_j$ is a utility valuation for $h_i \cup h_j$ defined by $(v_i \otimes v_j)(\mathbf{x}) = v_i(\mathbf{x}^{\downarrow h_i}) v_j(\mathbf{x}^{\downarrow h_j})$ for all $\mathbf{x} \in \mathscr{W}_{h_i \cup h_j}$. This method will not apply directly in problems where the joint utility valuation decomposes additively. In such problems, we will first have to combine all utility valuations before we apply the method described in this section. Thus the fusion method described in this section is unable to take computational advantage of an additive decomposition of the utility valuation. Shenoy (1992) describes a modification of the fusion algorithm that is able to take advantage of an additive decomposition of the utility function. The modification involves some divisions.

DELETION SEQUENCES

Since $>$ is only a partial order, in general, we may have many deletion sequences (sequences that satisfy the condition stated in Theorem 11.1). If so, which deletion sequence should one use? First, we note that all deletion sequences will lead to the same final result. This is implied in the statement of the theorem. Second, different deletion sequences may involve different computational efforts. For example, consider the VBS shown in Fig. 11.5. In this example, the deletion sequence $DPTS$ involves less computational effort than $PDTS$ as the former involves combinations on the frame of three variables only whereas the latter involves combination on the frame of all four variables. Finding an optimal deletion sequence is a secondary optimization problem that has been shown to be NP-complete (Arnborg et al., 1987). But there are several heuristics for finding good deletion sequences (Kong, 1986; Mellouli, 1987; Zhang, 1988).

One such heuristic is called **one-step look ahead** (Kong, 1986). This heuristic tells us which variable to delete next from among those that qualify. As per this heuristic, the variable that should be deleted next is one that leads to combination over the smallest frame. For example, in the VBS of Fig. 11.5, two variables qualify for first deletion, P and D. This heuristic would pick D over P since deletion of P involves combination over the frame of $\{S, D, P, T\}$ whereas deletion of D only involves combination over the frame of $\{T, P, D\}$. Thus, this heuristic would choose deletion sequence *DPTS*.

PROOFS

In this section we give proofs for all results in the chapter.

Proof of Lemma 11.1
The result follows directly from the definition of marginalization. ∎

Proof of Lemma 11.2
The result follows directly from the definition of marginalization. ∎

Proof of Lemma 11.3
The result follows directly from the definition of marginalization. ∎
    To prove Theorem 11.1, we need a lemma.

Lemma 11.4
Suppose $\Delta = \{\mathcal{X}_D, \mathcal{X}_R, \{\mathcal{W}_X\}_{X \in \mathcal{X}}, \{v_1\}, \{\rho_1, ..., \rho_n\}, \rightarrow\}$ is a well-defined decision problem. Suppose X is a minimal variable in $\mathcal{X} = \mathcal{X}_D \cup \mathcal{X}_R$ with respect to the partial order $>$, where $>$ is the transitive closure of $\rightarrow$. Then

$$(\otimes\{v_1, \rho_1, ..., \rho_n\})^{\downarrow(\mathcal{X} - \{X\})} = \otimes \operatorname{Fus}_X\{v_1, \rho_1, ..., \rho_n\}$$

Proof of Lemma 11.4
We will prove this result in two mutually exclusive and exhaustive cases. Suppose v_1 is a payoff valuation for h_1, and suppose ρ_i is a potential for g_i, $i = 1, ..., n$.

Case 1
Suppose X is a decision variable. Without loss of generality, assume that v_1, $\rho_1, ..., \rho_k$ are the only valuations that bear on X. Let $v = v_1 \otimes \rho_1 \otimes ... \otimes \rho_k$, let $h = h_1 \cup g_1 \cup .... \cup g_k$, and let $c \in \mathcal{W}_{\mathcal{X} - \{X\}}$. Then

$$(\otimes\{v_1, \rho_1, ..., \rho_n\})^{\downarrow(\mathcal{X} - \{X\})}(\mathbf{c})$$

$$= \max\{[v_1(\mathbf{c}^{\downarrow h_1}, \mathbf{x})\rho_1(\mathbf{c}^{\downarrow g_1}, \mathbf{x})...\rho_k(\mathbf{c}^{\downarrow g_k}, \mathbf{x})\rho_{k+1}(\mathbf{c}^{\downarrow g_{k+1}})...\rho_n(\mathbf{c}^{\downarrow g_n})] | \mathbf{x} \in \mathcal{W}_X\}$$

$$= \max\{[v_1(\mathbf{c}^{\downarrow h_1}, \mathbf{x})\rho_1(\mathbf{c}^{\downarrow g_1}, \mathbf{x})...\rho_k(\mathbf{c}^{\downarrow g_k}, \mathbf{x})] | \mathbf{x} \in \mathcal{W}_X\}[\rho_{k+1}(\mathbf{c}^{\downarrow g_{k+1}})...\rho_n(\mathbf{c}^{\downarrow g_n})]$$

$$= \max\{v(\mathbf{c}^{\downarrow h - \{X\}}, \mathbf{x}) | \mathbf{x} \in \mathcal{W}_X\}[\rho_{k+1}(\mathbf{c}^{\downarrow g_{k+1}})...\rho_n(\mathbf{c}^{\downarrow g_n})]$$

$$= v^{\downarrow h - \{X\}}(\mathbf{c}^{\downarrow h - \{X\}})[\rho_{k+1}(\mathbf{c}^{\downarrow g_{k+1}})...\rho_n(\mathbf{c}^{\downarrow g_n})]$$

$$= \otimes \operatorname{Fus}_X\{v_1, \rho_1, ..., \rho_n\}(\mathbf{c})$$

Case 2

Suppose X is a random variable. Without loss of generality, assume that $v_1, \rho_1, ..., \rho_k$ are the only valuations that bear on X. Let $v = v_1 \otimes \rho_1 \otimes ... \otimes \rho_k$, let $h = h_1 \cup g_1 \cup ... \cup g_{k'}$ and let $\mathbf{c} \in \mathcal{W}_{\mathcal{X} - \{X\}}$. Then

$$(\otimes \{v_1, \rho_1, ..., \rho_n\})^{\downarrow(\mathcal{X} - \{X\})}(\mathbf{c})$$

$$= \Sigma \{[v_1(\mathbf{c}^{\downarrow h_1}, \mathbf{x}) \rho_1(\mathbf{c}^{\downarrow g_1}, \mathbf{x}) ... \rho_k(\mathbf{c}^{\downarrow g_k}, \mathbf{x}) \rho_{k+1}(\mathbf{c}^{\downarrow g_{k+1}}) ... \rho_n(\mathbf{c}^{\downarrow g_n})] | \mathbf{x} \in \mathcal{W}_X\}$$

$$= \Sigma \{[v_1(\mathbf{c}^{\downarrow h_1}, \mathbf{x}) \rho_1(\mathbf{c}^{\downarrow g_1}, \mathbf{x}) ... \rho_k(\mathbf{c}^{\downarrow g_k}, \mathbf{x})] | \mathbf{x} \in \mathcal{W}_X\} [\rho_{k+1}(\mathbf{c}^{\downarrow g_{k+1}}) ... \rho_n(\mathbf{c}^{\downarrow g_n})]$$

$$= \Sigma \{v(\mathbf{c}^{\downarrow h - \{X\}}, \mathbf{x}) | \mathbf{x} \in \mathcal{W}_X\} [\rho_{k+1}(\mathbf{c}^{\downarrow g_{k+1}}) ... \rho_n(\mathbf{c}^{\downarrow g_n})]$$

$$= v^{\downarrow h - \{X\}}(\mathbf{c}^{\downarrow h - \{X\}}) [\rho_{k+1}(\mathbf{c}^{\downarrow g_{k+1}}) ... \rho_n(\mathbf{c}^{\downarrow g_n})]$$

$$= \otimes \text{Fus}_X\{v_1, \rho_1, ... \rho_n\}(\mathbf{c})$$

Proof of Theorem 11.1

By definition, $(\otimes \{v_1, \rho_1, ..., \rho_n\})^{\downarrow \varnothing}$ is obtained by sequentially marginalizing a minimal variable. A proof of this theorem is obtained by repeatedly applying the result of Lemma 11.4. At each step, we delete a minimal variable and fuse the set of all valuations with respect to the minimal variable. Using Lemma 11.4, after fusion with respect to X_1, the combination of all valuations in the resulting VBS is equal to $(\otimes \{v_1, \rho_1, ..., \rho_n\})^{\downarrow(\mathcal{X} - \{X_1\})}$. Again, using Lemma 11.4, after fusion with respect to X_2, the combination of all valuations in the resulting VBS is equal to $(\otimes \{v_1, \rho_1, ..., \rho_n\})^{\downarrow(\mathcal{X} - \{X_1, X_2\})}$. And so on. When all the variables have been deleted, there will be a single valuation left. Using Lemma 11.4, this valuation will be $(\otimes \{v_1, \rho_1, ..., \rho_n\})^{\downarrow \varnothing}$.

ACKNOWLEDGEMENTS

This work was supported in part by the National Science Foundation under grant IRI-8902444. I am grateful for discussions with, and comments from Dan Geiger, Steffen Lauritzen, Pierre Ndilikilikeśha, Anthony Neugebauer, Geoff Schemmel, Philippe Smets, Glenn Shafer and Po-Lung Yu.

REFERENCES

Arnborg, S., Corneil, D.G. and Proskurowski, A. (1987) Complexity of finding embeddings in a k-tree. *SIAM Journal of Algebraic and Discrete Methods*, **8**, 277–284.

Bertele, U. and Brioschi, F. (1972) *Nonserial Dynamic Programming*, New York: Academic Press.

Darroch, J.N., Lauritzen, S.L. and Speed, T.P. (1980) Markov fields and log-linear interaction models for contingency tables. *Annals of Statistics*, **8**(3), 522–539.

Dubois, D. and Prade, H. (1990) Inference in possibilistic hypergraphs, in *Proceedings of the Third International Conference on Information Processing and Management of Uncertainty in Knowledge-based Systems (IPMU-90)*, Paris, France, 228–230.

Edwards, D. and Kreiner, S. (1983) The analysis of contingency tables by graphical models. *Biometrika*, **70**, 553–565.

Ezawa, K.J. (1986) Efficient evaluation of influence diagrams. Ph.D. thesis, Department of Engineering-Economic Systems, Stanford University.

Howard, R.A. and Matheson, J.E. (1984) Influence diagrams, in *The Principles and Applications of Decision Analysis*, **2**, R.A. Howard and J.E. Matheson (eds), Menlo Park, CA: Strategic Decisions Group, pp. 719–762.

Jensen, F.V., Olesen, K.G. and Andersen, S.K. (1990) An algebra of Bayesian belief universes for knowledge-based systems. *Networks*, **20**, 637–659.

Kiiveri, H., Speed, T.P. and Carlin, J.B. (1984) Recursive causal models. *Journal of the Australian Mathematics Society*, A, **36**, 30–52.

Kong, A. (1986) Multivariate belief functions and graphical models. Ph.D. thesis, Department of Statistics, Harvard University.

Lauritzen, S.L. and Spiegelhalter, D.J. (1988) Local computations with probabilities on graphical structures and their application to expert systems (with discussion). *Journal of the Royal Statistical Society*, B, **50**(2), 157–224.

Mellouli, K. (1987) On the propagation of beliefs in networks using the Dempster–Shafer theory of evidence. Ph.D. thesis, School of Business, University of Kansas.

Olmsted, S.M. (1983) On representing and solving decision problems. Ph.D. thesis, Department of Engineering-Economic Systems, Stanford University.

Pearl, J. (1988) *Probabilistic Reasoning in Intelligent Systems*, San Mateo, CA: Morgan Kaufmann.

Shachter, R.D. (1986) Evaluating influence diagrams. *Operations Research*, **34**(6), 871–882.

Shafer, G. and Shenoy, P.P. (1988) Local computation in hypertrees. Working Paper No. 201, School of Business, University of Kansas.

Shafer, G. and Shenoy, P.P. (1990) Probability propagation. *Annals of Mathematics and Artificial Intelligence*, **2**, 327–352.

Shenoy, P.P. (1989) A valuation-based language for expert systems. *International Journal for Approximate Reasoning*, **3**(5), 383–411.

Shenoy, P.P. (1990) Valuation-based systems for propositional logic, in *Methodologies for Intelligent Systems*, Z.W. Ras, M. Zemankova and M.L. Emrich (eds), Amsterdam: North-Holland, pp. 305–312.

Shenoy, P.P. (1991a) On Spohn's rule for revision of beliefs. *International Journal of Approximate Reasoning*, **5**(2), 149–181.

Shenoy, P.P. (1991b) Valuation-based systems for discrete optimization, in *Uncertainty in Artificial Intelligence*, **6**, P.P. Bonissone, M. Henrion, L.N. Kanal and J. Lemmer (eds), Amsterdam: North-Holland, pp. 385–400.

Shenoy, P.P. (1991c) Valuation-based systems: A framework for managing uncertainty in expert systems. In L.A. Zadeh and J. Kacprzyk (eds), *Fuzzy Logic for the Management of Uncertainty* New York: John Wiley & Sons, pp. 83–104.

Shenoy, P.P. (1991d) Valuation networks, decision trees, and influence diagrams: A comparison. Working Paper No. 227, School of Business, University of Kansas.

Shenoy, P.P. (1992) Valuation-based systems for Bayesian decision analysis. *Operations Research*, **40**(3), 463–484.

Shenoy, P.P. and Shafer, G. (1986) Propagating belief functions using local computations. *IEEE Expert*, **1**(3), 43–52.

Shenoy, P.P. and Shafer, G. (1988) Constraint propagation. Working Paper No. 208, School of Business, University of Kansas.

Shenoy, P.P. and Shafer, G. (1990) Axioms for probability and belief-function propagation, in *Uncertainty in Artificial Intelligence*, **4**, R.D. Shachter, T.S. Levitt, J.F. Lemmer and L.N. Kanal (eds), Amsterdam: North-Holland, pp. 169–198.

Tatman, J.A. (1986) Decision processes in influence diagrams: Formulation and analysis. Ph.D. thesis, Department of Engineering-Economic Systems, Stanford University.

Wermuth, N. and Lauritzen S.L. (1983) Graphical and recursive models for contingency tables. *Biometrika*, **70**, 537–552.

Whittaker, J. (1990) *Graphical Models in Applied Multivariate Statistics*, Chichester: John Wiley & Sons.

Zadeh, L.A. (1979) A theory of approximate reasoning, in *Machine Intelligence*, **9**, J.E. Ayes, D. Mitchie and L.I. Mikulich (eds), Chichester: Ellis Horwood, pp. 149–194.

Zhang, L. (1988) Studies on finding hypertree covers of hypergraphs. Working Paper No. 198, School of Business, University of Kansas.

PART THREE
Learning

Inferring causal structure in mixed populations

12

C. Glymour, P. Spirtes and R. Scheines

INTRODUCTION

In this chapter we will examine the problem of reliably inferring causal relations from statistical data and fragmentary background knowledge. Such causal inference problems arise in many instances in statistics, sociology, economics and epidemiology, among other areas. The problem can also arise when building expert systems that use Bayes networks. In many cases such networks are constructed on the basis of some expert's background knowledge; in many other cases, however, our background knowledge is woefully inadequate for constructing a useful expert system.

The causal structure among a set V of n random variables can be represented by a directed graph over V, where there is an edge from A to B if and only if A is a direct cause of B relative to V. (We will say that A is a direct cause of B relative to V if and only if there is a causal chain from A to B that does not include any of the other variables in V.)

Given a joint distribution over V, the sheer number of different possible causal theories over V makes inferring which causal structure generated the joint distribution extremely difficult. There are $\binom{n}{2}$ pairs of variables in V, and for each pair of variables there are four possibilities: A causes B, B causes A, neither causes the other, or both cause the other (which we interpret as a feedback loop). Hence, there are $4\binom{n}{2}$ different causal structures over V. If $n = 6$, there are 1 073 741 824 different theories; if background knowledge eliminates the possibility of cycles, there are approximately 3 000 000 theories; and if background knowledge provides the time order for each pair of variables, then there are approximately 32 768 different theories. For 12 variables, the corresponding numbers are 5 444 517 870 735 015 415 413 993 718 908 291 383 296, 521 939 651 343 829 405 020 504 063, and 73 786 976 294 838 206 464. It is not uncommon for medical models or econometric models to include several hundred variables.

We will describe an algorithm that efficiently and reliably infers causal structures from statistical data (under the assumption that every common cause of a pair of

Artificial Intelligence Frontiers in Statistics: AI and statistics III. Edited by D.J. Hand. Published in 1993 by Chapman & Hall, London. ISBN 0 412 40710 8

measured variables is itself measured). In order to execute the algorithm (called PC), it is necessary to determine for certain pairs of variables a and b, and certain sets of variables C, whether a and b are independent conditional on C (in the discrete case), or whether the partial correlation $\rho_{ab.C}$ vanishes (in the linear case). (This is also true of a number of other causal inference algorithms.)

We will then examine the extra difficulties that are posed by populations that consist of mixtures of subpopulations in which each subpopulation has the same causal structure, but with quantitatively causal relationships. In these populations, conditional independence relations that hold in each subpopulation generally do not hold in the population as a whole. Also, partial correlations that vanish in each subpopulation generally do not vanish in the population as a whole. In such mixed populations, PC cannot be reliably employed.

Finally, we will show that, in the special case that the variables are linearly related, each unit in the population has the same causal structure, but the linear coefficients are independently distributed the partial correlations are the same as those in a population generated by the same causal structure in which the linear coefficients do not vary. In populations in which the linear coefficients are independently distributed, PC can be reliably employed.

PSEUDO-INDETERMINISTIC AND INDETERMINISTIC SYSTEMS

Consider any collection of finite structures in which there is a set of input variables whose values are independently and randomly distributed, a set of variables each of which is some function—linear, nonlinear, or whatever—of some subset of the input variables, a set of variables each of which is some function of some subset of the union of the set of input variables and the set of first-level variables, and so on. We call such a collection of random variables, functional relations and distributions over the independent variables a **causal system**.

In all of these cases there is a common formal structure. The causal structure can always be represented by a directed graph. In circuits without feedback and in most

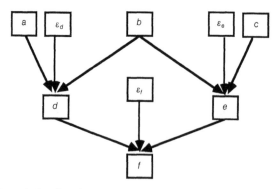

Figure 12.1 Causal structure I.

applied statistical cases, the directed graph is acyclic. Henceforth we will consider only acylic directed graphs. Figure 12.1 is an example of a graph of a causal structure.

Any directed acyclic graph represents a causal structure, the values of the variables represented as vertices of zero indegree (the inputs or exogenous variables) take their values randomly and independently of each other. (We assume that for any three disjoint sets A, B and C of exogenous random variables that A is independent of B conditional on C.) Assume, for the moment, that the value of each variable that is not an input variable is uniquely determined by its immediate causes. Then all variables are in fact random variables, and a graph of causal relations also represents a system of functional relations among these random variables: e.g.,

$$a \text{ input}$$
$$b \text{ input}$$
$$c \text{ input}$$
$$\varepsilon_d \text{ input}$$
$$\varepsilon_e \text{ input}$$
$$\varepsilon_f \text{ input}$$
$$d = r(a, b, \varepsilon_d)$$
$$e = s(b, c, \varepsilon_e)$$
$$f = t(d, e, \varepsilon_f)$$

For any three disjoint subsets X, Y, Z of $\{a, b, c, \varepsilon_d, \varepsilon_e, \varepsilon_f\}$ X is independent of Y conditional on Z. The graph specifies that the exogenous variables are independent of each other, and for each variable the graph determines what other variables it is a function of, but does not further specify the function. We say that the graph represents the **causal structure**. Many causal systems share the same causal structure and can be represented by the same graph.

A causal system S generates a probability distribution P in the following way. Since each endogenous variable is a function of the exogenous variables (or a function of variables that are themselves ultimately functions of the exogenous variables), once values for each of the exogenous variables have been specified, the values of all the variables in the system have been completely determined. Similarly, specifying a probability distribution over the exogenous variables completely determines a joint probability distribution over all the variables. We will extend this terminology to causal structures as well as causal systems: if causal system S generates distribution P, and S has causal structure C, we will also say that C generates P. Each causal system S generates a unique probability distribution. However, since many different causal systems share the same causal structure, many different distributions can be generated by one causal structure.

The connection between causality, directed acyclic graphs and probability distributions described above is tacitly assumed in many of the usual causal modelling formalisms in applied statistics, for example, in factor analysis, in linear structural equation models and in causal models with discrete variables. 'Recursive' structural equation models (sometimes called **linear causal models**), for example, specify a system of linear equations that can be viewed individually as regression equations with random regressors with non-zero variances. At least implicitly, there is a regression equation of this kind for each endogenous variable in the system. A system of such equations determines a directed acyclic graph, G, with its variables (omitting the error variables) as vertices in the obvious way. A joint probability distribution is imposed consistent with these assumptions.

In the complete set of variables, the value of an endogenous variable is always completely determined by its causal parents. However, in the example depicted in Fig. 12.1, if only the subset of variables $V = \{a, b, c, d, e, f\}$ is considered, the set of immediate causes in V of each endogenous variable does not uniquely determine the value of the variable. If we assume that for each endogenous variable X in the set V there is a unique exogenous variable of unit outdegree (i.e. at the tail of exactly one edge in the graph) and non-zero variance (an 'error' variable) not in V, that together with the causal parents of X in V completely determines the value of X, we call the causal structure of such a set of variables **pseudo-indeterministic**. The distribution generated by a pseudo-indeterministic causal system is the marginal of a distribution generated by a deterministic causal system. A particular kind of pseudo-indeterministic causal system is a **linear causal system**, in which all of the functions relating the variables are linear. In that case we assume that, for each variable, the variance of its distribution conditional on any set of variables not including error variables is non-zero, and that all partial correlations among non-error variables exist.

In the case of pseudo-indeterministic causal systems, the reason why the values of the endogenous variables are not completely determined by their causal parents in V is that the 'error' terms which partially determine the values of the endogenous variables are not in V. Another possible reason why the values of the endogenous variables are not completely determined by their causal parents in V is that there is a genuinely indeterministic relation between the endogeneous variables and the complete set of its causal parents. If that is the case, we say that the causal system is **indeterministic**.

CAUSAL GRAPHS AND BAYES NETWORKS

A Bayes network is a directed acyclic graph G and a distribution P that satisfies the following conditions:

1. *Markov condition*: In P, each variable in G is independent of all of its non-parental non-descendants conditional on its parents.
2. *Minimality condition*: No proper subgraph of G satisfies the Markov condition for P.

We will assume that any distribution P generated by a causal structure G satisfies the Markov and Minimality conditions with respect to P.

If every conditional independence relation that holds in P is entailed by satisfying the Markov condition for G, then we say that G is a **perfect representation** of P. We will also say that P is **faithful** to G. We will henceforth assume that each probability distribution generated by a causal structure is perfectly represented by the causal graph G of that structure; justification for this assumption is provided in Spirtes *et al.* (1991).

Pearl (1988) shows how to determine whether an atomic conditional independence statement is implied by the Markov boundary conditions for a graph G, using a graph-theoretic concept named d-separability.

In an acyclic graph $G = \langle V, E \rangle$ an **undirected path** from v_1 to v_n is an ordered n-tuple of vertices $\langle v_1, v_2, \ldots, v_{n-1}, v_n \rangle$ such that each vertex occurs only once, and for each pair of vertices v_k and v_{k+1}, either the edge $\langle v_k, v_{k+1} \rangle$ is in E, or the edge $\langle v_{k+1}, v_k \rangle$ is in E. If, in an undirected path $U = \langle v_1, v_2, \ldots, v_{n-1}, v_n \rangle$ there is a vertex v_k where the edge $\langle v_{k-1}, v_k \rangle$ is in E, and $\langle v_{k+1}, v_k \rangle$ is in E then v_k is a **collider** on U. If, in addition, there is no edge between v_{k-1} and v_{k+1} in G, then v_k is an **unshielded collider** on U.

A set of vertices X is d-separated from a set of vertices Y by a set of vertices Z in a graph $G = \langle V, E \rangle$ if and only if there is no undirected path U between a variable in X and a variable in Y such that

(a) for every vertex v_k on U that is a collider on U, either v_k is in Z or there is a directed path from v_k to some variable in Z; and

(b) for every vertex v_k on U that is not a collider on U, v_k is not in Z.

Theorem 12.1 (Pearl, 1988)
If G is a Bayes network of P, and X and Y are d-separated by Z in G, then X and Y are independent conditional on Z in P.

It follows that if P is faithful to G, then X and Y are independent conditional on Z in P if and only if X and Y are d-separated by Z in G.

INFERENCE OF CAUSAL STRUCTURE

Let us call a set of variables V **causally sufficient** if every common cause of any pair of variables in V is also in V. In Fig. 12.1 the set $V = \{a, b, c, d, e, f\}$ is causally sufficient, but the set $V' = \{a, c, d, e, f\}$ is not because b, a common cause of d and e, is not in V'.

Given a causally sufficient set of variables, and assuming that the graph of a causal structure is a perfect representation of any distribution generated by the causal structure, the following algorithm correctly constructs a pattern that represents a set of models that includes the true causal structure. The input to the algorithm is the set of conditional independence relations true of the probability P generated by a causal structure (or, under the assumption of faithfulness, a set of d-separation relations).

PC algorithm

Let $\mathbf{A}_C(A)$ denote the set of vertices adjacent to A or to B in graph C, except for A and B themselves. Let $\mathbf{U}_C(A, B)$ denote the set of vertices in graph C on (acyclic) undirected paths between A and B, except for A and B themselves. (Since the algorithm is continually updating C, $\mathbf{A}_C(A)$ and $\mathbf{U}_C(A, B)$ are constantly changing as the algorithm progresses.)

A. Form the complete undirected graph C on the vertex set $\mathbf{V}$.
B. $n = 0$.
    repeat
        For each pair of variables A, B adjacent in C, and for $X = A$ and $X = B$ do
           if $\mathbf{A}_C(X) \cap \mathbf{U}_C(A, B)$ has cardinality greater than or equal to n and A, B are

d-separated by any subset of $\mathbf{A}_C(X) \cap \mathbf{U}_C(A, B)$ of cardinality n, delete $A-B$ from C.

$n = n + 1$.

until for each pair of vertices A, B that are adjacent in C, $\mathbf{A}_C(A) \cap \mathbf{U}_C(A, B)$ and $\mathbf{A}_C(B) \cap \mathbf{U}_C(A, B)$ is of cardinality less than n.

C. Let F be the graph resulting from step B. For each triple of vertices A, B, C such that the pair A, B and the pair B, C are each adjacent in F but the pair A, C are not adjacent in F, orient $A - B - C$ as $A \Rightarrow B \Leftarrow C$ if and only if A and C are not d-separated by any subset of $\mathbf{A}_F(A) \cap \mathbf{U}_F(A, C)$ or any subset of $\mathbf{A}_F(C) \cap \mathbf{U}_F(A, C)$ containing B.

D. repeat

If there is a directed edge $A \Rightarrow B$, and undirected edge $B - C$, and no edge of either kind connecting A and C, then orient $B - C$ as $B \Rightarrow C$. If there is a directed path from A to B, and an undirected edge $A - B$, orientate $A - B$ as $A \Rightarrow B$.

until no more arrowheads can be added.

The output of this algorithm is a pattern that contains both directed and undirected edges and represents a set of directed acyclic graphs all of which have the same d-separation relations. A directed acyclic graph G is in the set of graphs represented by pattern Π if and only if:

1. G has the same adjacency relations as Π.
2. If the edge between A and B is oriented $A \Rightarrow B$ in Π, then it is oriented $A \Rightarrow B$ in G.
3. X is an unshielded collider on path U in G if and only if U is an unshielded collider on U in Π.

An algorithm that is similar in spirit but constructs undirected graphs has been independently suggested by Fung and Crawford (1990).

The complexity of the algorithm for a graph G is determined by $\max(|\mathbf{A}_G(a)|)$ over all pairs of vertices a, b, which is never more than the largest degree in G. Generally stage B of the algorithm continues testing for some steps after the correct undirected graph has been identified. The number of steps required before the true graph is found (but not necessarily until the algorithm halts) depends on the maximal number of **treks** between a pair of variables, say a, b, that share no vertices adjacent to a or b. (A trek is a pair of directed paths from some vertex z to a, b, respectively, intersecting only at z, or a directed path from a to b or a directed path from b to a.) If these maximal numbers are held constant as the number of vertices increases, so that k, the maximal order of the conditional independence relations that need to be tested, does not change, then the worst-case computational demands of the algorithm increase as

$$\binom{n}{2} \sum_{i=0}^{k} \binom{n-2}{i}$$

which is bounded by

$$\frac{n^2}{2} \cdot \frac{(n-1)^k}{k!}$$

It is possible to recover sparse graphs with as many as 100 variables. Of course, the computational requirements increase exponentially with k.

In many cases it is more efficient to perform conditional independence tests on all subsets of $\mathbf{A}_G(A)$ and $\mathbf{A}_G(B)$ rather than to compute $\mathbf{U}_G(A, B)$. We have not yet theoretically determined the trade-off.

The structure of the algorithm and the fact that it continues to test even after having found the correct graph suggest a natural heuristic for very large variable sets whose causal connections are expected to be sparse. If A and B are independent conditional on C, let us call the cardinality of C the **order** of the conditional independence. A natural heuristic is to set a fixed bound on the order of conditional independence relations that will be considered.

Theorem 12.2
If the causal graph G that generated a distribution P is causally sufficient and a perfect representation of P, then, given a list of the conditional independence relations true of P as data, the PC algorithm constructs a set of graphs that includes the true graph.

The proof for this can be found in Spirtes and Glymour (1991).

The PC algorithm has two major advantages over other algorithms that have been suggested for discovering causal structures. First, it takes advantage of the sparseness of the graph to reduce the number of conditional independence relations that need to be tested. Second, it can be reliably applied to large number of variables even if the sample size is only moderately large. For discrete variables, reliable tests of high-order conditional independence relations require huge sample sizes. Because the PC algorithm takes advantage of the sparseness of the graph to reduce the order of the conditional independence relations that need to be tested, it can be applied to large sets of variables with only moderate sample sizes.

We judge whether or not a pair of variables is independent conditional on a given set of variables by performing a statistical test. When statistical tests are applied to sample data, sampling error can lead to incorrect conclusions about whether or not a pair of variables is independent conditional on a given set of variables in the population. In order to determine how sensitive the PC algorithm is to sampling error we have applied the PC algorithm to simulated data generated from the causal network depicted in Fig. 12.2 taken from Beinlich *et al.* (1989).

The network represents a causal network in an intensive care unit. The graph and the conditional probability of each vertex given its parent were constructed by consulting an expert. Herskovitz and Cooper (1990) generated discrete data for the ALARM network, using variables with two, three and four values. Given their data, the TETRAD II program with the PC algorithm reconstructs almost all of the undirected graph (it omitted two edges in one trial; and in another also added one edge) and orients most edges correctly. The algorithm was run on a DecStation 3100 in about 10 minutes. In most orientation errors an edge was oriented in both directions. In each trial the output pattern omitted two edges in the ALARM network; in one of the cases it also added one edge that was not present in the ALARM network. The results were scored as follows:

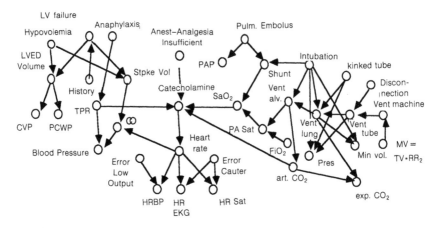

CO: cardiac output CVP: central venous pressure. LVED volume: left ventricular endiastolic volume, LV failure: left ventricular failure. MV: minute ventilation. PA Sat: pulmonary artery oxygen saturation. PAP pulmonary artery pressure. PCWP pulmonary capillary wedge pressure. Pres breathing pressure. RR: respiratory rate. TPR: total peripheral resistance. TV: tival volume.

Figure 12.2 Alarm network.

An **edge existence error of commission** (Co) occurs when any pair of variables are adjacent in the output but not in the pattern of the graph in Fig. 12.2. An **edge direction error of commission** occurs when any arrowhead not in the pattern of the graph in Figure 12.2 occurs in the output on an edge that is both in the output and an edge occurring in the pattern of the graph in Figure 12.2. **Error of omission** (Om) are defined analogously in each case. The results are tabulated as the average over the trial distributions of the ratio of the number of actual errors to the number of possible errors of each kind. With simulated data obtained from Herskovitz and Cooper at sample size 10 000 the results are described in Table 12.1.

In the linear case, we ran ten different simulation trials. In each trial, using the same directed graph, linear coefficients with values between 0.1 and 0.9 were randomly assigned to each directed edge in the graph. An 'error' variable was assigned to each variable in the graph. We simulated uncorrelated standard normal distributions on all exogenous variables (those with no parents) of sample size 2000. For each unit in the population, a pseudo-random number was generated as a value for each exogenous variable, including error variables. (We used the UNIX 'random' facility to generate pseudo-random numbers from a uniform distribution, and then transformed them into pseudo-random samples from a normal distribution.) Then the value of each

Table 12.1 Simulation results for discrete model

Trial	% Edge existence errors		% Edge direction errors	
	Commission	Omission	Commission	Omission
1	0	4.3	27.1	10.0
2	0.2	4.3	5.0	10.4

Table 12.2 Simulation results for linear model

# Trials	% Edge existence errors		% Edge direction errors	
	Commission	Omission	Commission	Omission
10	0.06	4.1	17	3.5

endogenous variable was calculated as a linear function of the exogenous variables. The covariance matrix and sample size were given to a version of the TETRAD II program with an implementation of the PC algorithm. This implementation takes as input a covariance matrix, and it outputs a pattern. No information about the orientation of the variables was given to the program. Run on a DecStation 3100, for each data set the program required less than 15 seconds to return a pattern. The results at sample size 10 000 are summarized in Table 12.2.

PARTIAL CORRELATIONS AND d-SEPARABILITY

The input to the PC algorithms requires determining when a given conditional independence relation holds in the distribution P. However, in the case of linear causal theories, rather than using facts about conditional independencies in P as input, we use facts about vanishing partial correlations. The following theorems justify using the results of statistical tests of vanishing partial correlations as input to the PC algorithm.

For a linear causal theory with graph G, let us say that a partial correlation $\rho_{xz.Y}$ is **strongly implied** to vanish if and only if it vanishes for every linear distribution generated by G. We assume that any partial correlation that vanishes in the population is strongly implied to vanish because of the following theorem:

Theorem 12.3
Let M be a linear model with n free linear coefficients $a_1, \ldots, a_n$ and k variances $v_1, \ldots, v_k$. Let $M(\mathbf{U})$ be the model obtained by specifying values $\mathbf{U} = \langle u_1, \ldots, u_n, u_{n+1}, \ldots, u_{n+k} \rangle$ for $a_1, \ldots, a_n$ and $v_1, \ldots, v_k$. Let $\mathbf{P}$ be the set of probability measures P on the space $\mathbb{R}^{n+k}$ of values of the parameters of model M such that for every subset S of $\mathbb{R}^{n+k}$ having Lebesque measure zero, $P(S) = 0$. Let Q be the set of vectors $\mathbf{U}$ of coefficient and variance values such that every multinormal probability distribution consistent with $M(\mathbf{U})$ has at least one statistical independence relation not represented in the directed acyclic graph of M according to d-separability. Then for $P \in \mathbf{P}$, $P(Q) = 0$.

The proof for this is described in Spirtes (1989) and in Spirtes *et al.* (forthcoming). We have also proved the following:

Theorem 12.4
In a linear causal system with graph G and distribution P, if x and z are distinct variables, and Y is a set of variables not including x and z, then Y d-separates x and z if and only if $\rho_{xz.Y}$ is strongly implied to vanish.

The proof for this is described in Spirtes (1989) and Spirtes *et al.* (forthcoming).

MIXED CAUSAL STRUCTURES

Consider a population that is a mixture of structures $\langle g, P_1 \rangle$ and $\langle g, P_2 \rangle$ where P_1 and P_2 are distinct and, we will suppose, both faithful to graph g. Let the proportions in the mixture be $n:m$. This sort of case appears to be the simplest and easiest sort of mixing, for we know in this case that the two distributions have the very same conditional independence relations. The unfortunate fact, however, is that even in this case the mixed population does not generally have those same conditional independence relations, and indeed unless special constraints are satisfied by P_1 and P_2, the mixed distribution will have no non-trivial conditional independence relations at all.

Yule (1903) concluded his fundamental paper on the theory of association of attributes is statistics with a section 'On the fallacies that may be caused by the mixing of distinct records' (where $|AB|$ is a measure of associate between A and B that vanishes when A and B are independent):

> It follows from the preceding work that we cannot infer independence of a pair of attributes within a sub-universe from the fact of independence within the universe at large.... The theorem is of considerable practical importance from its inverse application; i.e. even if $|AB|$ have a sensible positive or negative value we cannot be sure that nevertheless $|AB|C|$ and $|AB| \sim C|$ are not both zero. Some given attribute might, for instance, be inherited neither in the male line nor the female line; yet a mixed record might exhibit a considerable apparent inheritance.
>
> The fictitious association caused by mixing records finds its counterpart in the spurious correlation to which the same process may give rise in the case of continuous variables, a case to which attention was drawn and which was fully discussed by Professor Pearson in a recent memoir. If two separate records, for each of which the correlation is zero, be pooled together, a spurious correlation will necessarily be created unless the mean of one of the variables, at least, be the same in the two cases.

Let $P(XYZ) = nP_1(XYZ) + mP_2(XYZ)$, with $n + m = 1$, $n \neq 0$, $m \neq 0$. Elementary algebra shows that $P(XY|Z) = P(X|Z)P(Y|Z)$ if and only if

$$n^2 P_1(XYZ)P_1(Z) + nmP_2(XYZ)P_1(Z) + mnP_1(XYZ)P_2(Z) + m^2 P_2(XYZ)P_2(Z)$$
$$= n^2 P_1(XZ)P_1(YZ) + nmP_1(XZ)P_2(YZ) + mnP_2(XZ)P_1(YZ) + m^2 P_2(XZ)P_2(YZ)$$
$$(12.1)$$

If, in both distributions, X, Y are independent conditional on Z (that is, $P_1(XY|Z) = P_1(X|Z)P_1(Y|Z)$ and $P_2(XY|Z) = P_2(X|Z)P_2(Y|Z)$), then Equation (12.1) reduces to

$$P_2(XYZ)P_1(Z) + P_1(XYZ)P_2(Z) = P_1(XZ)P_2(YZ) + P_2(XZ)P_1(YZ) \quad (12.2)$$

which is not a function of the proportions n, m. Equation (12.2) can be put in the slightly more perspicuous form:

$$P_2(XY|Z) + P_1(XY|Z) = P_1(X|Z)P_2(Y|Z) + P_2(X|Z)P_1(Y|Z) \quad (12.3)$$

or, since we are assuming that X and Y are conditionally independent on Z in both

P_1 and P_2,

$$P_2(X|Z)P_2(Y|Z) + P_1(X|Z)P_1(Y|Z) = P_1(X|Z)P_2(Y|Z) + P_2(X|Z)P_1(Y|Z) \quad (12.4)$$

The rather surprising conclusion is that when we mix probability distributions we should expect to find all possible conditional **dependence** relations. Hence, in mixed populations, conditional independence and dependence will not be a reliable guide to causal structure. Applying the PC algorithm to such data will, for example, produce a complete undirected graph. This has a practical if informal moral for the significance we ought to give to inferences from non-experimental data. When from properly collected data sets with large sample sizes we find that the resulting undirected graph is not complete, we ought to be a little impressed. Either some constraints have been satisfied by chance, or over some variables almost all units in the sample have the same causal structure, and that structure does not include the missing connection.

An interesting question is therefore whether there is any means to infer causal dependence in mixed populations or samples. Suppose we have prior knowledge that excludes the complete undirected graph; it may even forbid particular edges. If we then obtain a probability distribution in which no non-trivial conditional independence relations hold, can we infer anything about the class of mixtures (consistent with the prior knowledge) from which the population distribution may have come? We have no idea.

In the case of linear structures the input to the PC algorithm is the set of zero partial correlations true of a distribution generated by some causal structure. When populations with two different distributions each associated with a linear structure are mixed, vanishing correlations in the mixed distribution will not mark independence in the mixed distribution, and vanishing partial correlations in the mixed distribution will not mark conditional independence in the mixed distribution. It is easy to verify that for any mixture of two distributions—based on linear structures or not—the covariance of two variables vanishes in the mixture if and only if

$$k_1 \text{Cov}_1(X, Y) + k_2 \text{Cov}_2(X, Y)$$
$$= k_1 k_2 [\mu_1 X \mu_2 Y + \mu_1 Y \mu_2 X] + k_1 (k_1 - 1)\mu_1 X \mu_1 Y + k_2 (k_2 - 1)\mu_2 X \mu_2 Y]$$
$$= k_1 k_2 [\mu_1 X \mu_2 Y + \mu_1 Y \mu_2 X] - k_1 k_2 [\mu_1 X \mu_1 Y + \mu_2 X \mu_2 Y] \quad (12.5)$$

where the proportion of population 1 to population 2 is $n{:}m$, $k_1 = n/(n+m)$, $k_2 = m/(n+m)$, and $\mu_1 X$ is the mean of X in population 1, etc.

Some of the uncertainty occasioned by mixed populations disappears when we can impose experimental controls, and this is one of the principal advantages of experimental procedures.

A population of systems sharing the same causal structure but considered with regard to a set V of variables that fails to include some common cause T of variables in V might be considered to have a mixed distribution. If, for simplicity, T is binary, the population can be viewed as a mixture of a subpopulation in which $T = 1$ and a subpopulation in which $T = 0$. Assuming for simplicity that there is no other causal connection between X and Y besides a trek with T as its source, in the subpopulation with $T = 0$ the variables X and Y will be independent and in the subpopulation with $T = 1$ X and Y will be independent, but in the mixed population X and Y will, of course, be dependent.

When we consider discrete data, the phenomena of mixtures show a fundamental limitation in our means of representing causal relations. Consider a simple switch. Suppose battery A has two states: charged and uncharged. A charge in battery A will cause bulb C to light up provided the switch B is on, but not otherwise. If A and B are independent random variables, then A and C are dependent conditional on B and on the empty set, and B and C are dependent conditional on A and the empty set, and A and B are dependent conditional on C. The causal structure therefore looks like the directed graph shown in Figure 12.3.

There is nothing wrong with this conclusion except that it is not fully informative. The dependence of A and C arises entirely through the condition $B = 1$. When $B = 0$, A and C are independent. The graph does not tell us that when $B = 0$ manipulating A will have no effect on C. Knowledge of the causal graph without this further information could lead to very mistaken expectations. Consider, for example, cases in which 'switch' variables analogous to B have off-values in the vast majority of the population. Then manipulating causes such as A will in most cases have no effect. Since in discrete data the conditional independence facts, if known, identify the switch variables, a better representation would identify certain parents of a variable as switches. But because in variables that take several discrete values, A may be a switch for B and B may be a switch for A, and several distinct values of B may be 'on'-values for A and conversely, a general representation of this sort would often not be very easy to grasp. A better practical arrangement might be a query system that, besides inferring the causal graph or graphs, responds to the user's questions about the effects of the manipulation of variables.

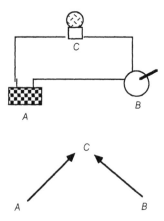

Figure 12.3 Switches.

RANDOM COEFFICIENT LINEAR STRUCTURES

Earlier in this chapter, we described linear causal structures in which each unit in the population has the same linear coefficients (i.e. they were constant random variables). Let us now call this a **constant-coefficient linear causal structure**. In a **random-coefficient linear causal structure**, the coefficients are non-constant random variables such that any set of coefficients is independent of any other disjoint set of coefficients or non-coefficient exogenous random variables. In both constant-coefficient and random-coefficient linear causal structures we assume that the partial correlation of each pair of variables on any set of variables exists. Note that this independence assumption is also true of constant-coefficient linear causal structures.

Theorem 12.5

For any two variables x and y in a random-coefficient linear causal structure RC, the covariance of x and y is equal to the covariance of the corresponding variables x' and y' in a constant-coefficient linear causal structure CC with the same graph, and in which the expected value of each linear coefficient in RC is equal to the constant value of the corresponding linear coefficient in CC.

The proof is given in Appendix 12A.

Note that this proof depends upon the independence of the linear coefficients from each other. This is true of both constant-coefficient and random-coefficient linear causal structures, but not true in general of linear causal structures in which large sub-populations share the same linear coefficients. (Of course, without *some* constraint upon the relationships between the linear coefficients for different members of a population, any population could be considered to arise from a linear model in which the linear coefficients depended upon the member of the population.)

The covariances completely determine the values of the partial correlations, and the partial correlations determine the output of the algorithm. Since the algorithm gives the correct output for non-random linear causal structures, and the covariance matrix of a random-coefficient linear causal structure is equal to the covariance matrix of a non-random-coefficient linear causal structure with the same graph, the algorithm gives the correct output for random-coefficient linear causal structures.

ACKNOWLEDGEMENTS

Research for this chapter was supported by the Naval Personnel Research and Development Center and the Office of Naval Research under contract number N00114-89-J-1964. We thank an anonymous reviewer for helpful suggestions and for correcting our analysis of the complexity of the PC algorithm.

APPENDIX 12A: PROOF OF THEOREM 12.5

The covariance of x and y is equal to $E(xy) - E(x)E(y)$. The formulae occurring in the following proof are correct for both models RC and CC.

If there is an edge from non-coefficient random variable a to b in the graph, then there is a non-zero coefficient of a in the equation for b. Label the edge from a to b by this non-zero coefficient. Label a directed path p by the product of the labels of the edges in the path. Let Ex be the set of exogenous variables, and P_{ab} be the set of paths from a to b.

The values of random variables x, y, and xy are, respectively:

$$x = \sum_{e \in Ex} \sum_{p \in P_{ex}} L(p)e$$

$$y = \sum_{f \in Ex} \sum_{q \in P_{fy}} L(q)f$$

$$xy = \sum_{e \in Ex} \sum_{p \in P_{ex}} \sum_{f \in Ex} \sum_{q \in P_{fy}} L(p)L(q)ef$$

The expected values of each of these variables is given below:

$$E(x) = E\left(\sum_{e \in Ex} \sum_{p \in P_{ex}} L(p)e \right) = \sum_{e \in Ex} \sum_{p \in P_{ex}} E(L(p)e)$$

$$E(y) = E\left(\sum_{f \in Ex} \sum_{q \in P_{fy}} L(q)f \right) = \sum_{f \in Ex} \sum_{q \in P_{fy}} E(L(q)f)$$

$$E(xy) = \left(\sum_{e \in Ex} \sum_{p \in P_{ex}} \sum_{f \in Ex} \sum_{q \in P_{fy}} L(p)L(q)ef \right) = \sum_{e \in Ex} \sum_{p \in P_{ex}} \sum_{f \in Ex} \sum_{q \in P_{fy}} E(L(p)L(q)ef)$$

Since the coefficients are independent of each other and the non-coefficient random variables, and the path labels are products of coefficients,

$$E(L(p)L(q)ef) = E(L(p))\, E(L(q))E(ef)$$

The label of a path is equal to the product of the labels of the edges

$$L(p) = \prod_{edge \in p} L(edge)$$

Substituting into the formula for $E(xy)$, we obtain

$$E(xy) = \sum_{e \in Ex} \sum_{p \in P_{ex}} \sum_{f \in Ex} \sum_{q \in P_{fy}} E\left(\prod_{edge \in p} L(edge) \right) E\left(\prod_{edge \in q} L(edge) \right) E(ef)$$

By the independence of the linear coefficients,

$$E\left(\prod_{edge \in p} L(edge) \right) = \prod_{edge \in p} E(L(edge))$$

It follows that

$$E(xy) = \sum_{e \in Ex} \sum_{p \in P_{ex}} \sum_{f \in Ex} \sum_{q \in P_{fy}} \prod_{edge \in p} E(L(edge)) \prod_{edge \in q} E(L(edge))E(ef)$$

Similarly,

$$E(L(p)e) = E(L(p))E(e)$$

and

$$E(L(q)f) = E(L(q))E(f)$$

Substituting these into the formula for $E(x)E(y)$, we obtain

$$E(x)E(y) = \left(\sum_{e \in Ex} \sum_{p \in P_{ex}} E(L(p))\, E(e) \right) \left(\sum_{f \in Ex} \sum_{q \in P_{fy}} E(L(q))E(f) \right)$$

Again, from the independence of the linear coefficients it follows that

$$E(x)E(y) = \left(\sum_{e \in Ex} \sum_{p \in P_{ex}} \prod_{edge \in p} E(L(edge))E(e) \right) \left(\sum_{f \in Ex} \sum_{q \in P_{fy}} \prod_{edge \in q} E(L(edge))E(f) \right)$$

In CC, since $L(edge)$ is a constant, $E(L(edge)) = L(edge)$. In RC, by hypothesis, $E(L(edge)) = L(edge)$ in CC. Hence the expression $E(xy) - E(x)E(y)$ is the same in both RC and CC.

REFERENCES

Beinlich, I., Suermondt, H., Chavez, R. and Cooper, G. (1989) The Alarm Monitoring System: Case study with two probabilistic inference techniques for belief network, in *Proceedings of the Conference on Artificial Intelligence in Medical Care*, London, 247–256.

Fung, R. and Crawford, S. (1990) Constructor: a system for the induction of probabilistic models, in *Proceedings of the American Association for Artificial Intelligence*, Boston, 762–769.

Herskovitz, E. and Cooper, G. (1990) Kutato: an entropy-driven system for construction of probabilistic expert systems from databases, in *Proceedings of the Sixth Conference on Uncertainty in Artificial Intelligence*, 54–63.

Pearl, J. (1988) *Probabilistic Reasoning in Intelligent Systems: Networks of Plausible Inference*, San Mateo, CA: Morgan Kaufmann.

Spirtes, P. (1989) Calculating Tetrad constraints implied by directed acyclic graphs. Report No. CMU-LCL-89-3, Laboratory for Computational Linguistics, Carnegie Mellon University.

Spirtes, P. and Glymour, C. (1991) An algorithm for fast recovery of sparse causal graphs. *Social Science Computer Reviews*, **9**, 62–72.

Spirtes, P., Glymour, C. and Scheines, R. (1991) From probability to causality. *Philosophical Studies*, **64**, 1–36.

Spirtes, P., Glymour, C. and Scheines, R., *Causality, Statistics, and Prediction*, Springer-Verlag, forthcoming.

Yule, G. (1903) Notes on the theory of association of attributes in statistics, *Biometrika*, **2**, 121–133.

A knowledge acquisition inductive system driven by empirical interpretation of derived results

13

K. Tsujino and S. Nishida

INTRODUCTION

Inductive learning is one of the most powerful techniques for constructing knowledge-based systems. It enables full (or semi-)automatic formalization of the knowledge base. In other words, if we specify proper knowledge representation, hypotheses and enough examples, i.e. proper inductive bias (Utgoff, 1986), we do not need to determine the details of the knowledge. Quinlan (1986) proposed an efficient induction algorithm named ID3 to generate classification knowledge in the form of a decision tree from examples represented in a vector form, which consists of a class to which an example belongs, the features that it has and the corresponding values that it satisfies. Although it provides quite reasonable knowledge statistically, the learned trees are not yet developed enough from the expert's point of view. That is because the system knows nothing about the target domain, so the induced knowledge sometimes lacks a lot of essential constraints required in the domain.

When a human expert inspects such a result, he can easily point out some 'improper' conditions (**improprieties**) in it. An impropriety is not an error or a fault statistically but something that seems strange to an expert. By interpreting how the improprieties arise, the expert can recommend new examples and new constraints according to his domain knowledge. Nevertheless, such knowledge is difficult to acquire beforehand by top-down reflection. This fact suggests that inductive learning can be an efficient mental stimulus for knowledge acquisition as well as a powerful technology for producing reasonable classifiers automatically.

This is the basic idea behind the proposed knowledge acquisition system named KAISER. As the acronym implies, KAISER (a Knowledge Acquisition Inductive System promoted by Explanatory Reasoning)—see Tsujino *et al.*, 1990; Dabija *et al.*, 1992— inductively learns classification knowledge in the form of a decision tree and analyses the result and process with domain- and task-specific knowledge to detect improprieties. Then it asks suggestive questions to eliminate the improprieties and

Artificial Intelligence Frontiers in Statistics: AI and statistics III. Edited by D.J. Hand. Published in 1993 by Chapman & Hall, London. ISBN 0 412 40710 8

acquires new domain knowledge, examples and various information for the next induction cycle.

The fundamental issues of KAISER include the following. What does the expert think improper in the result? Why does he feel so? What kind of knowledge is it based on? And how does he override the induction? What kinds of choices are there? By investigating these issues, we can derive the interview strategy that interprets the induction process and associates it with interviewing. The first issue is indispensable to clearing up the features of impropriety. The second issue is necessary to developing an interpretation and obtaining an explanation of the impropriety. The third issue makes it possible to provide considerate choices for resolving the impropriety.

ARCHITECTURE OF KAISER

Figure 13.1 illustrates the knowledge acquisition cycle of KAISER. KAISER consists of four sub-modules and three databases. The inductive learner adopts ID3 to generate a decision tree from the examples stored in the example database. The impropriety detector evaluates the decision tree based on the knowledge stored in the domain knowledge base and the impropriety knowledge base to detect the improprieties such as unreliable conditions in the tree and mismatching between the decision tree and domain knowledge. The domain knowledge base contains abstract causal knowledge that the target knowledge should satisfy. This knowledge base can initially be empty or specified by a human expert to the extent that he can recall. The impropriety knowledge base contains knowledge about the improprieties that are categorized into two types, structural improprieties that require no domain knowledge and semantic improprieties that refer to the domain knowledge. These types of impropriety knowledge can be prescribed beforehand, since neither depends on any domain knowledge, although the latter refers to the knowledge. Even if the domain knowledge base is empty, KAISER can acquire both examples and domain

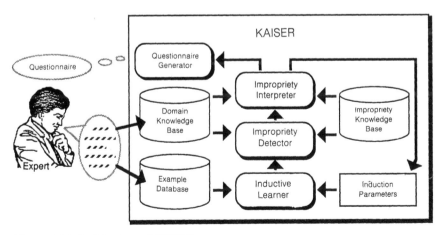

Figure 13.1 Architecture of KAISER.

knowledge by applying the domain-independent structural knowledge in the impropriety knowledge base. The impropriety interpreter analyses the detected improprieties and tries to suggest why they arise and how to eliminate them. The interpretation results are converted into queries and choices by the questionnaire generator. This questionnaire gives the human expert a mental stimulus that helps him recall missing examples, correct domain knowledge and proper induction parameters.

INDUCTION OF DECISION TREES

As mentioned above, KAISER adopts Quinlan's ID3 algorithm as its inductive learner. Although it provides simple and powerful methods for learning classifiers without any domain knowledge, it has several inductive problems including the following: it sometimes generates ignorant leaves that do not specify any class when multiple- (not binary-)valued attributes are used; the criterion for attribute selection has a strong tendency to overestimate detailed attributes and split the examples into too many sub-examples at once, which makes ID3 less resistant to noisy examples; and it cannot deal with numeric values. Much research (Quinlan, 1986; Tsujino et al., 1989) has been carried out to determine how to cope with these problems. KAISER thinks of these problems as structural improprieties and copes with them by heuristic detection and modification in a manner similar to the approach of Tsujino et al.'s (1989). This approach enables KAISER to use various kinds of domain knowledge to guide ID3.

STRUCTURAL IMPROPRIETIES

Improprieties of this type are task-specific and domain-independent, that is, they are not dependent on the target knowledge to be acquired but peculiar to the classification task or employed learning algorithm. They are detected as generalization failures or unreliable conditions, e.g., affected by the noisy examples. These improprieties give many kinds of clues for reasonable generalization according to the common knowledge about the classification task. To make the matter clear, let us think of the following two examples. First, when the price of T-shirts of size LL and M is $30 but that of size S is $28, how much do you pay for an L-size shirt? Second, when the price of a T-shirt is $32 while that of others on the same bargain rack is $30, do you pay $32 for that shirt without hesitation? Most of us will feel a kind of impropriety in these situations. The former example illustrates impropriety of missing information, and the latter of noisy examples. This sort of knowledge is common in all classification tasks. Currently, KAISER provides the following five types of structural impropriety. Some typical elimination methods recommended by each impropriety are also described.

1. *Noisy node impropriety.* A node was generated based on a few examples, e.g., less than twice the average number of examples in every leaf. This is a typical clue for pruning. It recommends changing the node to a leaf of its major class if there is no other impropriety at the node.
2. *Nil node impropriety.* The class of a leaf cannot be determined by induction because of the lack of examples. This often arises when we use multiple-valued (not binary)

attributes. A simple elimination method of this impropriety is to ask the correct class and/or to get new examples that will fit the leaf.

3. *Inseparable example impropriety.* Some examples cannot be separated only with given attributes. This impropriety possesses both features of nil node and noisy node improprieties.

4. *Similar node impropriety.* Two brother nodes refer to the same attribute, their entropies are near, and they consist of similar component classes. This is a structural clue to generalize the attribute of their father by merging the links to them. It recommends reconstructing an inferior tree by merging the examples at the brothers.

5. *Similar class impropriety.* More than one node tries to separate the same set of classes at different places in a decision tree. This is a clue for a new attribute. It recommends the induction of a new decision tree from the classes that may discriminate the confused examples efficiently as well as asking for such a new attribute.

REPRESENTATION OF DOMAIN KNOWLEDGE

In order to detect semantic improprieties, we must know about the domain and explain the decision tree by mapping the domain knowledge on to the nodes in the tree.

Abstract relationships between a class and attributes

Although it is difficult to describe exact rules, a human expert can prescribe explicit and abstract relationships between a class and the attributes. This knowledge is essential for the explanation-based reasoning of KAISER. It is formulated into the following three types of conditions: essential conditions (Econd); desirable conditions (Dcond); and permissible conditions (Pcond). For example, it is an Econd for insects that they have six legs. Their Dcond will be that they have four wings and the Pcond may be that they have eight legs.

Inter-attribute relationships

Some of the attributes can be derived from other attributes. For instance, the number of claws will be derived from the number of legs. Certain abstract and virtual attributes can be defined—for instance, density from mass and volume. KAISER accepts this knowledge in the form of Lisp function tables. For example, the *density* attribute can be defined as (*density*(/*mass volume*)). This knowledge is necessary to interpret the descriptors in the abstract domain knowledge described in the previous subsection, because *mass* and *volume* may appear in the decision tree instead of *density* used in the domain knowledge.

Ordinal relationships among attribute values

If an attribute represents an ordinal feature, the values have ordinal relationships. This information is necessary to identify the siblings of a node to eliminate many kinds of improprieties. For example, (*size-of-Tshirt* L LL M) will mean that the 'L' link of attribute *size-of-Tshirt* can be merged into another link if it has the rest values 'LL' and 'M'.

MAPPING OF DOMAIN KNOWLEDGE

KAISER examines two types of mapping of domain knowledge on to a decision tree. One is theory-based mapping (TBM), and the other is result-based mapping (RBM). The semantic improprieties are detected based on these mapping results. The TBM refers to the conditions on a path and those in the domain knowledge to predict the class for each leaf. We call the sets of class and the belief **support factors**. These are calculated by the following algorithm:

For all domain knowledge and paths in a decision tree,

1. if a path includes all of the Econd of a domain knowledge, then set the support factor of the path by the knowledge to 1.0;
2. if a path does not satisfy (1) but includes the conditions generalized by replacing some conditions in the Econd by Pcond, then set the support factor by the knowledge to 0.8^n, where n is the number of replaced conditions; and
3. if a path satisfies (1) or (2), and it also includes some conditions in Dcond, then multiply the support factor by 1.2^n, where n is the number of matched Dcond.

While the TBM infers the class of the path based on the domain knowledge without believing the induction results, the RBM verifies a path to a classified leaf based on the domain knowledge. The RBM is examined after the TBM for each path whose class is determined by induction but not supported by any TBM, that is, no domain knowledge could cover the path. If a path includes some of the Econd in a domain knowledge that supports the class of the path, then the path is marked as an insufficient path with the missing conditions. In many cases, these conditions become necessary when plenty of training examples are specified.

SEMANTIC IMPROPRIETIES

Ideally, all of the conditions in a decision tree will be interpreted in terms of the descriptors in the domain theory, in other words, the TBM succeeds. However, this mapping fails because of a lack of or fault in examples and domain knowledge. Consider the following example: You are trying to acquire knowledge for discriminating beetles from other creatures according to their features, i.e. number of legs, wings and eyes. You know that an insect has four wings and six legs, and beetles belong to insects. However, you get a decision tree that insists that if a creature has six legs and two wings, it is a beetle. (Although a beetle really has four wings, suppose that it has only two because the outer two look like shells.) This example suggests many improprieties for interviewing. Is there a fault in our domain theory, which insists on four wings, or in our examples? Are beetles really insects? A mosquito also seems to have only two wings. Is it a beetle? And so on. Currently, KAISER provides the following four types of semantic impropriety that have their own peculiar elimination methods.

1. *Explicable nil-node impropriety.* A nil leaf is explicable by the domain knowledge. Similar improprieties are also defined for inseparable nodes and noisy nodes. A simple elimination method is to adopt the derived class for their leaves.
2. *Contradictory explanation impropriety.* A leaf is explicable by domain knowledge of a different class. We can recommend many methods for eliminating this impropriety. One characteristic recommendation will be to specialize the Econd or Pcond of the mismatched domain knowledge, and/or to generalize the Econd or Dcond of the domain knowledge that should match and get higher support factors.
3. *Multiple explanation impropriety.* More than one explanation is suggested by domain knowledge. This impropriety recommends changing an Econd to a Dcond to lessen the support factor of a piece of knowledge that is not so important, and/or adding some conditions to Dcond to strengthen the factor of a preferable piece of knowledge.
4. *No explanation impropriety.* No explanation is given. It recommends generalizing the Econd and Pcond of a piece of knowledge that should match, and/or asking for new domain knowledge for the leaf.

IMPROPRIETY INTERPRETATION AND QUESTIONNAIRE GENERATION

If it is possible to offer the best recommendation based on the detected impropriety itself, we do not have to separate the impropriety detection and impropriety interpretation. With some simple improprieties this will be possible, but not with most. To understand this matter, consider previous examples again. In the T-shirt example, we discussed an unknown leaf problem. Since it was a quite simple problem that requires knowledge only about the order of the sizes LL, L and M, we could easily infer the price of size L. However, how about the following similar problem: If a creature has six legs, then it is an insect. If it has eight legs, then it is also an insect, e.g., a spider. But what is it if it has seven legs? The most common answer will be that there is no such creature, but a witty answer may be a crippled spider, but not a crippled beetle. To even deduce the common answer, we have to know much more about the domain. One major difference between these two examples is that, in the former example, all of the leaves, including the unknown L-size leaf, are covered by a single domain theory about the size of T-shirts, while the latter requires individual theories to explain each leaf. It is especially remarkable that there is no (common) knowledge that explains seven-legged creatures, which is a typical semantic impropriety.

This fact suggests that the combination of detected improprieties, especially a combination of structural and semantic ones, is an important clue for identifying an underlying fundamental impropriety. This is what we mean by impropriety interpretation. Because of the wide variety of interpretation knowledge, KAISER has not yet provided enough interpretation knowledge. Currently, it provides the following knowledge:

1. invalidation of the twin node impropriety when the twin nodes are explained by different domain knowledge;

2. cooperation of the inseparable node and multiple explanation improprieties to demand new attributes; and

3. cooperation of the contradictory and noisy or nil node improprieties to demand new domain knowledge.

Another point of issue is query generation. To achieve intelligent knowledge acquisition, sophisticated interviewing is indispensable. Since it is very hard to make human-like conversation in a natural language, KAISER does not try, but provides sophisticated questionnaires (i.e. queries and choices) according to the improprieties proposed by the interpreter. Although each impropriety has its own elimination method, it will not be a good idea to present all of them in the questionnaire at once and ask for answers to all of them. Currently, KAISER employs the following four criteria to select important improprieties. They are examined in this order.

1. Semantic improprieties take precedence over structural ones.
2. An impropriety that recommends tree modification takes precedence over those that require refinement of domain knowledge.
3. An impropriety that recommends structural tree reconstruction takes precedence over those that only prune or change the labels of nodes.
4. An impropriety related with an upper node (near the root) takes precedence.

AN EXAMPLE

EXPERIMENTAL PROBLEM

In this section, we will discuss KAISER by illustrating a simplified example of a famous computer game called TETRIS. Figure 13.2 shows a scene from TETRIS. A brick gradually drops from the top of the screen. We can move and rotate it right and left until it reaches the bottom. We can also let if fall immediately if we can determine a good place. When a row is filled with bricks, the row vanishes and the rows above it slide down. The object of this game is to pile up bricks neatly and crash as many bricks as possible until the peak of the stack reaches the top of the screen. Although this task of

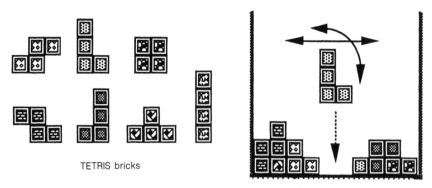

TETRIS bricks

Figure 13.2 A scene from TETRIS.

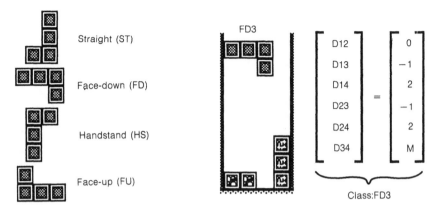

Figure 13.3 An example of a J-shaped brick.

piling bricks is a kind of planning, it can be handled as a classification task to classify a new brick to an adequate place according to the pattern of piled bricks.

To make the problem simple, we adopt a simplified version of TETRIS that has the following restrictions:

1. The width of the screen is four columns to make the problem space small.
2. A brick does not fall but stays at the top of the screen until we determine the place to put brick.
3. A brick cannot be brought under an overhanging row to fill the hollow under it.
4. Bricks of different shapes are dealt with separately.

Figure 13.3 shows a training instance of a J-shaped brick. An instance is described by six attributes, i.e. D12, D13, D14, D23, D24 and D34, where D12 represents the difference of height between the first and the second column from the left. Each attribute has one of seven values, i.e. S, -2, -1, 0, 1, 2 or M, where -2 means that the difference is two steps, S and M respectively mean that the difference is less than or more than two steps.

These six-dimensional vectors are classified into ten classes: FU3 (Face-Up at the third column, i.e. the flat side down and the right edge at the third column); FU4; FD3 (Face-Down at the third); FD4; ST2 (STraight at the second); ST3; ST4; HS2 (HandStand at the second); HS3; or HS4. Sixty-three examples of J-shaped brick are prepared out of 352 trials that consist of all types of brick.

DOMAIN KNOWLEDGE

Figure 13.4 shows the domain knowledge used for this problem. For example, an Econd of FD3 $-$ 1, (D12 0) (D23 $-$ 1), means that if D12 = 0 and D23 = $-$1, i.e. if the heights of the first and the second column are same, and the third column is one step lower than the second, then we can put the brick face-down at the third column. The Dcond (D34 1 2 M) means that if the fourth column is higher than the third column, it is more

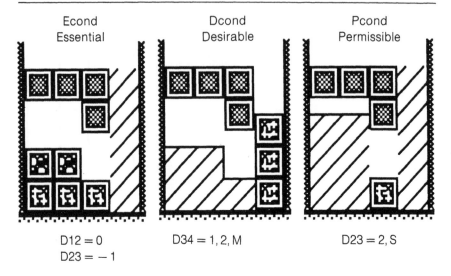

Econd   Dcond   Pcond
Essential   Desirable   Permissible

D12 = 0   D34 = 1, 2, M   D23 = 2, S
D23 = − 1

Figure 13.4 Domain knowledge for TETRIS.

desirable for FD3. The Pcond (D23 − 2 S) means that even if the third column is two or more steps below to the second, we may put the brick at FD3. In addition to the class-attribute relationships, inter-attribute and ordinal relationships are also defined for the TETRIS problem. For example, D13 attribute is defined as (D13 (calc-D-from-DD D12 D23)). This means that the first element D13 can be derived from D12 and D23 by applying the function calc-D-from-DD to these known attributes. Ordinal relationships are also defined as (D12 − 1 − 2 0).

EXPERIMENTAL RESULTS

Figure 13.5 shows a part of a decision tree for J-shaped bricks. The labels depicted under the leaves are the results of TBM. By checking the differences between the induced labels and the TBM results, i.e. support factors, KAISER recommends:

1. modifying the class of the leaf50 and leaf51 to FD4 according to the two support factors, FD4-1 and FD4-2;
2. choosing proper classes for the leaf53, leaf54 and leaf56 out of their support factors;
3. ignoring the result of the leaf55 and choosing the class out of the support factors, if there are few examples in the leaf, otherwise generalizing the Econd of FD3 rules.

In this figure, we can also see that the leaf57 could not get any support by domain knowledge although the class is specified by induction. This is the clue for RBM. Based on the domain knowledge of ST2, we can know that the path to leaf57 does not have a condition about D12, which is necessary (Econd) for ST2. In this case, KAISER marks leaf57 as an insufficient leaf that may lack the condition about D12, and propose

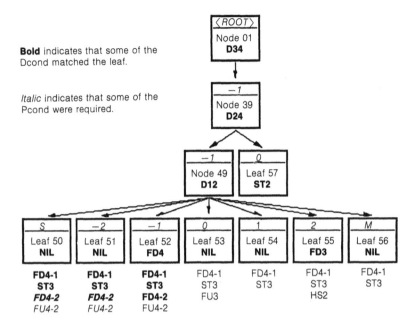

Figure 13.5 An example of a TBM.

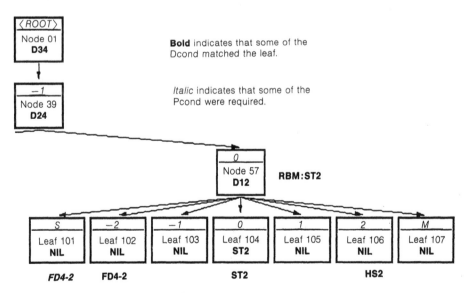

Figure 13.6 An example of an RBM.

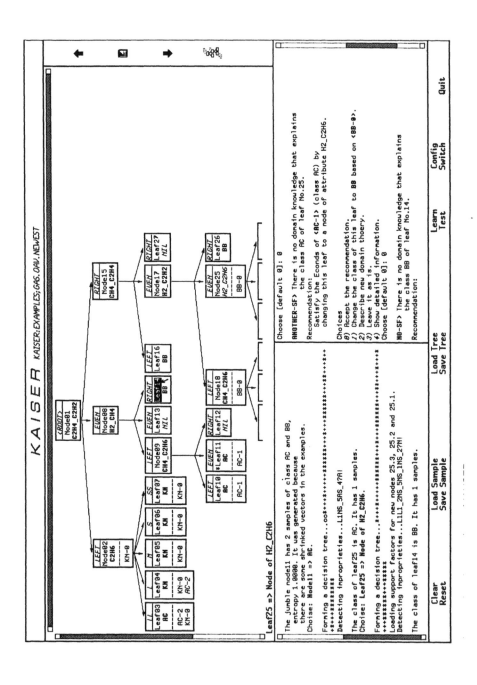

additional conditions as shown in Fig. 13.6. TBM is examined for the generated paths to the new leaves from 101 to 107 and the support factors are calculated for them.

IMPLEMENTATION

KAISER has been implemented on a Symbolics Lisp machine with Common-Lisp and OPS5 (Forgy, 1981). Impropriety detection and interpretation are performed by OPS5, which makes it easy to describe and modify improprieties. Figure 13.7 shows a snapshot of KAISER. The upper panel shows a decision tree under examination, and the lower left-hand panel displays some useful information about the tree and improprieties. The questionnaires are printed in the lower right-hand panel.

CONCLUSION

We have discussed an inductive and interactive knowledge acquisition system called KAISER. This system aims at integrating similarity-based inductive learning of classification knowledge and explanation-based knowledge evaluation to realize sophisticated knowledge acquisition and reliable knowledge formalization. Since KAISER is still at an initial stage of research, there are many pending issues; however, the proposed impropriety-based approach turned out to be efficient in carrying out intelligent interviewing based on the idea of improprieties and in supplementing the inductive power of inductive learning.

REFERENCES

Dabija, V.G., Tsujino, K. and Nishida, S. (1992) *Learning to learn decision trees. Proc AAAI'92*.

Forgy, C.L. (1985) OPS5 user's manual. CMU technical report, CMU-CS-81-135.

Quinlan, J.R. (1986) Induction of decision trees. *Machine Learning*, **1**, 81–106.

Tsujino, K., Takenouchi, S., Sakurai, T., Chigusa, S., Nomura, Y., Mizoguchi, R. and Kakusho, O. (1989) Adaptive Rule Induction System: ARIS (in Japanese). *Trans. of IEICE*, **J72–D2**, 121–131.

Tsujino, K., Takegaki, M. and Nishida, S. (1990) A knowledge acquisition system that aims at integrating inductive learning and explanation-based reasoning. *Proc. JKAW (the first Japanese Knowledge Acquisition for Knowledge-based Systems Workshop)*, pp. 175–190.

Tsujino, K., Dabija, V.G. and Nishida, S. (1992) Knowledge acquisition driven by constructive and interactive induction, in *Current Developments in Knowledge Acquisition: EKAW-92*, Wetter, T., Althoff, K.-D., Boose, J., Gaius, B., Linster, M. and Schmalhofer, F. (eds), Berlin/Heidelberg: Springer-Verlag.

Utgoff, P.E. (1986) *Machine Learning of Inductive Bias*, Boston, MA: Kluwer.

Incorporating statistical techniques into empirical symbolic learning systems

14

F. Esposito, D. Malerba and G. Semeraro

INTRODUCTION

Apart from connectionist approaches and genetic algorithms, for the most part the methods of inductive concept learning share the common objectives of classifying and producing predictive knowledge from observations. Although the rules produced are generally required to be intelligible and accurate, some problems arise due both to the complexity of the description languages and the noise and uncertainty in the initial observations.

A first problem is related to the **description language**: increasing the power and complexity of the description languages for inductive learning always implies an exponential growth of the concepts which are consistent with the training data. Thus, it raises the need for a bias or some heuristic rule to apply to the generalization process, i.e. a rule for selecting one among the many hypotheses which are consistent with the training set (Rendell, 1986). Statistical information concerning the information content or the relevance of a descriptor in the prediction task may be used to decide which concept descriptions are preferable. A form of statistical guidance is sometimes present in inductive methods which use measurements of information gain and entropy in order to point out the predictive utility of a single attribute or of configurations of attributes (Quinlan, 1986; Niblett, 1987). This approach is followed by those learning systems which use a representation language of the attribute–value pairs kind. The introduction of statistical techniques into such systems, in order to guide the inductive process and to learn quantitative relations, is feasible due to the possibility of considering the concept instance as a feature vector (Fisher and Chan, 1990). However, such an approach is not easily applicable when more powerful languages, which enable structural relations to be represented, are used.

Another problem is the **predictive accuracy**, which is related to the description language: the success in producing predictive knowledge often comes from the careful choice, based on a good knowledge of the domain, of the variables themselves. It is

Artificial Intelligence Frontiers in Statistics: AI and statistics III. Edited by D.J. Hand. Published in 1993 by Chapman & Hall, London. ISBN 0 412 40710 8

fundamental to determine which descriptors should be employed for the best learning results, through a process of selection and extraction. The capability of generating appropriate new features, that is, of inductively constructing new descriptions, is a way of increasing the performance of a learning system (Rendell, 1985; Muggleton, 1987). The task of constructing new derived descriptors may be accomplished, in symbolic learning too, by statistical data analysis methods which, by specifying relationships among attributes of objects, allow a constructive inductive learning to be realized.

Handling **continuous data** also constitutes a problem: in symbolic languages this is possible only by a discretization process, which is sometimes performed ignoring the variable distributions, so causing a loss in information and diminishing the predictive accuracy. Some authors have approached the problem by using different languages to represent the examples and the concepts and by applying partially defined predicates which contain parameters whose values are automatically settable (Bergadano and Bisio, 1988).

In symbolic methods the intrinsic difficulty in acquiring knowledge in **noisy environments** poses a further problem. The predictive accuracy is bound up with the hypothesis of a perfect class–concept consistency (Michalski, 1987), that cannot be guaranteed by covering algorithms (Schlimmer and Granger, 1986) due to the presence of noise: this results in a severe disadvantage, during the testing phase, as the classification accuracy is related to a strict matching between the new observations and the class descriptions. In order to increase the classification accuracy, flexible matching, based on a probabilistic interpretation of the matching predicate, has been proposed (Michalski *et al.*, 1986).

On the basis of these considerations, we have set up an approach integrating statistical data analysis and probability-based techniques with symbolic concept learning methods. The observations and concepts are described by a symbolic language which allows us to represent relations between objects. In contrast to what some authors affirm (Falkenheiner and Michalski, 1986), statistical data analysis is very useful when there is little prior knowledge and the trainer is not able to evaluate the relevance of the descriptors or their interrelationships. In this paper we present RES, a learning system that combines a data analysis technique for linear classification (discriminant analysis) with a conceptual method for generating disjunctive cover for each class based on the Star methodology (Michalski, 1983). The star is computed by INDUBI (Esposito, 1990), a radically modified version of the INDUCE algorithm: it uses a larger subset of the VL_{21} language, including full universal and numerical quantification of complex formulae, provides a statistical guidance in the generalization process (as a bias in selecting the candidate concepts) and uses flexible matching both to classify noisy data and to cope with structural deformations.

In this chapter, after a brief review of the representation language VL_{21}, two aspects of RES are presented at some level of detail, showing: how constructive induction may be realized through a form of statistical data analysis, in order efficiently to handle continuous descriptors in a symbolic description language where structural descriptions are possible; and how the classification accuracy is improved by flexible matching based on a probabilistic similarity measure between structural descriptions. The system has been applied to learning classification rules to recognize digitized documents, starting from the physical layout characteristics (Esposito *et al.*, 1990a).

THE REPRESENTATION LANGUAGE

The need to represent objects defined by intension has led to the introduction of symbolic objects (Diday and Brito, 1989), which extend the usual objects processed in statistical data analysis. In traditional notation, events or objects can be any kind of entity, defined by attributes, to which properties or variables are ascribed. Each object o can be identified by the vector of the observed values of the descriptive variables or features

$$\mathbf{X} = (X_1(o), \dots, X_p(o))$$

The input to trainable classifiers are data structured as vectors in the feature space; the classifier assigns each feature vector to a decision region in the feature space by a set of decision hypersurfaces, i.e. a set of decision rules partitioning the feature space into class regions.

In symbolic learning the objects are defined by a list of their properties or attributes:

$$o = (V_1, V_2, \dots, V_k)$$

where V_j is a subset of the attribute domain D_j for each $j = 1, 2, \dots, k, k \leqslant p$, and means that the variable X_j takes values in the power set $P(D_j)$. An event E, denoted as $E = [X_i = V_i]$, where $V_i \subseteq D_i$ means that X_i takes values in V_i, is a mapping on the boolean set such that an observed event $E(o)$ is true if and only if $X_i(o) \in V_i$.

Given a universe of objects, a class is a subset of the objects, and a concept is the description of the class (intension). The facts are descriptions of objects preclassified by the trainer and can be viewed as a collection of decision rules in the form

$$F = (E_{ik} ::> C_i), \qquad i \in I$$

where I is the number of classes, E_{ik} is the event concerning the kth object, that is, an example of the ith concept C_i.

The result of learning is the inductive assertion H of the inductive paradigm expressed as a set of rules of the type

$$H = (G_i ::> C_i), \qquad i \in I$$

where G_i are approximate descriptions of the ith class or concept.

Having a complex description language at one's disposal makes it possible to realize multi-level class descriptions, involving relations among objects and higher-level descriptors of concepts. An extension of the annotated predicate calculus suitable for structural descriptions has been proposed: the VL_{21} language (Michalski, 1980). Its basic component is the relational statement, or **selector**, written as

$$[L \# R]$$

where L, called the **referee**, is a function symbol with its arguments; R, called the

reference, is a set of values of the referee's domain; and **#** is a relational operator defining the relation between the referee and the reference.

The selector can be seen as a test to determine whether the predicate and the function values belong to a defined set or not. A domain is associated to each variable, predicate or function symbol; it can be nominal, linear or tree-structured. Furthermore, we distinguish between **numerical** and **conceptual** descriptors: the former are considered by discriminant analysis while the latter are exploited by the symbolic method. Obviously, only interval or ratio level measurements will be specified as numerical descriptors, while nominal, ordinal and partially ordered level measurements will be treated as conceptual descriptors.

Selectors may be combined by applying different operators, such as AND, OR (V), decision operators (::>) and logic implication (⇒) in order to define: decision rules representing facts; inference rules representing the background knowledge; and generalization rules defining the transformations applicable to facts in the generation of hypotheses.

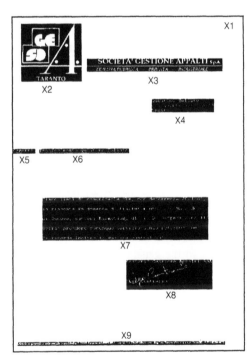

[WIDTH(X2)=medium__large]
[WIDTH(X3)=large]
[WIDTH(X4)=medium__large]
[WIDTH(X5)=medium__small]
[WIDTH(X6)=medium__large]
[WIDTH(X7)=very__large]
[WIDTH(X8)=medium__large]
[WIDTH(X9)=very__large]
[HEIGHT(X2)=medium__large]
[HEIGHT(X3)=small]
[HEIGHT(X4)=small]
[HEIGHT(X5)=very__very__small]
[HEIGHT(X6)=very__very__small]
[HEIGHT(X7)=medium]
[HEIGHT(X8)=medium__small]
[HEIGHT(X9)=very__very__small]
[TYPE(X2)=mixture]
[TYPE(X3)=mixture]
[TYPE(X4)=text]
[TYPE(X5)=text]
[TYPE(X6)=text]
[TYPE(X7)=text]
[TYPE(X8)=mixture]
[TYPE(X9)=text]
[TO__RIGHT(X5,X6)=true]
[TO__RIGHT(X2,X3)=true]
[ALIGN(X5,X6)=both__row]
[ALIGN(X3,X9)=ending__column]

[CONTAIN_IN_POS(X1,X2)=north__west]   [CONTAIN_IN_POS(X1,X3)=north]
[CONTAIN_IN_POS(X1,X4)=north__east]   [CONTAIN_IN_POS(X1,X5)=west]
[CONTAIN_IN_POS(X1,X6)=west]   [CONTAIN_IN_POS(X1,X7)=centre]
[CONTAIN_IN_POS(X1,X8)=south]   [CONTAIN_IN_POS(X1,X9)=south]

Figure 14.1 An example of VL$_{21}$ structural description of a document page layout.

Figure 14.1 shows an example of VL_{21} structural description concerning a document page layout. All the selectors are implicitly connected by ANDs and each variable denotes a single block of the layout or the whole document.

In addition to the descriptors originally used in order to describe an instance, the generalizations produced for each concept could also contain new descriptors which have been constructed during the generalization process itself. More formally, in concept learning the process of feature construction, often referred to as **constructive induction** (Matheus and Rendell, 1989), is the application of a set of constructive operators

$$\{op_1, op_2, \ldots, op_n\}$$

to all or a subset of initial features $\{X_1, X_2, \ldots, X_p\}$ resulting in the construction of one or more new features $\{X'_1, X'_2, \ldots, X'_q\}$ intended for use in describing the target concept. A constructive operator is a function mapping a tuple of existing variables into a new descriptor. A constructive operand is the tuple of initial variables to which an operator is applied. The term **constructor** is used to refer to both a constructive operand and an operator, while the term **metadescriptor** refers to a constructed descriptor.

A first step towards the integration between statistical and symbolic learning methods involves the possibility of expressing the results of the statistical data analysis in the symbolic language. In this case, only the numerical descriptors or a subset of these will be used as constructive operands or constructors.

STATISTICAL DATA ANALYSIS AS A FORM OF CONSTRUCTIVE INDUCTION

The success of empirical learning methods depends on the selection of good descriptions and appropriate features for the target concept. When a large number of examples is available the computational complexity grows as the number of the concepts consistent with the training set increases: statistical data analysis can be used in order to reduce the dimensionality of the problem. We propose the use of the discriminant function as a constructive operator: this linear combination of the initial continuous features (all or the most significant of them), allows us to construct a powerful metadescriptor, to define a heuristic in the generalization process and to increase the computational efficiency.

In order to evaluate the suitability of our approach, we will try to examine the aspects inherent to the problem of feature construction as proposed by Matheus and Rendell (1989):

1. detection of the need;
2. selection of constructors;
3. generalization of selected constructors;
4. evaluation of the new features.

As to the first point, the weakness of the symbolic learning methods in handling real numbers leads to a need to use the discriminant function as a constructive operator when working with continuous-valued attributes. The process of dichotomizing or discretizing continuous variables in order to generate disjunctive covers for each class may be arbitrary (without hypotheses on the distributions), and results in a serious loss of information. Given that this technique is able to synthesize several continuous descriptors in a unique function, the constructive process raises the computational efficiency of the whole learning process because it limits the space of the descriptions. On the other hand, the representation efficacy diminishes since the final metadescriptor appearing in the classification rules is not 'understandable', but then, for the final user, how understandable are very long complex rules, with many or-atoms, expressed as ranges of numerical attributes? Moreover, discriminant analysis may also be applied to data which have a categorical structure (Goldestein and Dillon, 1978): it may be used in the same way as other methods of statistical data analysis (regression, cluster and factor analysis), on the complete initial data set, as a technique for detecting the influence of various features on the classification results. There are statistical tests for measuring the success with which the discriminating variables actually discriminate when combined into the discriminant functions. Letting the variables with nominal domains and the structural descriptions be managed by the symbolic method, the weight with which each initial descriptor (discrete variable) contributes to the definition of the discriminant function may be used as a bias for the conceptual method in preferring, during the generalization process, those hypotheses involving the most discriminating descriptors.

As to the second point, selecting constructors deals with the reduction of the set of potential operators and operands. For the operators, there are a certain number of different discriminant functions (linear, quadratic, piecewise, etc.) and an increasing number of discriminant analysis procedures, from the simple k-nearest-neighbour classification methods to very sophisticated robust kernel and estimators. Selection may be done at runtime, for example, by test hypotheses in order to choose between linear or quadratic classifiers (training set bias). We select the operator: a linear pattern classifier that seems to be sufficiently robust as regards different aspects (lack of multivariate normality, initial misclassification). All these operators are domain-independent. The selection of constructive operands is similar to variable selection in parametric methods of classification for which formal statistical approaches are adopted. Most of these methods begin with a large number of variables measured on the design set and try to find an effective subspace of the complete set of variables. We propose a stepwise method on the whole space of the numerical features.

As to the generalization of constructors, discriminant analysis may be used for prospective classification. The generalization is performed during the construction of the metadescriptor *class__disc__analysis* which specifies the class assigned to an object by the discriminant analysis. A linear classifier can predict regions of positive and negative instances that have not been observed using a measurement, the discriminant score, based on the results of the analysis on the training set. The bias in constructive generalization of *class__disc__analysis* is the probability density function of the set of continuous features (normal multivariate).

The problem of evaluating features seems irrelevant when only one constructor is proposed. Using discriminant analysis, the choice of an 'optimal' model evaluation criterion is made according to the specific problem. If the aim is prospective classification, criteria based on the error rate are preferred; if the aim is to find out the best discriminators, distance functions are recommended. Many error rates are of interest: optimal error rate, actual error rate, 'resubstitution' error rate, 'leaving-one-out' error rate.

In the integrated approach, the numerical descriptions of the objects are processed by statistical analysis in order to compute the coefficients of the discriminant functions and the posterior probabilities of the class membership of each object. The 'classification expertise' of the statistical method is encoded in the following VL_{21} metadescriptor:

$$[class\_\_disc\_\_analysis(variable) = value]$$

where *class__disc__analysis* represents the value of the discriminant function mapping the feature vector associated with the object to the class identifier; *variable* is the name of the object as a whole; and *value* is the class identifier.

This metadescriptor is appended to the symbolic description of each object and will be successively considered by the conceptual method INDUBI in order to generate the class descriptions (concepts). The value of the metadescriptor is computed for each observation by applying the results of the discriminant analysis (discriminant scores) only to those descriptors selected as 'constructors' for their discriminating power. Once acquired by INDUBI, this metadescriptor may be selected or rejected, according to its provisional capacity, which can be measured by testing the discriminant scores on the same training examples. The evaluation of the predictive power is the criterion which guides the inductive process: those descriptions of concepts containing the *class__disc__analysis* metadescriptor will be preferred among all the possible generalizations, if the error rate reveals it to be significant.

INCREASING THE CLASSIFICATION ACCURACY BY A PROBABILISTIC FLEXIBLE MATCHING PROCESS

Symbolic concept learning from examples has the advantage of supplying results even with a limited number of significant examples, allowing rules to be incrementally updated when new events are added to the training data set. The predictive accuracy may be guaranteed only by a complete set of examples, but the utility of a method which can also work with a partial data set cannot be denied. The rules initially produced are often already significant, but classification is not feasible: in fact, optimal learning requires canonical matching between the new observations and the class descriptions, which is difficult because of the initial noise, thus reducing the possibility of classifying new objects. The problem of noisy data could be faced by using consolidated statistical techniques if the events were described by feature vectors. This is no longer possible when extensions of the FOPL (First-order Predicate Logic) are used as representation languages, since the multiple unification of formulae makes even the simplest statistics hard to compute. The problem of missing data may be solved, during the learning phase, by considering the unknown value as a 'don't care' value, but such a solution is useless for the classification task, when we have to match the event described by a VL_{21} formula, for example:

$$E3: [on\_top(x1, x2)][length(x1) = ?][length(x2) = 28]$$

against the left part of the following rule concerning a concept:

$$G: [on\_top(s1, s3)][length(s1) = 13]::> [class = first]$$

When noise is present or the training sample is insufficient, the predictive accuracy cannot be guaranteed: a classifier working in an all-or-nothing fashion is useless and it is necessary to rely on a measure of **similarity** between objects, taking on values in a continuous range and not in a boolean set. We propose its complementary measure, the **distance**, based on a probabilistic interpretation of the matching predicate (Esposito *et al.*, 1992): it is used to classify examples which do not present exactly all the regularities appearing in the corresponding recognition rule.

The distance measure, Δ, suitable for structural deformations, is defined on the space of the well-formed formulae (WFFs) of a variable-valued logic, the VL_{21}. Let F1 and F2 be two VL_{21} WFFs. Then:

$$\Delta(F1, F2) = 1 - Flex\_Match(F1, F2) \tag{14.1}$$

where $Flex\_Match(F1, F2)$ is a function taking on values in $[0, 1]$ and representing the probability that F1 perfectly matches F2. Therefore:

$$Flex\_Match(F1, F2) = P(Match(F1, F2)) \tag{14.2}$$

where *Match* is the canonical matching predicate defined on the same space. The definition (14.2) marks the transition from deterministic to probabilistic matching.

The main application of the distance measure is concept recognition in noisy environments; therefore, from now on, F1 will denote the description of a concept and F2 the observation to be classified.

Flex_Match is computed according to a top-down evaluation scheme:

1. F1 and F2 are disjunctions of conjuncts:

$$Flex\_Match(F1, F2) = P(Match(Or\_atom_1, F2) \vee \ldots \vee Match(Or\_atom_N, F2)) \tag{14.3}$$

where Or_atom_i is the generic conjunction of selectors in F1 and $Match(Or\_atom_i, F2)$, with $i \in [1, N]$, are independent events;

2. F1 and F2 are conjunctions of selectors (VL_{21} literals):

$$Flex\_Match(F1, F2) = \begin{cases} \max_{\sigma_j} \prod_i w_i \, Flex\_Match_j(Sel_i, F2) & \text{if exists a substitution } \sigma_j \\ & \text{such that } F2 \Rightarrow \sigma_j(F'1) \\ 0 & \text{otherwise} \end{cases} \tag{14.4}$$

where σ_j is one of the possible substitutions among variables; Sel_i is the generic selector in F1; $Flex\_Match_j$ is the measure of fitness between Sel_i and F2 when the variable substitutions are fixed (by σ_j); w_i denotes the weight of the function in Sel_i; and $F'1$ is the shortest conjunction of selectors in F1 such that all the distinct variables in F1 are also in $F'1$;

3. *F1* and *F2* are selectors:

$$Flex\_Match_j(F1, F2) = \max_{i \in [1, s]} P(EQUAL(g_i, e)) \qquad (14.5)$$

where g_i is one of the s values of the reference of *F1*; e is the only element in the reference of the observation *F2*; and $EQUAL(x, y)$ denotes the matching predicate defined on any two values x and y of the same domain.

The term $P(EQUAL(g_i, e))$ is defined as the probability that an observation e may be considered a distortion of g_i, that is:

$$P(EQUAL(g_i, e)) = P(\delta(g_i, X) \geqslant \delta(g_i, e)) \qquad (14.6)$$

where X is a random variable assuming values in the domain of the function contained in *F1* (or *F2*); and δ is the distance defined on the domain itself.

The definition of $P(EQUAL(g_i, e))$ takes into account both the type of function and the probability distribution of its domain.

The function *Flex__Match(F1, F2)* actually computes the probability that any observation of the concept described by *F1* would be further from the centroid *F1* than the case *F2* being considered. If *Flex__Match(F1, F2)* is too small, it signals the possibility that *F2* is not an instance of the class described by *F1*, even though it is the closest. *Flex__Match(F1, F2)* can be considered from a theoretical point of view. In fact the probability that the referee's value changes is a deformation probability, so finding the most similar concept description is equivalent to detecting the most probable deformation.

Decision-making based on a distance measure is more expensive than a true/false matching procedure, therefore a multi-layered framework is more suitable (Esposito *et al.*, 1992). Moreover, two distinct methods have been applied to reduce the computational complexity, a branch-and-bound algorithm and a heuristic approach (Esposito *et al.*, 1991).

To sum up, the proposed distance measure aims to cope with the problem of noise and uncertainty in the classification process by computing a numeric degree of membership of an event description *F2* to a class described by a generalization rule *F1*. This is different from approximate matching in fuzzy set theory since we do not define any membership function for the elements of a set, but we do need the probability function of each feature (Cheeseman, 1986).

AN APPLICATION TO PAGE LAYOUT RECOGNITION

RES has been used to produce automatically the rules of a knowledge-based system for the recognition of optically scanned documents. A document is intended as a related collection of printed objects (characters, columns, paragraphs, titles, figures, etc.), on paper or microform, for example, technical journals or reports. Provided there is a set of documents with common page layout features, an optically scanned document can be classified in the early phase of its processing flow, by using a defined set of relevant and invariant layout characteristics, **the page layout signature**. Based upon this geometrical knowledge, it is possible to build a system which can classify any document from a perceptional point of view, without using optical character recognition or syntactic descriptions of the document given by the user. In fact, a printed page is treated by dealing only with automatically detected and constructed characteristics of the document, namely the geometrical characteristics of the blocks (height, width, spacing and alignment), and the document structure, which is described in a symbolic notation.

In order to produce the classification rules, some significant examples of document classes, that may be of interest in a specific environment, are used as a training sample to discover the layout similarities within each class.

The first experiment was realized on a set of 161 single-page documents, namely printed letters and magazine indexes, belonging to eight different classes (the last one was a 'reject' class, representing 'the rest of the world'). Fifty-one instances were selected as training cases, leaving the remaining 110 documents for the testing phase. All the sample documents were real letters received or sent by firms or copies of the indexes of international magazines, so that several forms of noise actually affected them.

Once a document has been digitized, the numerical output of the layout analysis is automatically translated into VL_{21} descriptions: these are structural descriptions where spatial relationships between cohesive blocks within the same reference frame (the document) are expressed. Seven descriptors are used: WIDTH (X), HEIGHT (X), TYPE (X), CONTAIN_IN_POS (Y, X), ON_TOP (X1, X2), TO_RIGHT (X1, X2), and ALIGN (X1, X2).

Working on the layout structural characteristics, the conceptual method INDUBI produces seven maximally specific classification rules, expressed as VL_{21} WFFs, satisfying both the completeness and consistency conditions. Most of these rules point out the invariant parts of the physical layout such as logos and titles of fixed sizes in fixed positions, and their relations with other parts of the documents, such as alignment with other blocks. However, due to the limited number of training cases, the predictive accuracy is low. In fact, only 76 of the 110 testing cases were correctly classified, although recognition was improved using the distance measure. Notice, for example, in

Table 14.1 Results of the application of the conceptual method by a strict matching and a flexible matching (training set, 51 cases; testing set, 110 cases)

Class	No. of cases	Correct classif.	Strict match.	Flex match.
Olivetti letter	27	20	74%	96%
Sifi letter	20	12	60%	90%
Sims letter	15	7	46%	93%
Sogea letter	10	8	80%	90%
PAMI index	21	17	81%	86%
Spectrum index	8	4	50%	75%
Computer index	9	8	89%	89%

Table 14.1, how the use of flexible matching improves the recognition rate for all the classes.

Increasing the number of the training examples should provide a better performance. A training set of 121 examples, including the original 51 instances and still independent from the 110 testing samples, was selected. The predictive accuracy for the whole set increased to 92%. However we have to report the presence of disjunctive and trivial rules. Two disjunctive rules with three or-atoms were generated both for *IEEE Trans. on PAMI* indexes and for *IEEE Trans. on Computers* indexes, because of the variability of the page layout and the presence of noise. Table 14.2 illustrates the results of the testing process. The learning process in the training phase took about 4 hours 30 minutes on a Sun 3/280 minicomputer. Half of the training time was spent in generating these rules.

With a consistent number of training cases, it is possible to use and to evaluate the performance of the numerical classifier on the numerical descriptors concerning the layout: these can be derived from the coordinates of two opposite corners of a block, such as height, length, area and eccentricity, as well as their means, standard deviations, maximum and minimum values, while other measurements related to the number of black pixels per block can be directly provided by the segmentation process. Such a quantity of numerical information cannot be dealt with by the symbolic method alone, but once synthesized by discriminant analysis into the metadescriptor *class_disc_analysis*, will be considered together with the symbolic descriptors.

Initially, 93 features were picked out to describe each document but only six of them were selected by the stepwise variable selection process minimizing Wilks's lambda:

1. maximum eccentricity of image blocks;
2. standard deviation of the number of black pixels;
3. standard deviation of the length of text blocks;
4. minimum eccentricity of image blocks;
5. symmetry along the vertical direction;
6. percentage of the textual part.

It is fairly evident that image blocks, such as logos and magazine headings, are important in the discrimination process. Another interesting feature which is selected is

Table 14.2 Conceptual method: confusion matrix (training set, 121 cases; testing set, 110 cases)

	Olivetti	Sifi	Sims	Sogea	PAMI	Spectrum	Computer	Reject
Olivetti	96%							4%
Sifi		100%						
Sims			100%					
Sogea				100%				
PAMI					90%			10%
Spectrum						75%		25%
Computer	11%						67%	22%

the standard deviation of the length of text blocks, which is important in order to distinguish justified letters from those not justified, and magazine indexes with a high variability in the length of their entries from others which are better organized. The discriminant rules generated by discriminant analysis are not sufficient to recognize the documents: in fact, some misclassifications occurred when the same training documents were reclassified. This means that the eight classes are not linearly separable. When using the features selected, the resubstitution error rate is about 8%, nearly the same as that obtained using the leaving-one-out criterion.

The throughput time for selecting the variables and computing the discriminant functions was about 4 minutes on a Sun 3/280. The time needed to classify a single document was less than 1 second. Table 14.3 shows the results of the testing phase when only the statistical method is applied.

Using the integrated approach, the new metadescriptor, [class–disc–analysis(doc) = class], resulting from the application of discriminant analysis and representing the class membership of each document, was appended to each training/testing example. Moreover, a lower cost was assigned to *class_disc_analysis* than to the other descriptors due to the high discriminant power revealed by the statistical method.

With the application of INDUBI the hypotheses generated often changed significantly, especially for those classes characterized by disjunctive rules. This was due

Table 14.3 Statistical method:confusion matrix (training set, 121 cases; testing set, 110 cases)

	Olivetti	Sifi	Sims	Sogea	PAMI	Spectrum	Computer	Reject
Olivetti	96%							4%
Sifi		95%	5%					
Sims		7%	93%					
Sogea				100%				
PAMI					95%		5%	
Spectrum						100%		
Computer							100%	

Table 14.4 Integrated method:confusion matrix (training set, 121 cases; testing set, 110 cases)

	Olivetti	Sifi	Sims	Sogea	PAMI	Spectrum	Computer	Reject
Olivetti	96%							4%
Sifi		100%						
Sims			100%					
Sogea				100%				
PAMI					95%		5%	
Spectrum						100%		
Computer							100%	

to the fact that only the new appended metadescriptor was able to characterize some classes. The only rules which remained unchanged were those concerning two classes of letters, while all the others changed to include the new selector in the condition part. The results of the testing phase are reported in Table 14.4. The overall predictive accuracy of the combined approach is 98%. It is interesting to notice how complementary the symbolic and the statistical methods are, which justifies the robustness of the integrated approach. It is also interesting to notice that the application of the distance measure is necessary when the training sample is poor, while it becomes less important when the training sample rises due to the presence of the new metadescriptor which summarizes the class regularities and is robust against the noise.

Finally, the time needed to train the system was nearly halved, since it took only 2 hours 30 minutes, including the statistical analysis. This is a good result, bearing in mind that the learning system handles about 9000 selectors in the complete training set. Once the rules have been generated, the knowledge-based system is able to recognize a digitized document in 8 seconds.

A complete discussion of the experimental results is given in Esposito *et al.* (1990b).

CONCLUSIONS

The idea of combining data analysis and conceptual methods in order to reduce the amount of knowledge which is required from the trainer in selecting examples, and to realize a more powerful learning, seems attractive. Statistical techniques may be used combined with symbolic learning methods in order efficiently to handle qualitative and quantitative features, reduce the effects of noise, accelerate the learning process and realize a form of constructive induction.

The approach proposed is able to use exceptions to improve selective induction and to deal with uncertainty, by means of a distance measure between structural descriptions based on a probabilistic interpretation of the matching predicate. Discriminant analysis already expresses classes in terms of the probabilities of belonging to each class or degrees of belief. By extending this approach to the conceptual method it is possible to realize a faster covering algorithm and a more efficient learning process.

REFERENCES

Bergadano, F. and Bisio, R. (1988) Constructive learning with continuous-valued attributes, in *Uncertainty and Intelligent Systems*, B. Bouchon, L. Saitta and R.R. Yager (eds), Lecture Notes in Computer Science, Berlin: Springer-Verlag, pp. 154–162.

Cheeseman, P. (1986) Probabilistic versus fuzzy reasoning, in *Uncertainty in Artificial Intelligence*, L.N. Kanal and J.F. Lemmer (eds), Amsterdam: North-Holland, pp. 85–109.

Diday, E. and Brito, P. (1989) Introduction to symbolic data analysis, in *Conceptual and Numerical Analysis of Data*, O. Opitz (ed.), Berlin: Springer-Verlag, pp. 45–84.

Esposito, F. (1990) Automated acquisition of production rules by empirical supervised learning, in *Knowledge, Data and Computer-Assisted Decisions*, M. Schader (ed.), Berlin: Springer-Verlag, pp. 35–48.

Esposito, F., Malerba, D., Semeraro, G., Annese, E. and Scafuro, G. (1990a) Empirical learning methods for digitized document recognition: an integrated approach to inductive generalization, in *Proceedings of the Sixth Conference on Artificial Intelligence Applications*, Santa Barbara, California, pp. 37–45.

Esposito, F., Malerba, D., Semeraro, G., Annese, E. and Scafuro, G. (1990b) An experimental page layout recognition system for office document automatic classification, in *Proceedings of the 10th International Conference on Pattern Recognition*, Atlantic City, New Jersey, pp. 557–562.

Esposito, F., Malerba, D. and Semeraro, G. (1991) Flexible matching for noisy structural descriptions, *Proceedings of the 12th International Joint Conference on Artificial Intelligence*, Sydney, pp. 658–664.

Esposito, F., Malerba, D. and Semeraro, G. (1992) Classification in noisy environments using a distance measure between structural symbolic descriptions. *IEEE Trans. PAMI*, **14**, 390–402.

Falkenheiner, B.C. and Michalski, R.S. (1986) Integrating quantitative and qualitative discovery: the ABACUS system. *Machine Learning*, **1**, 367–402.

Fisher, D.H. and Chan, P.K. (1990) Statistical guidance in symbolic learning, in *Annals of Mathematics and Artificial Intelligence 2*, D. Hand (ed.), Basel: J.C. Baltzer, pp. 135–148.

Goldestein, M. and Dillon, W. (1978) *Discrete Discriminant Analysis*, New York: Wiley.

Matheus, C.J. and Rendell, L. (1989) Constructive induction on decision trees, in *Proceedings of the 11th International Joint Conference on Artificial Intelligence*, pp. 645–650.

Michalski, R.S. (1980) Pattern recognition as rule-guided inductive inference. *IEEE Trans. PAMI*, **2**, 349–361.

Michalski, R.S. (1983) A theory and methodology of inductive learning, in *Machine Learning: an Artificial Intelligence Approach*, R.S. Michalski, J.G. Carbonell, T.M. Mitchell (eds), Palo Alto, CA: Tioga.

Michalski, R.S. (1987) How to learn imprecise concepts, in *Proceedings of the Fourth International Workshop on Artificial Intelligence*, Irvine, CA: Morgan Kaufmann, pp. 50–58.

Michalski, R.S., Mozetic, I., Hong, J. and Lavrač, N. (1986) The AQ15 inductive learning system: an overview and experiments. Technical Report, Intelligent System Group, Department of Computer Science, University of Illinois at Urbana-Champaign.

Muggleton, S. (1987) DUCE, an oracle based approach to constructive induction, in *Proceedings of the 10th International Joint Conference on Artificial Intelligence*, pp. 287–292.

Niblett, T. (1987) Constructing decision trees in noisy domains, in *Progress in Machine Learning*, Bratko, I. and Lavrač, N. (eds), Wilmslow: Sigma Press, pp. 67–78.

Quinlan, J.R. (1986) Induction of decision trees. *Machine Learning*, **1**, 81–106.

Rendell, L. (1985) Substantial constructive induction using layered information compression, in *Proceedings of the 9th International Joint Conference on Artificial Intelligence*, pp. 650–658.

Rendell, L. (1986) A general framework for induction and a study of selective induction. *Machine Learning*, **1**, 177–226.

Schlimmer, J.C. and Granger, R.H. (1986) Incremental learning from noisy data. *Machine Learning*, **1**, 317–354.

Learning classification trees

15

W. Buntine

INTRODUCTION

A common inference task consists of making a discrete prediction about some object given other details about the object. For instance, in financial credit assessment as discussed by Carter and Catlett (1987) we wish to decide whether to accept or reject a customer's application for a loan given particular personal information. This prediction problem is the basic task of many expert systems, and is referred to in artificial intelligence as the **classification problem** (where the prediction is referred to as the classification). The task is to learn a classifier given a **training sample** of classified examples. In credit assessment, classes are 'accept' and 'reject' (the credit application). A training sample in this case would be historical records of previous loan applications together with whether the loan went bad. This generic learning task is referred to as **supervised** learning in pattern recognition, and **induction** or **empirical** learning in machine learning (see, for instance, Quinlan, 1986). The statistical community uses techniques such as discriminant analysis and nearest neighbour methods, described, for instance, by Ripley (1987).

This prediction problem is often handled with **partitioning classifiers**. These classifiers split the example space into partitions; for instance, ID3 (Quinlan, 1986) and CART (Breiman *et al.*, 1984) use classification trees to partition the example space recursively, and CN2 (Clark and Niblett, 1989) and PVM (Weiss *et al.*, 1990) use disjunctive rules (which also partition a space, but not recursively in the manner of trees). Tree algorithms build trees such as the ones shown in Fig. 15.1. The medical tree shown on the left has the classes *hypo* (hypothyroid) and *not* (not hypothyroid) at the leaves. This tree is referred to as a **decision tree** because decisions about class membership are represented at the leaf nodes. Notice that the first test, $TSH > 200$, is a test on the real-valued attribute *TSH*. An example is classified using this tree by checking the current test and then falling down the appropriate branch until a leaf is reached, where it is classified. In typical problems involving noise, class probabilities are

Artificial Intelligence Frontiers in Statistics: AI and statistics III. Edited by D.J. Hand. Published in 1993 by Chapman & Hall, London. ISBN 0 412 40710 8

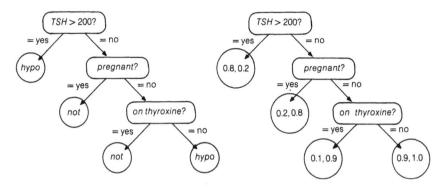

Figure 15.1 A decision tree and a class probability tree from the thyroid application.

usually given at the leaf nodes instead of class decisions, forming a **class probability tree** (where each leaf node has a vector of class probabilities). A corresponding class probability tree is given in Fig. 15.1 on the right. The leaf nodes give predicted probabilities for the two classes. Notice that this tree is a representation for a conditional probability distribution of class given information higher in the tree. This chapter will only be concerned with class probability trees since decision trees are a special case. Tree-based approaches have been pursued in many areas such as applied statistics, character recognition and information theory for well over two decades. Perhaps the major classical statistics text in this area is Breiman *et al.* (1984), and a wide variety of methods and comparative studies exist in other areas (see, for example, Quinlan, 1988; Mingers, 1989b; Buntine, 1991b; Bhal *et al.*, 1989; Quinlan and Rivest, 1989; Crawford, 1989; and Chou, 1991).

The standard approach to building a class probability tree consists of several stages: growing, pruning, and sometimes smoothing or averaging. A tree is first **grown** to completion so that the tree partitions the training sample into terminal regions of all one class. This is usually done from the root down using a **recursive partitioning** algorithm. Choose a test for the root node to create a tree of depth 1 and partition the training set among the new leaves just created. Now apply the same algorithm recursively to each of the leaves. The test is chosen at each stage using a greedy one-ply lookahead heuristic. Experience has shown that a tree so grown will suffer from over-fitting, in the sense that nodes near the bottom of the tree represent noise in the sample, and their removal can often increase predictive accuracy (for an introduction, see Quinlan, 1986). To help overcome this problem, a second process is subsequently employed to **prune** back the tree, for instance using resampling or hold-out methods (see, for example, Breiman *et al.*, 1984; or Crawford, 1989), approximate significance tests (Quinlan, 1988; or Mingers, 1989a), or minimum encoding (Quinlan and Rivest, 1989). The pruned tree may still have observed class probabilities at the leaves with zeros for some classes, an unrealistic situation when noisy data is being used. So **smoothing** techniques described by Bahl *et al.* (1989) and Chou (1991), explained later, are sometimes employed to make better class probability estimates. A final technique from Kwok and Carter (1990) is to build multiple trees and use the benefits of **averaging** to arrive at possibly more accurate class probability estimates.

This chapter outlines a Bayesian approach to the problem of building trees that is related to the minimum encoding techniques of Wallace and Patrick (1991) and Rissanen (1989). These two encoding approaches are based on the idea of the 'most probable model', or its logarithmic counterpart, the minimum encoding. But the approaches here average over the best few models using two techniques, the simplest is a Bayesian variant of the smoothing technique of Bahl *et al.* (1989). A fuller discussion of the Bayesian methods presented here, including the treatment of real values and extensive comparative experiments, is given by Buntine (1991b). The current experiments are described in more detail by Buntine (1991a). Source code and manual for the implementations and reimplementations are available in some cases as the IND tree package by Buntine and Caruana (1991).

THEORY

This section introduces the notation, the priors and the posteriors for the prediction task just described. This develops the theoretical tools on which the Bayesian tree learning methods are based. The section largely assumes the reader is familiar with the basics of Bayesian analysis, that is, the application of Bayes's theorem, the notion of subjective priors, etc. (see, for instance, Press, 1989; or Lee, 1989). In what follows, the term 'prior' is an abbreviation for 'prior probability distribution'; likewise for 'posterior'. The prior and the posterior are both subjective Bayesian probabilities or measures of belief, and so do not correspond to measurable frequencies.

If the loss function to be used is minimum errors in prediction, we need to determine, given a new unclassified example, which class has maximum posterior probability of being true. In the Bayesian framework with trees used to represent knowledge about making predictions, this task involves determining posterior probabilities for different tree structures and class proportions, and then returning the posterior class probability vector for the new example based on posterior probability averaged over all possible trees. The mathematics of this process is outlined in this section.

BASIC FRAMEWORK

To reason about posterior probabilities of class probability trees conditioned on a training sample, we need to separate out the discrete and the continuous components of a class probability tree. These need to be modelled by probability functions and probability density functions, respectively.

A class probability tree partitions the space of examples into disjoint subsets, each leaf corresponding to one such subset, and associates a conditional probability rule with each leaf. Denote the tree structure that defines the partition by T; this is determined by the branching structure of the tree together with the tests made at the internal nodes. Suppose there are C mutually exclusive and exhaustive classes $d_1, \ldots, d_C$. The probabilistic rule associated with each leaf can be modelled as a conditional probability distribution. Suppose example x falls to leaf l in the tree structure T. Then the tree gives a vector of class probabilities $\phi_{j,l}$ for $j = 1, \ldots, C$, which give the proportion of examples at the leaf that have class d_j. A class probability tree then corresponds to a (discrete) tree structure T, together with the (continuous) matrix of class proportions $\Phi_T = \{\phi_{j,l} : j = 1, \ldots, C, l \in leaves(T)\}$. If the choice of a test at a node requires selecting a 'cut-point', real value, as does the test at the root of the tree in Fig. 15.1, then this framework needs to be modified. The 'cut-points' chosen are continuous, not discrete parameters, so their specification involves additional continuous parameters for the tree structure T. A tree T, Φ_T therefore represents a conditional probability distribution for class c given example x of the form

$$\phi_{j,l} = P(c = d_j | x, T, \Phi_T)$$

where example x falls to leaf l in the tree structure T.

For the learning problem, we are given a training sample x, c consisting of N examples x_i with known classification given by class values c_i. Examples and their classes are assumed to be generated identically and independently. The distribution of a single classified example x, c can be specified by a probability distribution on the example, about which we have little concern, together with a conditional probability distribution on the class c given the example x, which corresponds to the class probability tree. Given a sample consisting of N examples x with known classification c, we are interested in determining the posterior distribution of class probability trees given the training sample. This distribution can be found using Bayes's theorem:

$$P(T, \Phi_T | x, c) \propto P(T, \Phi_T | x) \prod_{i=1}^{N} P(c_i | x_i, T, \Phi_T)$$

$$= P(T, \Phi_T | x) \prod_{l \in leaves(T)} \prod_{j=1}^{C} \phi_{j,l}^{n_{j,l}}$$

where $n_{j,l}$ is the number of examples of class d_j falling in the lth leaf of the tree structure T. The probability $P(T, \Phi_T | x)$ I refer to as the 'prior', because even though it is conditioned on the unclassified examples x, below I assume independence from these unclassified examples.

PRIORS

There are many different priors that could be used for trees. Entropic arguments like that of Rodriguez (1990) suggest that leaves should have more extreme probabilities (closer to 0 and 1) when the leaf is more infrequent (fewer examples occur at the leaves), for instance, lower down the tree. Trees tend to fracture the example space unnecessarily, which Pagallo and Haussler (1990) refer to as the 'replication problem', because some tests higher in the tree are unnecessary for some (but not all) of the lower leaves. For this reason, we might expect class probabilities to be similar across different leaf nodes. Smoothness arguments also suggest that class probabilities for neighbouring leaves in the tree might be more similar on average than for unrelated leaves. We therefore might use a hierarchical prior that relates class probabilities across leaves.

Rather than these, I use priors that have a simple conjugate (that is, the prior is the same functional form as the likelihood function; see Berger, 1985) and multiplicative form. Generic representation-independent techniques for forming priors, such as Rodriguez's entropic priors and smoothing priors described by Buntine and Weigend (1991), yield priors that lack the sample form I use. But although our motivation for choice of priors is expediency, I believe they are reasonable 'generic' priors for many circumstances.

Consider a prior consisting of a term for the tree structure and a term for the class proportions at each leaf given by

$$P(T, \Phi_T | x) = P(T)P(\Phi_T | T) = P(T) \prod_{l \in leaves\,(T)} \frac{1}{B_C(\alpha_1, \ldots, \alpha_C)} \prod_{j=1}^{C} \phi_{j,l}^{\alpha_j - 1} \quad (15.1)$$

This assumes that different class probability vectors at the leaves are a priori independent. Although this contradicts our intuitions described above, it does not add any incorrect information a priori. B_C is the C-dimensional beta function defined by

$$B_C(x_1, \ldots, x_C) = \frac{\Pi_{j=1}^{C} \Gamma(x_j)}{\Gamma(\Sigma_{j=1}^{C} x_j)}$$

and Γ is the gamma function. Each term indexed by l in Equation 15.1 is a Dirichlet distribution (a C-dimensional variant of the two-dimensional beta distribution). Let $\alpha_0 = \Sigma_{j=1}^{C} \alpha_j$. The Dirichlet terms ensure that the class probability vectors at each leaf l centre about a common mean

$$\left(\frac{\alpha_1}{\alpha_0}, \ldots, \frac{\alpha_C}{\alpha_0} \right)$$

with standard deviation proportional to $(\alpha_0 + 1)^{-1/2}$. I take an empirical Bayes approach (see Berger, 1985) and set the common mean to the observed common mean in the training sample. Because of the sample size, this should be close to the population base-rate. The 'prior weight' parameter α_0 is then set according to how much we expect a priori the class probabilities at leaves to vary from the base-rate class probabilities. A value of $\alpha_0 = 1$ means we expect class probabilities to differ strongly from the base-rate, and a value of $\alpha_0 = C$ means we expect mild difference.

I use several choices for the prior on the tree structure T, $P(T)$. Each of these priors is multiplicative on the nodes in the tree, so the posterior, given later, is also multiplicative on the nodes in the tree. Also, the priors I–III presented are not normalized as their later use does not require it, and prior I is a special case of prior II.

I. Each tree structure is equally likely, $P(T)$ is constant.
II. Give a slight preference to simpler trees, such as $P(T) \propto \omega^{|nodes(T)|}$, for $\omega < 1$. This means every node in the tree makes the tree a factor of ω less likely a priori.
III. The tree shapes (tree structure minus choice of tests at internal nodes) are

equally likely.

$$P(T) \propto \prod_{n \in nodes(T) - leaves(T)} \frac{1}{\# \, possible \, tests \, at \, n}$$

This means we have a uniform prior on the number of tests that will be made on an example to determine its class probability.

IV. Tree structures are coded using bits for 'node', 'leaf', and 'choice of test'. This is a combination of type II and III priors together with 'encoding' of cut-points. (For details, see Rissanen, 1980, p. 165; or Wallace and Patrick, 1991.)

The priors have been given in increasing strength, in the sense that type IV priors represent an extreme preference for simpler trees, but type I priors represent no such preference. In medical data, for instance, where many of the attributes supplied for each example may well be noise, we would expect the stronger priors to be more reasonable. In problems like the classic faulty LED problem of Breiman *et al.* (1984), we would expect all faulty LED indicators to be somewhat informative about the actual digit being represented, so large trees seem reasonable and the type I prior should be used.

Clearly, many variations on these priors are possible without departing from the basic conjugate and multiplicative form. Certain attributes could be penalized more than others if it is believed a priori that those attributes are not as effective in determining class. The prior weight α_0 could vary from node to node, for instance becoming smaller lower down the tree. Our claim, however, is that the priors described above are an adequate family of mildly informative tree priors for the purposes of demonstrating a Bayesian approach to learning tree classifiers.

POSTERIORS

Using standard properties of the Dirichlet distributions, given by Buntine (1991b, Section 2.5), posteriors conditioned on the training sample can now be computed as follows:

$$P(T, \Phi_T | x, c) = P(T | x, c) \cdot P(\Phi_T | x, c, T)$$

$$P(T | x, c) \propto P(T) \prod_{l \in leaves(T)} \frac{B_C(n_{1,l} + \alpha_1, \ldots, n_{C,l} + \alpha_C)}{B_C(\alpha_1, \ldots, \alpha_C)} \tag{15.2}$$

$$P(\Phi_T | x, c, T) \propto \prod_{l \in leaves(T)} \frac{1}{B_C(\alpha_1, \ldots, \alpha_C)} \prod_{j=1}^{C} \phi_{j,l}^{n_{j,l} + \alpha_j - 1}$$

One important property of these posteriors is that they are multiplicative on the nodes in the tree whenever the prior is also multiplicative on the nodes. This means posteriors for a collection of similar trees can be efficiently added together using an extended distributive law as described in the next section. A second important property is that the discrete space of tree structures is combinatoric, and only partially ordered. This means for smaller data sets we might expect to have very many different trees with a similar high posterior. In contrast, consider the space of polynomials of a single variable. This is a linearly ordered discrete space, so we can expect polynomials with a high posterior to be polynomials of a similar order.

Posterior expected values and variances for the proportions Φ_T can be deduced from the posterior using standard properties of the Dirichlet distribution. I use the notation $E_{x|y}(z(x, y))$ to denote the expected value of $z(x, y)$ according to the distribution for x conditioned on knowing y. For instance, the posterior expected class proportions given a particular tree structure are:

$$E_{\Phi_T|x, c, T}(\phi_{j,l}) = \frac{n_{j,l} + \alpha_j}{n_{.,l} + \alpha_0} \tag{15.3}$$

where $n_{.,l} = \sum_{j=1}^{C} n_{j,l}$ and $\alpha_0 = \sum_{j=1}^{C} \alpha_j$.

The log-posterior for the tree structure can be approximated when the sample size at each leaf $(n_{.,l})$ is large using Sterling's approximation.

$$-\log P(T|x, c) \approx -\log P(T) + N I(C|T) + constant \tag{15.4}$$

where the dominant term $I(C|T)$ represents the expected information gained about class due to making the tests implicit in the tree structure T. That is:

$$I(C|T) \equiv \sum_{l \in leaves(T)} \frac{n_{.,l}}{N} I\left(\frac{n_{1,l} + \alpha_1}{n_{.,l} + \alpha_0}, \dots, \frac{n_{C,l} + \alpha_C}{n_{.,l} + \alpha_0}\right)$$

The function I used here is the standard information or entropy function over discrete probability distributions. The information measure $I(C|T)$ when applied to a tree of depth 1 is used by Quinlan (1986) as a splitting rule when growing trees. If some leaves have small numbers of examples, then approximation 15.4 is poor, and the beta function of Formula 15.2 really needs to be used. The beta function then has the effect of discounting leaves with small example numbers.

METHODS

To implement a full Bayesian approach, we need to consider averaging over possible models, as, for instance, approximated by Kwok and Carter (1990), or done by Stewart (1987) using Monte Carlo methods. This section introduces methods closer in spirit to Henrion (1990) who collects the dominant terms in the posterior sum. Fuller details of the methods described below are given by Buntine (1991b).

Intuitively, this full Bayesian approach involves averaging all those trees that seem a reasonable explanation of the classifications in the training sample, and then predicting the class of a new example based on the weighted predictions of the reasonable trees. This estimates the posterior probability conditioned on the training sample that a new example x will have class c by the formula:

$$E_{T, \Phi_T x, c}(P(c = d_j|x, T, \Phi_T)) = \frac{\sum_{T, l = leaf(x, T)} P(T, x, c) \frac{n_{j,l} + \alpha_j}{n_{.,l} + \alpha_0}}{\sum_T P(T, x, c)} \tag{15.5}$$

where the summations are over all possible tree structures, and $leaf(x, T)$ denotes the unique leaf in the tree T to which the example x falls. Notice that the denominator of the right-hand side of Equation (15.5) can be simplified. However, when approximating the summations with a reduced set of tree structures as done below, this full form is required to normalize the result.

Given the combinatoric number of tree structures, such a calculation is not feasible. There are three computationally reasonable approximations to this general formula that can be made by restricting the summations to a reduced subset of tree structures. The first corresponds to a maximum a posteriori approximation, and the second and third, described below, collect more dominant trees from the a posteriori sum. To collect these dominant trees however, we first have to find high a posteriori trees. The growing method presented in the next subsection describes how.

TREE GROWING

Formula 15.2 suggests a heuristic for growing a tree in the standard recursive partitioning algorithm described in the introduction. When expanding the (single) current node, for each possible test grow a tree of depth 1 at that node by extending it one more level. Then choose the new test yielding a tree structure with the maximum posterior probability. Because the posterior is multiplicative, we only need look at that component of the posterior contributed by the new test and its leaves. The one-ply lookahead heuristic for evaluating a test then becomes:

$$Quality_1(test) = P(test) \prod_{l \in outcomes(test)} \frac{B_C(n_{1,l} + \alpha_1, \ldots, n_{C,l} + \alpha_C)}{B_C(\alpha_1, \ldots, \alpha_C)} \qquad (15.6)$$

where $P(test)$ is the contribution to the tree prior, $outcomes(test)$ is the set of test outcomes, and $n_{j,l}$ is the number of examples in the training sample at the current node with class j and having test outcome l. So the $n_{j,l}$ are also a function of the test. The subscript 1 is there in $Quality_1(test)$ to remind us that this is a one-ply lookahead heuristic. Computation should be done in log-space to avoid underflow. A test should be chosen to maximize Formula 15.6.

If the test being evaluated contains a cut-point, as does the test at the root of the tree in Fig. 15.1, then this too should be averaged over. Formula 15.6 in this case represents an evaluation of the test conditioned on knowing the cut-point. Since we have the freedom to choose this as well, we should calculate the expected value of Formula 15.6 across the full range of cut-points. Suppose we assume all cut-points are a priori equally likely between a minimum and maximum value, then for a real-valued attribute R, the quality of a cut-point test on R, denoted $cut\text{-}point(R)$ is given by,

$$Quality_1(cut\text{-}point(R)) = \frac{1}{max(R) - min(R)} \int_{cut = min(R)}^{max(R)} Quality_1(R < cut) \, dcut \qquad (15.7)$$

This test adds a penalty factor, given by $\dfrac{Quality_1(cut\text{-}point(R))}{Quality_1(R < best\text{-}cut)} \ll 1.0$, to the quality of the best cut, *best-cut*. This penalty has the same effect as the 'cost of encoding the cut-point', added by Quinlan and Rivest (1989) in their minimum encoding approach. While this evaluates the average quality of the cut-point test, we also need to select a cut-point. I use the cut that maximizes the quality test $Quality_1(R < cut)$.

Experiments show the quality heuristic of Equation 15.6 is very similar in performance to the information gain measure of Quinlan (1986) and to the GINI measure of CART (Breiman *et al.*, 1984). Approximation 15.4 explains why. The quality measure has the distinct advantage, however, of returning a measure of quality that is a probability. This can be used to advantage when growing trees from a very large training sample, growing trees incrementally, or for a stopping or pre-pruning rule.

When determining the test to make at a node using a one-ply lookahead heuristic, we need to do $O(AN)$ operations, or $O(AN \log N)$ if a sort is involved, where N is the size of the sample at that node, and A is the number of potential tests. To reduce computation for very large N, we could evaluate the tests on a subset of the sample (i.e. reduce N in the computation), as suggested by Breiman *et al.* (1984) and Catlett (1991). Because the measure of quality is in units of probability, one can readily determine if one test is significantly better than another according to the measure simply by taking their ratio. This can be used to determine if evaluation on the current subsample is sufficient, or if we need to view a larger subsample.

A related problem occurs when growing trees incrementally. In this regime, a tree needs to be updated given some additions to the training sample. Crawford (1989) shows that if we take the naive approach as done by Utgoff (1989) and attempt to update the current tree so the 'best' split according to the updated sample is taken at each node, the algorithm suffers from repeated restructuring. This occurs because the best split at a node vacillates while the sample at the node is still small. To overcome this problem Crawford suggests allowing restructuring only if some other test is currently significantly better than the current test at the node. We also need a test to determine whether the test at a node should be fixed and not changed thereafter, despite new data. Crawford uses a relatively expensive resampling approach to determine significance of splits, but the probabilistic quality measure of Formula 15.6 can be used instead without further modification.

The quality heuristic can also be improved using an N-ply lookahead beam search instead of the one-ply. In this, I use the one-ply quality test to obtain a few good tests to make up the search 'beam' at this level. I currently use a beam (search width) of 4. I then expand these tests, do an $(N-1)$-ply lookahead beam search at their children to find the best tests for the children, and finally propagate the quality of the children to calculate a quality measure corresponding to Equation 15.2 for the best subtree grown. This means that, at the cost of computation, we can do more extensive searches.

TREE SMOOTHING

The simplest pruning approach is to choose the maximum a posteriori tree. Of all those tree structures obtained by pruning back the complete grown tree, pick the tree structure with maximum posterior probability, $P(T|x, c)$. This pruning approach was tried with a range of different priors and the approach was sometimes better in performance than Quinlan's C4 or Breiman *et al.*'s CART, and sometimes worse. With careful choice of prior, this 'most probable model' approach was often better; however, it was unduly sensitive to the choice of prior. This suggests a more thorough Bayesian approach is needed.

The first approximation to the sum of Equation 15.5 I call **Bayesian smoothing**. The standard approach for classifying an example using a class probability tree is to send the example down to a leaf and then return the class probability at the leaf. In the smoothing approach, I also average all the class probability vectors encountered at interior nodes along the way (see Bahl *et al.*, 1989, p. 1005). Given a particular tree structure T', presumably grown using the algorithm described in the previous subsection, consider the space of tree structures, *pruned* (T'), obtained by pruning the tree structure T' in all possible ways. If we restrict the summation in Equation 15.5 to this space and the posterior on the tree structure is a multiplicative function over nodes in the tree, then the sum can be recursively calculated using a grand version of the distributive law. The sum is computable in a number of steps linear in the size of the tree. The sum takes the form of an average calculated along the branch traversed by the new example.

$$E_{T, \Phi T | x, c}(P(c = d_j | x, T, \Phi_T))$$

$$\simeq \frac{\sum_{T \in pruned(T'), \, l = leaf(x, T)} P(T, x, c) \frac{n_{j,l} + \alpha_j}{n_{.,l} + \alpha_0}}{\sum_{T \in pruned(T')} P(T, x, c)}$$

$$= \sum_{n \in traversed(x, T')} P(n \text{ is leaf} \,|\, x, c, \text{ pruning of } T') \frac{n_{j,n} + \alpha_j}{n_{.,n} + \alpha_0} \qquad (15.8)$$

where $traversed(x, T')$ is the set of nodes on the path traversed by the example x as it falls to a leaf, and $P(n \text{ is leaf} | x, c, \text{ pruning of } T')$ is the posterior probability that the node n in the T' will be pruned back to a leaf given that the 'true' tree is a pruned subtree of T'. It is given by

$P(n \text{ is leaf} | x, c, \text{ pruning of } T')$

$$= CP(leaf(n), x, c)/SP(T', x, c) \prod_{O \in ancestors(T', n)} CP(node(O)) \prod_{\substack{B \in children(O) \\ B \neq child(O, x)}} SP(B, x, c)$$

where $ancestors(T', n)$ in the set of ancestors of the node n in the tree T', $child(O, x)$ is the child tree of the node O taken by the example x during traversal, $children(O)$ is the set of children trees of the node O,

$$SP(T, x, c) = \sum_{ST \in pruned(T)} P(ST, x, c)$$

$$= CP(leaf(T), x, c) + 1_{T \text{ is not leaf}} CP(node(T)) \prod_{B \in branches(T)} SP(B, x, c) \qquad (15.9)$$

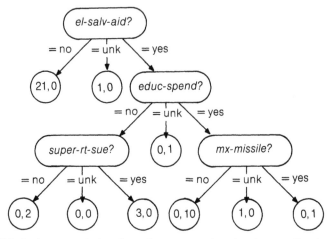

Figure 15.2 A class counts tree from the congressional voting application.

and $CP(node(T))$ is the component of the tree prior due to the internal node at the root of T, for instance 1 for type I priors and $\frac{1}{\#\ possible\ tests\ at\ node\ (T)}$ for type III priors. For $n_{1,l}, \ldots, n_{C,l}$ the class counts at the node T,

$$CP(leaf(T), x, c) = CP(leaf(T)) \frac{B_C(n_{1,l} + \alpha_1, \ldots, n_{C,l} + \alpha_C)}{B_C(\alpha_1, \ldots, \alpha_C)}$$

where $CP(leaf(T))$ is the multiplicative component of the tree prior due to the leaf T, for instance 1 for type I priors and ω for type II priors. Cut-points are currently handled by multiplying in the penalty factor described in the previous subsection with $CP(node(T))$.

For example, consider the tree given in Fig. 15.2. This is grown from data about voting in the US Congress. The numbers at the nodes represent class counts of Democrats and Republicans, respectively. To smooth this counts tree, we first compute the class probabilities for each node, and compute a node probability indicating how strongly we believe that node n gives appropriate class probabilities, $P(n$ is leaf $| x, c,$ *pruning of T'*). This was done with $\alpha_1 = \alpha_2 = 0.5$ and a type II tree prior with $\omega = 0.5$, intended to bias against larger trees. The intermediate tree is given in Fig. 15.3. The probability in brackets at each node represents the node probability. Notice these sum to 1.0 along any branch. The two probabilities below represent the class probabilities for that node for Democrats and Republicans, respectively. This intermediate tree allows smoothing as follows. Suppose we have a politician voting 'yes' on *el-salv-aid*, *educ-spend* and *mx-missile*. Then the probability the politician is Republican is given by running down the rightmost branch and computing the weighted sum of class probabilities:

$$0.000 \times 0.35 + 0.355 \times 0.76 + 0.348 \times 0.88 + 0.296 \times 0.75 = 0.80$$

The final tree after performing this smoothing process is given in Fig. 15.4. Notice

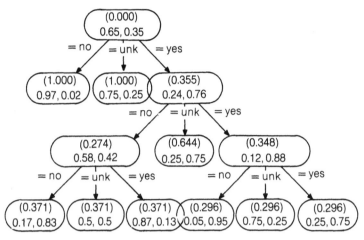

Figure 15.3 The intermediate calculation tree.

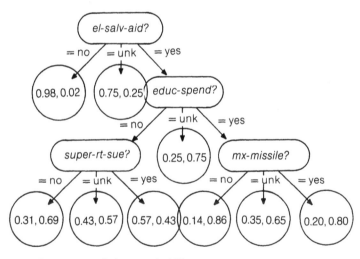

Figure 15.4 The averaged class probability tree.

the difference in probabilities for the leaf nodes of the intermediate class probability tree and the averaged class probability tree. In particular, notice that for the averaged tree the bottom right test on *mx-missile* has all its leaf nodes predict Republican. Rather than pruning these three leaf nodes we can keep them separate because the probabilities for the middle leaf are quite different from the probabilities for the other leaves. Of course, since the class counts are all quite small, a change in the prior parameters α_j and ω would change the final class probabilities quite a bit.

In experiments with smoothing, sometimes nodes will make so little contribution to the final averaged probabilities that they can be pruned without much affecting class probabilities of the resultant tree. For instance, this would happen if the cluster of leaves at the bottom right of Fig. 15.3 under the test *mx-missile* all had leaf probabilities of 0.001 instead of 0.296. This means the contribution to the sum in Equation 15.8 by a traversed node n and all its descendants will be so small that they will have no effect on the sum. In this case nodes n and below can be pruned.

Experiments reported below showed smoothing often significantly improved class probability estimates for a class probability tree (for instance, as measured using the half-Brier score), and sometimes made no significant difference. This happened regardless of the pruning and growing approach used to construct the original tree. In some cases smoothing is an adequate replacement for tree pruning compared with the standard pruning approaches such as pessimistic pruning or cost-complexity pruning. However, for really noisy data using a weak prior, this form of pruning was not strong enough, whereas, for strongly structured data with a strong prior, the pruning was too severe due to the choice of prior. The smoothing approach gives predictions still quite sensitive to the prior, although it was generally better than or at least as good as finding the most probable model.

OPTION TREES AND AVERAGING

The second approximation to the sum of Equation 15.5 involves searching for and compactly storing the most dominant (i.e. highest posterior) tree structures in the sum. The approach involves building **option trees** and then doing **Bayesian averaging** on these trees. Option trees are a generalization of the standard tree where options are included at each point. At each interior node, instead of there being a single test and subtrees for its outcomes, there are several optional tests with their respective subtrees. The resultant structure looks like an and–or tree. This is a compact way of representing many different trees that share common prefixes.

A class probability option tree built from Fisher's iris data is represented in Fig. 15.5. These trees have leaf nodes and test nodes as before, but optional test nodes are sometimes XORed together in a cluster. For instance, the very top node of the option tree XORs together two test nodes, *petal-width* < 8 and *petal-length* < 24.5. These two tests in the small ovals are referred to as **options** because, when selecting a single tree, we are to choose exactly one. Each node is labelled in brackets with the probability with which the node should be a leaf, which corresponds to the first probability in Equation 15.8. These determine the probability of selecting any single tree. Each node is also labelled with the estimated class probabilities at the node, used when averaging. The cluster of two tests at the top is referred to as a node of degree 2. So for the top node of degree 2, we should treat it as a leaf with probability 0.0 and otherwise choose either the test *petal-width* < 8 or the test *petal-length* < 24.5. Both options have subtrees with high probability leaves, so these two optional trees are about as good as each other. When parts of the option tree are not included in the figure, the word 'etc.' appears. The bottom cluster of options is a node of degree 4. This has its subtrees excluded from the figure. That this bottom cluster and its children contain

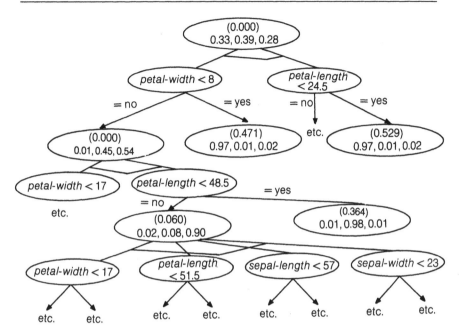

Figure 15.5 A class probability option tree from the iris application.

optional tests on all four attributes indicates that this cluster gives little indication as to which test is more appropriate, or if any test is appropriate at all.

Classification on the resultant structure is done using Equation 15.5 in a similar style to Bayesian smoothing. The same formulae apply accept that Equation 15.9 repeats the second term for every option at the node. Because this no longer involves dealing with a single tree, I refer to the process of growing and smoothing as **tree averaging**.

The current method for growing option trees is primitive yet adequate for demonstration purposes. Option trees are grown using an N-ply lookahead as described earlier in this section. Rather than only growing the best test as determined by the lookahead, the best few tests are grown as optional tests at the node. I currently use one-ply or two-ply lookahead and allow a maximum of four optional tests at each node, retaining only those within a factor of 0.005 of the best test. A depth bound is used to stop growing the tree after a fixed depth, and this is chosen a priori, although it sometimes had to be decreased due to a memory or time overrun. Note that these parameters affect the search, and more search usually leads to better prediction accuracy at the expense of a larger option tree. For nodes that have large counts, usually only one test is expanded because all others are insignificant according to the quality measure. With smaller samples, or in domains where trees are a poor representation (such as noisy DNF concepts) many tests may be almost as good according to the quality measure so many options will be expanded. This means option trees tend to have more options at the lower nodes where more uncertainty lies.

COMPARISON

Comparative trials were run with the Bayesian algorithms discussed above and reimplementations of CART, C4, and a generic minimum encoding method. The versions of CART and C4 were reimplementations by the auther, and are known to perform comparably to the original algorithms on the data sets used. With the CART reimplementation, cost-complexity pruning was done with tenfold cross-validation using the 0-SE rule. Breiman *et al.* (1984) suggest this gives the most accurate predictions, and this was confirmed experimentally. For the Bayesian algorithms, either type I or type II priors with $\omega = 0.5$ were used, depending on whether there were believed to be many irrelevant attributes. The prior weight parameter α_0 was set depending on how strongly it was believed class probabilities would vary from the base-rate. These prior choices were fixed before running the algorithms. To standardize the experiments, all methods used the tree growing method of the previous section. This behaves similarly to the standard GINI and information-gain criteria on binary splits.

Data sets came from different real and simulated domains, and have a variety of characteristics. They include Quinlan's (1988) hypothyroid and XD6 data, the CART digital LED problem, medical domains reported by Cestnik *et al.* (1987), pole balancing data from human subjects collected by Michie *et al.* (1990), and a variety of other data sets from the Irvine Machine Learning Database such as 'glass', 'voting records', 'hepatitis' and 'mushrooms'. The 'voting' data has had the attribute 'physician-fee-freeze' deleted, as recommended by Michie. These are available via **ftp** at **ics.uci.edu** in **"/pub"**.

For the LED data, $\alpha_0 = 1$ and the type I tree prior was used. This was because all attributes were believed to be relevant and a high accuracy was expected. The type II tree prior was used for the pole balancing, voting and hepatitis domains because attributes were believed to be only partly relevant. For pole balancing a type I prior was used out of personal inclination, however it turned out not to make much difference. For the hepatitis domain, $\alpha_0 = 2$ because class probabilities were expected to lie around the population base-rate, rather than to be near 0 or 1. For many of these domains depth bounds on option trees were set at about 6.

Data sets were divided into training/test pairs, a classifier was built on the training sample and the accuracy, predicted accuracy, and half-Brier score taken on the test sample. The half-Brier score is an approximate measure of the quality of class probability estimates, similar to mean-square error, where smaller is better. This was done for ten random trials of the training/test pair, and significance of difference between two algorithms was checked using the paired *t*-test. Several different training-set sizes were also tried for each data set to test small and large sample properties of the algorithms.

Algorithms are part of the IND tree package and options used on Version 1.1; more details of the data sets, and acknowledgements to the sources are given by Buntine (1991a).

RESULTS

A somewhat random selection of the results is presented in Table 15.1. The MSE column refers to half-Brier score. The 'Bayes trees' method corresponds to one-ply lookahead growing with Bayesian smoothing.

In general, Bayesian option trees and averaging yielded superior prediction accuracy. It was always competitive and usually superior to the other algorithms. In several cases it was pairwise significantly better than each of the non-Bayesian approaches at the 5% level. Bayesian option trees and averaging with a one-ply and two-ply lookahead during growing yielded improvement in predictive accuracy, averaged over 10 trials, often as high as 2–3%, sometimes more. Bayesian smoothing on either trees or option trees also yielded superior half-Brier scores. This is highly significant in most cases, even, for instance, when the prediction accuracy was worse. These results were consistent across most data sets and sizes, including those not shown here.

Bayesian option trees and averaging are also quite resilient to the setting of the prior

Table 15.1 Performance statistics

Data	Method	Set size (train + test)	Timing (train + test)	Error ± std. dev.	MSE
LED	CART-like	100 + 2900	1.2 s + 1.0 s	35.7 ± 4.0	0.57
LED	C4-early	100 + 2900	0.8 s + 1.0 s	36.1 ± 4.6	0.59
LED	Bayes trees	100 + 2900	0.8 s + 1.8 s	35.5 ± 2.6	0.55
LED	1-ply option trees	100 + 2900	14.0 s + 62.2 s	33.9 ± 2.5	0.51
LED	2-ply option trees	100 + 2900	15.3 s + 71.4 s	33.8 ± 2.5	0.51
LED	CART-like	900 + 2100	2.3 s + 0.7 s	28.8 ± 0.6	0.45
LED	C4-early	900 + 2100	1.0 s + 0.7 s	29.3 ± 0.5	0.46
LED	2-ply option trees	900 + 2100	12.4 s + 12.0 s	28.0 ± 0.6	0.41
pole	CART-like	200 + 1647	15.0 s + 3.8 s	15.7 ± 1.0	0.26
pole	C4-early	200 + 1647	3.4 s + 3.4 s	15.6 ± 1.6	0.27
pole	1-ply option trees	200 + 1647	171.2 s + 174.8 s	15.2 ± 0.6	0.22
pole	CART-like	1200 + 647	38.0 s + 1.0 s	12.2 ± 1.7	0.21
pole	MDL-like	1200 + 647	4.0 s + 0.7 s	13.0 ± 1.3	0.20
pole	Bayes trees	1200 + 647	3.9 s + 0.8 s	12.3 ± 0.9	0.18
pole	1-ply option trees	1200 + 647	269.8 s + 18.3 s	10.6 ± 0.8	0.16
glass	CART-like	100 + 114	4.8 s + 0.3 s	37.3 ± 4.7	0.60
glass	C4-early	100 + 114	1.5 s + 0.3 s	36.2 ± 4.3	0.59
glass	Bayes trees	100 + 114	3.0 s + 0.3 s	36.5 ± 5.8	0.56
glass	2-ply option trees	100 + 114	469.9 s + 102.1 s	30.5 ± 6.0	0.43
voting	CART-like	200 + 235	1.4 s + 0.1 s	12.3 ± 1.7	0.22
voting	C4-early	200 + 235	0.9 s + 0.1 s	12.9 ± 1.5	0.22
voting	MDL-like	200 + 235	1.0 s + 0.2 s	12.9 ± 1.4	0.21
voting	1-ply option trees	200 + 235	162.5 s + 18.0 s	10.4 ± 1.5	0.16
hepatitis	CART-like	75 + 70	4.2 s + 0.1 s	19.8 ± 3.7	0.35
hepatitis	Bayes trees	75 + 70	1.5 s + 0.1 s	23.1 ± 4.9	0.32
hepatitis	2-ply option trees	75 + 70	131.0 s + 23.1 s	18.8 ± 3.6	0.26

parameters. Use of either type $\overline{\text{I}}$ or II tree prior often made little difference. Simple smoothing, by contrast, was quite sensitive to the prior. The effects of averaging are clearly important.

DISCUSSION

With this simple experimental set-up, it is difficult to say with conviction that any one algorithm is 'better' than the others, since there are several dimensions on which learning algorithms can be compared, and there are combinations of algorithms which were not tried. Prediction accuracy and half-Brier scores of the Bayesian methods are impressive, however.

Several factors contribute here. The option trees are more than just a single decision tree, they effectively involve an extension of the model space, so we are not comparing like with like. The growing of option trees sometimes involved an extra order of magnitude in time and space, partly perhaps because of the primitive search used. Option trees do not have the comprehensibility of normal trees, although this could probably be arranged with some post-processing.

While option trees were often significantly better in accuracy by several percent, it is unclear how much of this is due to the smoothing/averaging process and how much is due to the improved multi-ply lookahead search during growing. Initial experiments combining multi-ply lookahead growing and CART-style cost-complexity pruning produced erratic results, and it is unclear why.

A final point of comparison is the parameters available when driving the algorithms. CART and C4 have default settings for their parameters. With CART, heavy pruning can be achieved using the 1-SE rule rather than the 0-SE rule. The number of partitions to use in cross-validation cost-complexity pruning can also be changed, but the effect of this is unclear. The minimum encoding approaches are (according to the purist) free of parameters. However, these approaches often strongly overprune, so Quinlan and Rivest (1989) introduce a parameter that allows lighter pruning. So all approaches, Bayesian and non-Bayesian, have parameters that allow more or less pruning that can be chosen depending on the amount of structure believed to exist in the data. In the fuller Bayesian approach with option trees and Bayesian averaging, choices available also allow greater search during growing and fuller elaboration of the available optional trees. These parameters have the useful property that predictive accuracy (or some other utility measure) and computational expense are on average monotonic in the value of the parameter. The parameter setting allows improved predictive accuracy at computational expense.

CONCLUSION

Bayesian algorithms for learning class probability trees were presented and compared empirically with reimplementations of existing approaches like CART (Breiman *et al.*, 1984), C4 (Quinlan, 1988) and minimum encoding approaches. The Bayesian option trees and averaging algorithm gave significantly better accuracy and half-Brier score on predictions for a set of learning problems, but at computational expense. Bear in mind the Bayesian algorithms had settings of mild prior parameters made and undertook considerably more search, whereas the other algorithms were not tuned in any such way.

First, this work has a number of implications for Bayesian learning. Simple maximum posterior methods and minimum encoding methods (which here would choose the single maximum posterior tree) may not perform well in combinatorial discrete spaces like trees if the prior is not well matched to the problem. Considerable improvement can be obtained by averaging over multiple high posterior models. With trees and a multiplicative posterior, efficient averaging over multiple models is possible. Standard computational techniques for performing averaging such as importance sampling and Gibbs sampling are therefore avoided. More sophisticated priors could help here, but it is surely just as important to consider more versatile classification models such as the decision trellises suggested by Chou (1991).

Second, the Bayesian methods derived here corresponded to a variety of subtasks previously done by a collection of disparate *ad hoc* approaches honed through experience. The splitting rule derived here suggested improvements such as multi-ply lookahead search, penalty factors for cut-points, and a modification for doing learning in an incremental rather than a batch mode. A comparison with previous pruning and smoothing methods is difficult because the derived Bayesian methods are highly parametric, although cost-complexity pruning is in some ways comparable with use of the type II prior. Cross-validation is difficult to interpret from a Bayesian perspective.

More research is needed on these Bayesian methods. Multi-ply lookahead and smoothing could be combined with CART-style methods. It is unknown how much smoothing, option trees and multi-ply lookahead each contribute to the observed gain in prediction accuracy and half-Brier score. Further priors need to be developed. For instance, the current tree structure priors are difficult to conceptualize, and the whole Bayesian framework becomes dubious when priors are not somehow 'meaningful' to the user. More advanced any-time best-first searches could be developed for option trees, and an importance sampling approach might also compare favourably.

ACKNOWLEDGEMENTS

Thanks to Peter Cheeseman, Stuart Crawford and Robin Hanson for their assistance and to the Turing Institute and Brian Ripley at the University of Strathclyde who sponsored me while much of this research was taking shape. I am particularly indebted to Ross Quinlan, whose grasp of the tree field and insistence on experimental evaluation helped me enormously during the thesis development.

REFERENCES

Bahl, L., Brown, P., de Souza, P. and Mercer, R. (1989) A tree-based language model for natural language speech recognition. *IEEE Transactions on Acoustics, Speech and Signal Processing*, **37**, 1001–1008.

Berger, J.O. (1985) *Statistical Decision Theory and Bayesian Analysis*, New York: Springer-Verlag.

Breiman, L., Friedman, J., Olshen, R. and Stone, C. (1984) *Classification and Regression Trees*, Belmont: Wadsworth.

Buntine, W. (1991a) Some experiments with learning classification trees. Technical report, NASA Ames Research Center. In preparation.

Buntine, W. (1991b) A theory of learning classification rules. PhD thesis, University of Technology, Sydney.

Buntine, W. and Caruana, R. (1991) Introduction to IND and recursive partitioning. Technical Report FIA-91-28, RIACS and NASA Ames Research Center, Moffett Field, CA.

Buntine, W. and Weigend, A. (1991) Bayesian back-propagation. *Complex Systems*, **5**, 603–643.

Carter, C. and Catlett, J. (1987) Assessing credit card applications using machine learning. *IEEE Expert*, **2**, 71–79.

Catlett, J. (1991) Megainduction: machine learning on very large databases. PhD thesis, University of Sydney.

Cestnik, B., Kononenko, I. and Bratko, I. (1987) ASSISTANT86: A knowledge-elicitation tool for sophisticated users, in *Progress in Machine Learning: Proceedings of EWSL-87*, Bratko, I. and Lavrač, N. (eds), Wilmslow: Sigma Press, pp. 31–45.

Chou, P. (1991) Optimal partitioning for classification and regression trees. *IEEE Transactions on Pattern Analysis and Machine Intelligence*, **13**.

Clark, P. and Niblett, T. (1989) The CN2 induction algorithm. *Machine Learning*, **3**, 261–283.

Crawford, S. (1989) Extensions to the CART algorithm. *International Journal of Man–Machine Studies*, **31**, 197–217.

Henrion, M. (1990) Towards efficient inference in multiply connected belief networks, in *Influence Diagrams, Belief Nets and Decision Analysis*, Oliver, R. and Smith, J. (eds), New York: Wiley, pp. 385–407.

Kwok, S. and Carter, C. (1990) Multiple decision trees, in *Uncertainty in Artificial Intelligence 4*, Schachter, R., Levitt, T., Kanal, L. and Lemmer, J. (eds), Amsterdam: North-Holland.

Lee, P. (1989) *Bayesian Statistics: An Introduction*, New York: Oxford University Press.

Michie, D., Bain, M., and Hayes-Michie, J. (1990) Cognitive models from subcognitive skills, in *Knowledge-based Systems for Industrial Control*, McGhee, J., Grimble, M. and Mowforth, P. (eds), Stevenage: Peter Peregrinus.

Mingers, J. (1989a) An empirical comparison of pruning methods for decision-tree induction. *Machine Learning*, **4**, 227–243.

Mingers, J. (1989b) An empirical comparison of selection measures for decision-tree induction. *Machine Learning*, **3**, 319–342.

Pagallo, G. and Haussler, D. (1990) Boolean feature discovery in empirical learning. *Machine Learning*, **5**, 71–99.

Press, S. (1989) *Bayesian Statistics*, New York: Wiley.

Quinlan, J. (1986) Induction of decision trees. *Machine Learning*, **1**, 81–106.

Quinlan, J. (1988) Simplifying decision trees, in *Knowledge Acquisition for Knowledge-based Systems*, Gaines, B. and Boose, J. (eds), London: Academic Press, pp. 239–252.

Quinlan, J., Compton, P., Horn, K., and Lazarus, L. (1987) Inductive knowledge acquisition: A case study, in *Applications of Expert Systems*, Quinlan, J. (ed.), Wokingham: Addison-Wesley.

Quinlan, J. and Rivest, R. (1989) Inferring decision trees using the minimum description length principle. *Information and Computation*, **80**, 227–248.

Ripley, B. (1987) An introduction to statistical pattern recognition, in *Interactions in Artificial Intelligence and Statistical Methods*, Aldershot: Gower Technical Press, pp. 176–187.

Rissanen, J. (1989) *Stochastic Complexity in Statistical Enquiry*, Teaneck, NJ: World Scientific, Section 7.2.

Rodriguez, C. (1990) Objective Bayesianism and geometry, in *Maximum Entropy and Bayesian Methods*. Fougère, P. (ed.), Dordrecht: Kluwer.

Stewart, L. (1987). Hierarchical Bayesian analysis using Monte Carlo integration: computing posterior distributions when there are many possible models. *The Statistician*, **36**, 211–219.

Utgoff, P. (1989) Incremental induction of decision trees. *Machine Learning*, **4**, 161–186.

Wallace, C. and Patrick, J. (1991) Coding decision trees. Technical Report 151, Monash University, Melbourne. To appear in *Machine Learning*.

Weiss, S., Galen, R., and Tadepalli, P. (1990) Maximizing the predictive value of production rules. *Artificial Intelligence*, **45**, 47–71.

An analysis of two probabilistic model induction techniques

16

S.L. Crawford and R.M. Fung

INTRODUCTION

In this chapter, we compare the properties of two algorithms for the induction of probabilistic models: CONSTRUCTOR (Fung and Crawford, 1990; Fung *et al.*, 1990); and CART (Breiman *et al.*, 1984).

Both CART and CONSTRUCTOR may be viewed loosely as concept formation algorithms in a vein similar to earlier machine learning work in this area. From this perspective, CART and CONSTRUCTOR are similar with respect to their goals: they are designed 'to help one better understand the world and to make predictions about its future behavior' (Gennari *et al.*, 1990). There are, however, significant differences between CART and CONSTRUCTOR, For example, CART requires that each observation in the training set have an associated class label (it is a **supervised** learning system), whereas CONSTRUCTOR makes no such demands. The two algorithms also differ markedly in the way the induced models are represented, since CART generates binary classification trees, and CONSTRUCTOR generates probabilistic networks.

In the sections to follow, we briefly describe each algorithm and then broadly compare the two approaches to model induction. This broad comparison raises a number of important questions and serves to motivate the more detailed analyses that follow.

CART

CART is useful when one has access to a training set consisting of a set of measurement vectors (observations), each labelled according to known class membership, and wishes to obtain rules for classifying future observations of unknown class. CART forms its classification trees via a recursive partitioning of the attribute space. A classification tree initially consists only of the root node, containing all of the observations in the training set. A search is made through the set of possible binary splits of the root until an optimal

Artificial Intelligence Frontiers in Statistics: AI and statistics III. Edited by D.J. Hand. Published in 1993 by Chapman & Hall, London. ISBN 0 412 40710 8

split is found. The optimal split is one that divides the attribute space into that pair of regions most homogeneous with respect to class distribution. The training set is then partitioned using this split and a pair of leaves is generated to represent the newly partitioned training data. For real-valued attributes A, splits take the form $A \leqslant r$, where r is a midpoint of any adjacent pair found in the set of unique values attained by A. For categorical attributes which can attain possible values $D_1, \ldots, D_M$, splits take the form $A \in \{D\}$ where $\{D\}$ is one of the subsets of $\{D_1, \ldots, D_M\}$. The partitioning process is applied recursively to each leaf, continuing until all leaves contain observations from a single class only, or contain identical attribute vectors. When partitioning ends, the initial tree has grown to become a terminal tree, T_{max}.

Associated with T_{max} is an error rate computed by using T_{max} to classify the observations in the training set. It is well known, however, that this error rate (known as the **apparent** error rate) is optimistic (downwardly biased) in that we cannot expect T_{max} to classify *new* observations as well—in fact, it is likely that a smaller tree will actually perform better on new data. One advantage of the CART approach to tree induction is that T_{max} is pruned to a nested sequence of successively smaller trees, allowing one to ask which of these subtrees will yield the lowest misclassification rate when tasked with classifying new observations. CART typically makes use of a statistical resampling technique known as cross-validation (other methods are available—for details, see Crawford, 1989) to deliver estimates of the *true* error rate of each subtree. The tree with the best estimated performance is then selected as the optimal tree. Subsequent observations are dropped down the selected tree and classified according to the class membership of the plurality of observations populating the leaf into which they fall.

CONSTRUCTOR

Probabilistic networks (Pearl, 1988) are a new knowledge-based approach for reasoning under uncertainty. Like other knowledge-intensive approaches, acquiring the qualitative and quantitative information needed to build these networks is highly labour-intensive. In an effort to address this problem, techniques for network induction (Rebane and Pearl, 1987) have been explored. However, these techniques are restricted to the recovery of tree structures and these structures are often not expressive enough to represent real-world situations.

The CONSTRUCTOR algorithm is designed to address this 'knowledge acquisition bottleneck' more fully. CONSTRUCTOR induces discrete, undirected, probabilistic networks (Lauritzen, 1982) of arbitrary topology from a training set of examples. The induced networks contain a quantitative (i.e. probabilistic) characterization of the data but, perhaps more importantly, also contain a qualitative structural description of the data. By 'qualitative structure' we mean, loosely, the positive and negative *causal* relationships between factors as well as the positive and negative *correlative* relationships between factors in the processes under analysis. CONSTRUCTOR has as a primary focus the recovery of qualitative structures not only because these structures determine which quantitative relationships are recovered but also because such structures are readily interpretable and thus are valuable in explaining the real-world processes under analysis.

The CONSTRUCTOR algorithm builds a probabilistic network from a training set in three stages. The first stage involves instantiating a network and adding to the network a node for each attribute in the training set. In the second and most crucial stage, the topology of the network is found by iteratively identifying the neighbours of each node in the network. Identifying the neighbours of a particular node is operationalized by heuristically searching for the smallest set of nodes (i.e. attributes) that makes that node conditionally independent of all other nodes in the network. Conditional independence between three sets of nodes is tested by building a contingency table of the nodes' corresponding attributes and then applying the χ^2 test of independence. Fung and Crawford (1990) have identified several powerful heuristics that help make searching efficient. These heuristics choose the order for the nodes to be processed as well as which nodes to hypothesize as neighbours. In the third stage, the joint distributions of the nodes in each clique of the network are estimated using frequencies from the training set.

When CONSTRUCTOR is used to generate a network that will be used as a classifier, then CONSTRUCTOR can generate several possible alternative networks and, like CART, choose the best by estimating the error rate for each network through the technique of cross-validation.

Fung *et al.* (1990) have successfully applied CONSTRUCTOR to a number of synthetic training sets as well as a real-world task that involved building a probabilistic network that was used to perform document retrieval.

MOTIVATION

In this section we compare CART and CONSTRUCTOR qualitatively in an effort to assess their potential strengths and weaknesses.

Both CART and CONSTRUCTOR take a training set as input and deliver a probabilistic model as output. For CART, the training set consists of a collection of observations, each of which consists of a *known* class assignment and a vector of attributes. The attribute values may be boolean, integer, real or categorical and, in addition, may be both noisy and incomplete. The output from CART is a probabilistic model represented as a binary classification tree. Unlike CART, no strict 'class attribute' needs to be defined for CONSTRUCTOR. Although we are presently working on a new version of CONSTRUCTOR that will be able to process real-valued attributes, the current version of CONSTRUCTOR is restricted to measurements on attributes that are categorical-valued. The output from CONSTRUCTOR is a discrete, undirected network of arbitrary topology.

The probabilistic network induced by CONSTRUCTOR represents the full joint probability distribution of the training set and is used for classification purposes by the application of maximum likelihood estimation. In network terms, the simplest type of tree induced by CART (i.e. one without surrogate splits resident at the nodes) can be thought of as a representation of the network neighbours of the attribute that acts as the class identifier. Selection of a single attribute as a class identifier limits CART to use as a simple classifier, whereas CONSTRUCTOR may be used to address more general inference tasks. For example, given a training set, CONSTRUCTOR will generate a probabilistic network and, from this network, it is possible to compute the probability of *any* attribute given *any* subset of attribute values present in the network. In contrast, each CART tree is useful only for a single inference (classification) task. The generality of CONSTRUCTOR is not without cost, however. The computational requirements of CART grow log-linearly with training-set size whereas, in the worst case, CONSTRUCTOR grows exponentially (search heuristics provide better average-case performance).

The algorithms also differ in the manner in which observations with missing values are handled. For both model-building and subsequent estimation purposes, CART makes use of a mechanism based upon finding the best predictors (surrogates) of decision points in the classification tree. CONSTRUCTOR performs estimation based upon the partial information present in incomplete observations and, during model-building, makes use of partial information in the construction of the contingency tables used for assessing attribute independence. More detailed descriptions of the way in which both CART and CONSTRUCTOR handle missing values may be found in a later section. The experiments described in this chapter are intended to investigate how these differences between CART and CONSTRUCTOR manifest themselves in terms of both the performance and the expressiveness of the models they are able to induce.

THE FAULTY CALCULATOR DATA

We illustrate the two algorithms using variations on a problem originally described in Breiman *et al.* (1984). This problem involves an LED display connected to a numeric keypad. The display is *faulty*, however, since the output of the display may not always match the key that was depressed. Figure 16.1 shows the digit display unit and

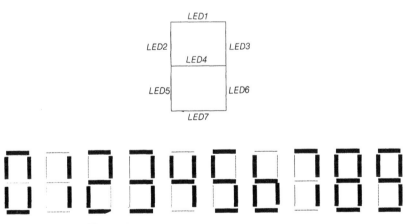

Figure 16.1 The LED display.

Table 16.1 $P(LED1 = \text{ON})$ given states of *LED2* and *LED3*

	LED3 ON	LED3 OFF
LED2 ON	0.9	0.3
LED2 OFF	0.7	0.1

illustrates the components of the display that must be illuminated to generate all ten digits.

The model described in Breiman *et al.* (1984) was very simple in that each component of the display was defined to have a 0.9 probability of illuminating in error. We introduce an additional source of random error by constraining the behaviour of one component (*LED1*) to be determined by the states of *LED2* and *LED3*, both of which depend upon the state of the depressed key. The dependencies are summarized in Table 16.1. The entries in Table 16.1 indicate the probability that *LED1* will illuminate, given the states of *LED2* and *LED3*. For example, when *LED2* and *LED3* are both on, the probability that *LED1* will also be on is 0.9.

Using this model, a training set of 200 observations (20 from each of the ten-digit classes) was generated. Each observation consists of an attribute value representing the key that was depressed, as well as a vector of attribute values representing the actual state of each LED. The task is to use both CART and CONSTRUCTOR to recover probabilistic models that can be used both *qualitatively* to understand the workings of the LED display and *quantitatively* to predict the intended digit given new LED displays.

EXPLANATORY POWER

In this section, we compare the two algorithms from a strictly explanatory perspective, asking only how the results obtained from each algorithm help us to understand the mechanism underlying the observed phenomena. In terms of the faulty calculator training set, we wish to construct a model from data that is able to explain the inner workings of the mechanism controlling the illumination of the LED segments.

THE INDUCED CALCULATOR TREE

Here, we present an analysis of the faulty calculator training set from the perspective of an investigator who does not know the details of the probability model used to generate the data.

When CART is applied to the training set, the tree shown in Fig. 16.2 is selected as optimal. Inspection of the interior nodes of the tree provides little insight into the nature of the calculator's faulty behaviour, illustrating only the importance of *LED5* for distinguishing digits 0, 2, 6, and 8 from the remaining digits. A look at the leaves, however, reveals the surprising absence of a leaf for digit 1—of course, once one

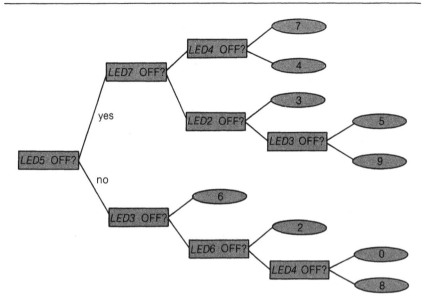

Figure 16.2 Classification tree for calculator data.

knows the probability model used to generate the training set, the absence of this leaf from the CART tree is hardly surprising! This means that, regardless of the LED configuration, no observations will be classified as digit 1.

A more detailed feel for the classification behavior of this tree can be obtained by examination of the misclassification matrix shown in Table 16.2. For each digit, the matrix displays the number of training observations classified correctly (the entry along the main diagonal) as well as the number misclassified (the other entries in the row). For example, we see that there are three instances in which the correct class was digit 4 and the resulting LED configuration was classified as digit 9. A glance at Table 16.2 immediately leads us to notice the column of zeros for digit 1 and to observe that 16 of the 20 digit-1 observations were actually classified as digit 7. The observation that the characteristic distinguishing digit 7 from digit 1 is the illumination of *LED1* leads us to believe that *LED1* is behaving strangely.

Since the mere observation that an attribute is present or absent from a tree is no indicator of its power to discriminate among classes (one attribute may act as a proxy for others, thus masking their discriminatory ability), CART maintains a list of surrogate attributes at each node—those attributes most predictive of the way in which the most discriminating attribute actually partitions the training set. By computing a weighted sum (summing over all of the nodes in the optimal tree weighted by the proportion of observations resident at a node) of the degree to which each attribute predicts the behaviour of the optimal split, a list of the most discriminating attributes can be readily calculated. This list (not shown) reveals that *LED1* has little ability to discriminate among the ten digit classes—further evidence that *LED1* is behaving strangely.

Table 16.2 Misclassification matrix for the CART tree

	9	8	7	6	5	4	3	2	1	0
9	12	1	0	0	4	1	1	1	0	0
8	1	15	0	3	0	0	0	0	0	1
7	0	1	16	0	0	2	1	0	0	0
6	1	1	0	18	0	0	0	0	0	0
5	0	0	0	2	18	0	0	0	0	0
4	3	1	0	0	0	16	0	0	0	0
3	2	1	0	0	0	2	15	0	0	0
2	0	1	0	3	0	0	1	15	0	0
1	0	1	16	0	0	1	1	0	0	1
0	0	2	1	1	0	0	0	2	0	14

To summarize, analysis of the CART results reveals that it is difficult to distinguish between observations of digits 1 and 7 in the training data. Furthermore, the only difference between such observations is the illumination of *LED1*—an LED with very low discriminatory power. Although *LED1* is a poor discriminator, it appears to have *some* predictive abilities, thus weakening any hypothesis that it might be illuminating purely at random. A best guess at this point might be that *LED1* is dependent on the key that is depressed, but illuminates far less frequently than it should. If so, one would expect some instances in which digit 5 is misclassified as digit 6, and in which digit 8 is misclassified as digit 6. This expectation is borne out in Table 16.2, thus supporting this hypothesis.

THE INDUCED CALCULATOR NETWORK

Expressed as a probabilistic network, the model used to generate the faulty calculator training set is shown in Fig. 16.3. A probabilistic network represents a joint state space and a probability distribution on that space. Each node represents one component of the joint state space (i.e. a mutually exclusive and exhaustive set of states). For example, the *Depressed Key* node in Fig. 16.3 represents the set of states {0 1 2 3 4 5 6 7 8 9}. Each arc of a probabilistic network represents a relationship between the nodes it connects. For example, the arc between the *Depressed Key* node and the *LED2* node indicates that the state of *LED2* is probabilistically related to the key that is depressed. The primary innovation of probabilistic networks is the explicit representation of conditional

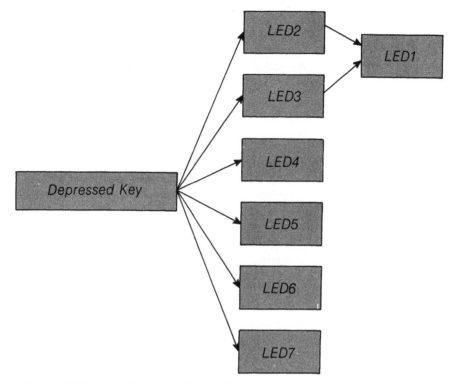

Figure 16.3 Faulty calculator network with loop.

independence relations. The representation of conditional independence relations in a probabilistic network is encoded in the topology of the network and is clearly illustrated in Fig. 16.3. In the figure, the *Depressed Key* node separates LED segments 2 through 7 from one other. This topology implies that the state of any one of those LED segments is independent of the state of any other *given* that it is known what key has been depressed. Furthermore, the figure clearly illustrates that the state of *LED1* is dependent only upon the states of *LED2* and *LED3*—the depressed key affects *LED1* only through its effect on *LED2* and *LED3*.

CONSTRUCTOR can uncover this structure when provided with a set of training examples generated from the probability model. Our application of CONSTRUCTOR generated the undirected analogue of the network shown in Fig. 16.3. Note that the induced network generated by CONSTRUCTOR is the *exact* network that would result from translating the directed network shown in Fig. 16.3 into an undirected network, clearly demonstrating that CONSTRUCTOR is able successfully to recover models with loops.

Useful inferences can be made given a probabilistic network that represents a situation and evidence about the situation. For example, given the evidence that *LED2*, *LED4*, *LED6*, and *LED7* are illuminated, and the network shown in Fig. 16.3, one can infer updated beliefs about which key was pressed. As one would expect for this example, the result is strong belief for '5' or '6' and weak beliefs for the remaining keys. Chang and Fung (1989), Lauritzen and Spiegelhalter (1988), and Pearl (1988) all describe algorithms for inference. While each of these algorithms employs significantly different methods for inference, they are equivalent in that given a probabilistic network and a particular query, they will infer identical results.

PREDICTIVE POWER

In this section we compare the predictive performance of the CART classification tree and the CONSTRUCTOR network through three experiments. In the first experiment, we compare the classification error rates as measured against an independently derived test set of size 1000. In the second experiment, the effect of missing values on error rate is determined. In the third experiment, we compare the performance of the two algorithms with respect to classifying the 128 possible configurations of the LED display. In each experiment, the CART and CONSTRUCTOR results are compared with the theoretical Bayes error rate.

EXPERIMENT 1

In this experiment, a test set containing 1000 observations is generated from the calculator model described earlier. Since the probability model is known, the Bayes error rate can be computed, and is found to be 0.349.

When CART is applied to the faulty calculator training set, a tree with 32 leaves (not shown) is obtained. This tree exhibits an error rate of 0.22. As already described, however, this error rate is optimistic (downwardly biased) in that we cannot expect this large tree to classify *new* examples as well. Cross-validation is used to obtain unbiased estimates of true error rate and the subtree with the best estimated performance is selected as the optimal tree. For this experiment, the tree selected as optimal has nine leaves, and the cross-validation estimate of true error rate is 0.375. This tree is shown in Fig. 16.2. When used to classify the 1000 observations in the independent test sample, the subtree with nine leaves is again selected as optimal with an error rate of 0.359, thus confirming the cross-validation results.

When the induced CONSTRUCTOR network is used to classify the 200 observations in the training set, an error rate of 0.285 is observed. Once again, this apparent error rate is downwardly biased. In order to gain a more realistic estimate of the predictive power of the network, it was used to classify the 1000 observations in the independent test sample, and was found to have an error rate of 0.354. Note that both the CART and CONSTRUCTOR models have misclassification rates remarkably close to the Bayes rate.

EXPERIMENT 2

It is often the case in data analysis that some of the available data contains observations with missing values. Missing values cause problems both at model-building time and at subsequent estimation (classification) time. CART and CONSTRUCTOR take rather different approaches to the handling of missing values.

As described earlier, the CART algorithm builds trees by examining candidate splits based upon the existing attribute values for each attribute in the training set. For each attribute, only those observations with values for the attribute in question are used when searching for the optimal split. Once the optimal split has been found, those observations with missing values for that split are directed towards the appropriate partition based upon their value for a different split—the one with the greatest predictive association with the optimal split. This 'alternate' split is known as a **surrogate** split; each node in the tree will often have an associated set of surrogate splits, ranked by the degree to which they are predictive of the optimal split. These surrogate splits are also used when new observations are classified—if the observation lacks a value for an optimal split, then the best surrogate is used.

As described earlier, the CONSTRUCTOR algorithm builds probabilistic networks by constructing contingency tables for subsets of attributes and assessing their independence via a χ^2 test. The contingency tables are filled using observations complete for the attributes in question. During subsequent estimation, CONSTRUCTOR makes use of whatever partial information is available in the observations to be classified, making use of prior distributions for the missing attributes.

We address only the second part of the missing value problem here, assuming that the model in question has been generated from a collection of complete observations, and is being used to classify new observations with missing values. We leave the analysis of the other half of the problem (building models using incomplete observations) for a later study. For this experiment, 20 independent test sets of 1000 observations were generated from the calculator model. Each test set had some proportion of its attribute values randomly removed, and these proportions ranged from 0% to 95% in 5% increments. The induced CART tree and CONSTRUCTOR network were then used to classify each of the 20 test sets. In Fig. 16.4, we illustrate the

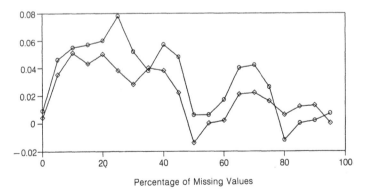

Figure 16.4 CONSTRUCTOR error rate (diamonds) and CART error rate (circles) versus percentage of missing values.

difference in predictive performance between CART and CONSTRUCTOR measured against the baseline performance of the Bayes rate (the Bayes error rate starts out at 0.349 for no missing values and monotonically increases to 0.877 for 95% missing values). In other words, the Bayes error rate for each missing value level is subtracted from the observed error rates for both the CART and CONSTRUCTOR trees and the resulting differences are plotted. As shown in Fig. 16.4, we find that CONSTRUCTOR has a lower error rate than CART over most of the range of missing values.

EXPERIMENT 3

In this experiment, the CART tree and CONSTRUCTOR network were used to classify each of the 128 possible LED configurations. The CART tree classified a total of 16 LED configurations in a different manner than the optimal Bayes classifier. The 16 LED configurations, however, make up only 3.1% of the probability mass of distribution of LED configurations. In addition, of the 16 differences, only one of the LED configurations was actually present in the 200 observation training set.

The CONSTRUCTOR network classified only eight LED configurations in a different manner from the Bayes classifier. These eight configurations make up only 1.1% of the probability mass associated with the complete distribution. Furthermore, none of these eight LED configurations was found in the training sample.

SUMMARY

In this section, we summarize the results of our experiments and gather together what we know of the CART and CONSTRUCTOR algorithms in an attempt to formulate some recommendations about their use.

Perhaps the most interesting observation to arise from our study is the degree to which the performance of both CART and CONSTRUCTOR approximated that of the Bayes classifier. The poorest performing model obtained in this analysis exhibited an error rate only 0.08 worse than the optimal Bayes rule! Similarly, it is interesting to note that, although CONSTRUCTOR delivered overall a better error rate performance than CART, the differences were slight. A more important distinction between the performance of CART and CONSTRUCTOR was in the interpretability of the induced models. The CART tree provided clues regarding the faulty nature of the calculator, but the CONSTRUCTOR network clearly illustrates what was wrong with the calculator.

As mentioned above, CONSTRUCTOR is computationally more complex than CART. In the worst case, CONSTRUCTOR grows exponentially with the size of the training set, whereas CART grows only log-linearly. The availability of the good search heuristics identified by Fung and Crawford (1990), along with some evidence (Fung et al., 1990) that models of real-world situations are rather sparsely connected, suggests that the average-case performance of CONSTRUCTOR may be far from exponential. We are currently exploring new techniques to reduce the computational complexity of CONSTRUCTOR by allowing 'approximate' networks in which the maximum number of neighbours a node can have is limited. In any event, when all that is required of an analysis is a high-performance classifier, then the results that can be obtained by CONSTRUCTOR may not warrant the CPU demands. However, if interpretability is important, CONSTRUCTOR is an appealing choice.

For some induction tasks, *causal* explanations are very important. For example, path analysis (Blau and Duncan, 1967) is a technique much used in the social sciences to extract causal relationships from data. Freedman (1987) has, however, criticized path analysis largely because of the often unrealistic distributional and independence assumptions that the technique demands. Used along with new techniques for recovering arc directions (Pearl and Verma, 1990; Verma and Pearl, 1990), CONSTRUCTION can, in certain circumstances, be used to extract causal relationships from data, and thus has potential for use as a tool for path analysis.

As mentioned earlier, both CART and CONSTRUCTOR take quite different approaches to handling missing values. Preliminary evidence, obtained from analysis of half of the missing-value problem (missing values in new data) suggests that CONSTRUCTOR may have a very slight edge. We believe this is because the surrogate splits that CART uses are essentially an approximation to the posterior probabilities that CONSTRUCTOR computes explicitly. Since CONSTRUCTOR builds a *full* model of the data, missing values can be more readily estimated.

CART trees are easy to use as classifiers—one can simply look at a CART tree and manually use it to classify a new observation. With CONSTRUCTOR, this is not possible since classification is only possible via application of an inference algorithm. On the other hand, CART trees are *restricted* to use as classifiers, whereas CONSTRUCTOR networks may be used to handle arbitrary queries about probabilistic relationships among the training set attributes.

CART can make use of training sets containing real-, integer-, and categorical-valued attributes. At this time, CONSTRUCTOR is able to handle real-valued attributes only by first discretizing the attribute values. However, CONSTRUCTOR is currently being extended so that real-valued attributes can be handled directly.

Before summarizing this section we must stress the fact that the results obtained in this chapter are from analysis of a single training set. The evidence suggesting that CONSTRUCTOR can outperform CART in certain situations is interesting, but preliminary. More analyses must be done to explore the performance of both algorithms. We conclude by stating that the results of the experiments described here strongly suggest that networks induced by the CONSTRUCTOR algorithm may be roughly comparable, in terms of predictive performance, with the classification trees generated by the CART algorithm. Evidence suggests that CONSTRUCTOR networks can, however, provide considerably more insight into the underlying model responsible for generation of the training set than do the CART trees. In addition, CONSTRUCTOR can be applied to inference tasks beyond the scope of the simple classification capabilities of CART. Although CART is a more mature, better-understood, and less computationally intensive approach, we feel that CONSTRUCTOR shows considerable promise as a powerful tool for model induction.

ACKNOWLEDGEMENT

This work was funded by an ADS Internal Research and Development Program.

REFERENCES

Blau, P.M. and Duncan, O.D. (1967) *The American Occupational Structure,* New York: John Wiley.

Breiman, L., Friedman, J.H., Olshen, R.A. and Stone, C.J. (1984) *Classification and Regression Trees,* Belmont: Wadsworth.

Chang, K.C. and Fung, R.M. (1989) Node aggregation for distributed inference in Bayesian networks, in *Proceedings of the 11th International Joint Conference on Artificial Intelligence,* Detroit.

Crawford, S.L. (1989) Extensions to the CART algorithm. *International Journal of Man–Machine Studies,* **31**, 197–217.

Freedman, D.A. (1987) As others see us: A case study in path analysis. *Journal of Educational Statistics,* **12**(2), 101–128.

Fung, R.M. and Crawford, S.L. (1990) CONSTRUCTOR: A system for the induction of probabilistic models, in *Proceedings of the Eighth National Conference on Artificial Intelligence,* Boston, pp. 762–769.

Fung, R.M., Crawford, S.L., Appelbaum, L.A. and Tong, R.M. (1990) An architecture for probabilistic concept-based information retrieval, in *Proceedings of the 13th International Conference on Research and Development in Information Retrieval,* Brussels.

Gennari, J., Langley, P. and Fisher, D. (1990) Models of incremental concept formation. *Artificial Intelligence,* **40**(1–3), 11–61.

Lauritzen, S.L. (1982) *Lectures on Contingency Tables,* Aalborg, Denmark: University of Aalborg Press.

Lauritzen, S.L. and Spiegelhalter, D.J. (1988) Local computations with probabilities on graphical structures and their application in expert systems. *Journal of the Royal Statistical Society B,* **50**, 157–224.

Pearl, J. (1988) *Probabilistic Reasoning in Intelligent Systems: Networks of Plausible Inference,* San Mateo, CA: Morgan Kaufmann.

Pearl, J. and Verma, T.S. (1990) A formal theory of inductive causation. Technical report, Cognitive Systems Laboratory, UCLA.

Rebane, G. and Pearl, J. (1987) The recovery of causal poly-trees from statistical data, in *Proceedings of the 3rd Conference on Uncertainty in Artificial Intelligence.*

Verma, T.S. and Pearl, J. (1990) Equivalence and synthesis of causal models, in *Proceedings of the 6th Conference on Uncertainty in Artificial Intelligence,* Cambridge, MA, pp. 220–227.

PART FOUR
Neural networks

A robust back-propagation learning algorithm for function approximation

17

D.S. Chen and R.C. Jain

INTRODUCTION

Function approximation from a set of input–output pairs has numerous applications in the fields of signal processing, pattern recognition and computer vision. Recently, feedforward neural networks containing hidden layers have been proposed as a tool for nonlinear function approximation as in Hornik *et al.* (1989), Poggio and Girosi (1990), and Ji *et al.* (1990). Parametric models represented by such networks are a cascade of nonlinear functions of affine transformations. In particular, approximation using a three-layer network is closely related to projection-based approximation, a parametric form first used in projection pursuit regression and classification studied by Friedman and Stuetzle (1981) and Friedman (1985). Diaconis and Shahshahani (1984) have shown that the projection-based approximation is dense in any real-valued continuous function. Recently, Hornik *et al.* (1989) extended this result by showing that a three-layer network is actually capable of approximating any Borel measurable function, provided sufficiently many hidden layer units are available. Due to these results, multi-layer networks have been referred to as **universal approximators** (Hornik *et al.*, 1989).

Learning algorithms are methods for estimating parameters in a neural network from a set of training data. Frequently, these parameters are connection weights between neural units of adjacent layers, whereas network architectures and activation functions are often chosen prior to learning. In the 1960s, Widrow and Hoff (1960) developed the Least Mean Squares (LMS) algorithm to train two-layer networks (with no hidden layer) where input and output are linearly related. The LMS algorithm is a kind of recursive linear regression method which produces optimal solutions only when output noise deviations have Gaussian distributions. The back-propagation (BP) learning algorithm (Rummelhart *et al.*, 1986) successfully advances the LMS algorithm to train multi-layer networks with hidden layers. The highly nonlinear modelling flexibility of multi-layer networks equipped with the powerful BP learning algorithm has made neural computation increasingly popular for nonlinear function approximation.

Artificial Intelligence Frontiers in Statistics: AI and statistics III. Edited by D.J. Hand. Published in 1993 by Chapman & Hall, London. ISBN 0 412 40710 8

However, the BP algorithm is extremely sensitive to noise in the training data. Using the least squares principle, the BP algorithm iteratively adjusts the network parameters to minimize an objective (energy) function which is defined as the sum of squares of the residuals. Due to the nonlinear modelling capability of multi-layer networks, it is possible for the network parameters to converge to a set of values such that the objective function reaches 0. At this stage, the learned function interpolates all the training data points. This interpolating phenomenon is undesirable in situations where training data contain noise. Such a phenomenon is also known as **overfitting** or **tuning** to the noise; it causes the learned function to oscillate abruptly among noise-contaminated training points. Even when training data are noise-free, the learned function can still oscillate wildly among discrete training data in cases where a network architecture has too many hidden layer units (Ji *et al.*, 1990). The interpolating effect worsens when in the presence of gross errors such as outliers. In general, poor approximation results occur in the vicinity of outliers.

In this chapter we propose a robust BP learning algorithm for training multi-layer networks, that is stable under small noise perturbation and robust against gross errors in the training data. The new algorithm extends the BP algorithm by employing a new form of objective function whose scale varies with learning time. It is known from statistical estimation theory that the shape of an objective function depends upon the noise probability distribution. Since noise distribution cannot be reliably estimated without any knowledge about the underlying function, an iterative strategy is adopted. The new objective function takes the shape of Hampel *et al.*'s (1981) hyperbolic tangent estimator in order to reject outliers. The shape of the objective function dynamically shrinks during learning; its rate of shrinking is iteratively estimated based upon the increasing knowledge about the underlying function and the noise statistics. During the process of learning, the function approximation and the noise estimation are continually being concomitantly refined. Although the spirit of this new algorithm comes from the pioneering work in robust statistics by Huber (1981) and Hampel *et al.* (1986), our work is different from the M-estimators in two aspects: first, the shape of the objective function changes with the iteration time; and second, the parametric model of functional approximator is highly nonlinear. Three advantages of the robust BP algorithm are: first, that it approximates an underlying mapping rather than interpolating training samples; second, that it is robust against gross errors; and third, that its rate of convergence is improved since the influence of incorrect samples is gracefully suppressed.

NEURO-COMPUTING FOR FUNCTION APPROXIMATION

In this section we first show how a multi-layer network can be utilized to approximate nonlinear functions and what kinds of parametric model are represented by such networks. We then briefly review the BP learning algorithm which is widely used for estimating parameters (weights) in a multi-layer network. Finally, we discuss two problems associated with BP learning, namely, non-smooth interpolation and noise sensitivity.

COMPUTATIONAL FRAMEWORK

Numerous engineering problems can be abstracted into the task of approximating an unknown function from a training set of input–output pairs, say $\mathcal{T} = \{(X_p, Y_p);\ p = 1, \ldots, P\}$. For example, in signal processing, unknown continuous signals are estimated from a finite set of signal values collected at known time instances. The inputs are the time instances, and the outputs are the corresponding signal values. Likewise, in three-dimensional machine vision applications, unknown analytical surface representations are reconstructed from a set of sparse surface depth values collected at irregularly spaced grid points. The spatial grid locations and the corresponding depth values form the input–output pairs. Another example is statistical pattern recognition in which objects of interest are encoded by vectors of feature attributes. Complex boundaries of decision regions that partition the feature space are approximated from a set of training samples consisting of input feature vectors along with their class labels as output. We notice that pattern classification is a special case of function approximation in which an unknown function takes values on a finite set of class labels. In all these examples, it is hypothesized that the input vector $X_p = (x_{p1}, \ldots, x_{pn})^{\mathrm{T}}$ and the output vector $Y_p = (y_{p1}, \ldots, y_{pm})^{\mathrm{T}}$ are related by an unknown function f such that $Y_p = f(X_p) + e_p$. The output noise deviation, e_p, is a random variable due to the imprecise measurements made by physical devices in real-world environments. The task of function approximation is to find an estimator $\hat{f}$ of f such that some metric of approximation error is minimized.

Multi-layer networks are highly nonlinear models for function approximation. In general, an L-layer feedforward network with fixed nonlinear activation functions can be parametrized by a set of $L - 1$ weight matrices, $\mathcal{W} = \{W^l, l = 0, 1, \ldots, L - 2\}$. The weight matrix W^l relates the lth layer output vector X^l to the $(l + 1)$th layer activation vector A^{l+1} by means of an affine transformation, $A^{l+1} = W^l X^l$, $0 \leqslant l < L - 1$. Notice that the zeroth layer output is the network input X. For an input X, the network estimated output $\hat{Y}$ is determined by the following forward propagating recursive equations which characterize the network dynamics:

$$X^0 = X$$

$$X^l = G_l(A^l), \quad \text{where } A^l = W^{l-1} X^{l-1} \qquad l = 1, \ldots, L - 1 \qquad (17.1)$$

$$\hat{Y} = \hat{f}(X; \mathcal{W}) = X^{L-1}$$

Here X^0 and X^{l-1} are the input vector and network output, respectively, and G_l is a vector function which applies a nonlinear activation function $g_l(\ )$ to each component of its vector argument. From Equations (17.1), it is clear that $\hat{f}(X, \mathcal{W})$ is uniquely determined by the set of weight matrices $\mathcal{W}$ provided the G_l functions are fixed. When Equations (17.1) are written in a non-recursive form, it is easy to see that the parametric model for a network is a cascade of nonlinear functions of affine transformations of the input. For example, the parametric form represented by a five-layer network illustrated in Fig. 17.1(a) is $\hat{f}(X; \mathcal{W}) = G_4(W^3 G_3(W^2 G_2(W^1 G_1 (W^0 X))))$. Figure 17.1(b) shows a three-layer network with n nodes in the input layer,

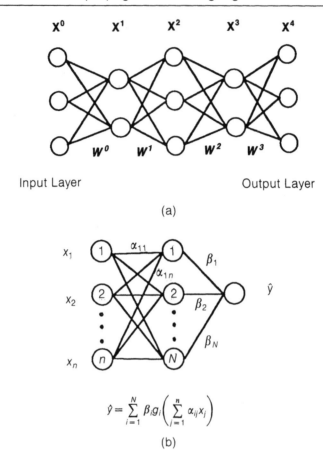

Figure 17.1 (a) A five-layer network. (b) A single-valued n-variate function represented by a three-layer network.

a single-output node with a linear (identity) activation function for a single-valued n-ary function, and N nodes in the hidden layer. The parametric model represented by this network is of the form

$$\hat{f}(x_1, x_2, \ldots, x_n; \mathcal{W}) = \sum_{j=1}^{N} \beta_j g_j \left(\sum_{i=1}^{n} \alpha_{ji} x_i \right) \qquad (17.2)$$

where $\beta_j, 1 \leqslant j \leqslant N$, are weights connecting N hidden units to the output unit; α_{ji}, $1 \leqslant i \leqslant n$ are weights connecting n input layer units to the jth hidden layer unit; and g_i are the hidden layer activation functions. This representation is closely related to the projection-based approximation which has been used in projection pursuit regression and classification by Friedman and Stuetzle (1981) and Friedman (1985). It can be shown that any polynomial function can be represented exactly as in Equation (17.2). Thus, a three-layer network can approximate any real-valued continuous functions following the Stone–Weierstrass theorem (Diaconis and Shahshahani, 1984). In most applications of neural networks, activation functions are often chosen to be sigmoid (logistic) functions, and the network architecture, i.e. the number of hidden layer units, is chosen by users based upon their knowledge about the problem. Thus, unlike linear regression models, multi-layer networks can learn arbitrary nonlinear mappings. The degree of nonlinearity is dependent upon the network architecture.

The class of parametric models provided by neural networks with at least one hidden layer is much more general than many traditional function approximation techniques. In classical approximation theory, arbitrary nonlinear functions are approximated by a linear combination of nonlinear basis functions such as in the techniques of spline approximation, polynomial approximation and Fourier series expansion. This type of approximation can rather naturally be embedded in a three-layer feedforward network (with one hidden layer), in which each hidden layer node simply implements one basis function, and the output node generates a linear combination of the hidden layer's basis functions. McKay (1989) demonstrates how a class of spline basis functions can be implemented with hidden layer nodes. His work relates function approximation by layered networks to spline fitting. Hecht-Nielsen (1989) showed that a simple neural node with a sigmoid activation function can approximate sinusoidal functions in L^2 norm. This result relates neuro-computing by layered networks to Fourier series approximation. Poggio and Girosi (1990) and Hartman et al. (1990) showed that radial basis functions can be implemented by a single neural node. Their work establishes a relationship between layered network computation and kernel-based approximation and regularization theory. All these research results demonstrate the fact that the class of parametric models provided by neural networks includes those of many other well-established techniques.

REVIEW OF THE BP LEARNING ALGORITHM

The BP algorithm is a generalization of Widrow's LMS algorithm (Widrow and Hoff, 1960). The LMS algorithm is a recursive solution to linear regression using a gradient descent technique. It has been used to train adaptive linear elements, i.e. networks with no hidden layers. Weights connecting input layer to output layer are iteratively updated upon each training input presentation. The updating formula, which is often referred to as the **Widrow–Hoff delta rule**, obeys the Hebbian learning law, i.e. the amount of weight adjustment at each input presentation is proportional to the product (correlation) of the error signal and the input. The error signal is defined to be the residuals, i.e. the difference between the training output and network estimated output. A major limitation of the LMS algorithm is that it only conducts learning for two-layered networks for linear approximation.

Historically, it has proven difficult to generalize the LMS algorithm to train networks having hidden layers. The fundamental obstacle is the determination of error signals for hidden layer nodes. The error signals are essential in weight adjustment in accordance with the Hebbian learning law. The BP learning algorithm successfully generalizes the delta learning rule by means of error back-propagation of variational calculus. The algorithm iteratively adjusts the network weights to minimize the least squares objective function (the sum of squares of the residuals), $E_{LS}(\mathcal{W}, \mathcal{T}) = \sum_{p=1}^{P} \sum_{j=1}^{m} (y_{pj} - \hat{y}_{pj})^2$, where y_{pj} is the jth component of the training output vector $\mathbf{Y}_p$, and $\hat{y}_{pj}$ is the estimated output at the jth output node obtained by propagating the training input $\mathbf{X}_p$ forward through the network using Equations (17.1). Clearly, this objective function depends upon the network weights $\mathcal{W}$ and the training set $\mathcal{T}$. We now introduce some notation to facilitate the algorithm description. Upon the presentation of the pth training sample, for the jth unit in layer l, $1 < l \leqslant L - 1$, let a_{pj}^l be its activation value, o_{pj}^l be its output, $g_l()$ be its activation function, and θ_j^l be its threshold. Let $w_{ji}^l(t), 0 \leqslant l < L - 1$, be the weight from the ith unit in layer l to the jth unit in layer $l + 1$ at time t, and let δ_{pj}^l be the error signal at the jth unit. The activation value and output value for each unit in layer l are computed by the forward-propagation equations:

$$a_{pj}^l = \sum_{i=1}^{n_{l-1}} w_{ji}^{l-1} o_{pi}^{l-1} + \theta_j^{l-1} \qquad l = 1, 2, \ldots, L - 1 \qquad (17.3)$$

$$o_{pj}^l = \begin{cases} x_{pj} & \text{if } l = 0 \\ g_l(a_{pj}^l) & \text{otherwise} \end{cases} \qquad (17.4)$$

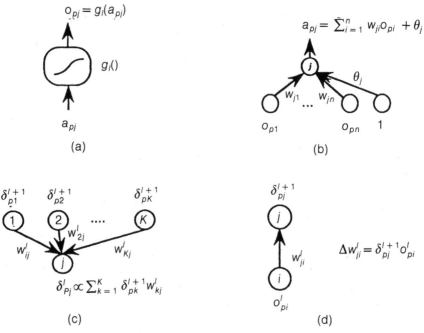

Figure 17.2 (a) Relation between activation and output values. (b) The forward-propagation equation. (c) The back-propagation equation. (d) The weight adaptation equation.

These equations are also graphically illustrated in Figs 17.2(a) and 17.2(b). The central part of the BP algorithm is the method of error back-propagation, i.e. the method of computing error signals for the hidden layer units. Similar to the LMS algorithm, the error signals for the output layer nodes are computed to be proportional to the differences between the training and estimated output values, or the residuals. Error signals at the output layer are then propagated back to lower (hidden) layers in the following way. For each hidden node, its error signal is computed to be proportional to the sum of weighted error signals at the nodes to which it has connections in the layer above. This rule of back-propagation can be derived analytically by variational calculus, as in Rummelhart *et al.* (1986), or by the method of Lagrangian multipliers for constrained optimization, as in Le Cun (1988). The solution is expressed by the following two equations:

$$\delta_{pj}^{L-1} = \dot{g}_{L-1}(a_{pj}^{L-1})(y_{pj}^{L-1} - \hat{y}_{pj}^{L-1}) \tag{17.5}$$

$$\delta_{pj}^{l} = \dot{g}_l(a_{pj}^{l}) \sum_{k=1}^{n_{l+1}} \delta_{pk}^{l+1} w_{kj}^{l} \qquad l = 1, \dots, L-2 \tag{17.6}$$

where $\dot{g}()$ denotes the derivative of $g()$. Once the error signal for each unit is computed using a gradient descent technique, the weights are adjusted by the adaptation equation:

$$w_{ji}^{l}(t+1) = w_{ji}^{l}(t) + \eta \delta_{pj}^{l+1} o_{pi}^{l} \qquad 0 \leqslant l < L-1 \tag{17.7}$$

where η specifies the learning rate. The back propagation equation 17.6 and the weight adaptation equation (17.7) are graphically illustrated in Figs 17.2(c) and 17.2(d). Equations (17.3)–(17.7) summarize the BP learning algorithm.

PROBLEMS WITH THE BP LEARNING ALGORITHM

Non-smooth interpolation

It is easy to see that the objective function $E_{LS}(\mathcal{W}, \mathcal{T})$ reaches 0 when the network interpolates all the training points, i.e. $Y_p = \hat{f}(X_p)$. However, interpolation is not a desirable property in nonlinear function approximation. For the sake of illustration, let us consider using only two data points to train a multi-layer $(l > 2)$ network. While there is only one line that can go through these two points, there are infinitely many nonlinear curves that can connect them. Each such nonlinear curve is a solution for minimizing the objective function. This example illustrates the existence of multiple minima of the objective function E_{LS} for networks having more than two layers. Mathematically, $E_{LS}(\mathcal{W}, \mathcal{T})$ is no longer a convex function of the weights for networks with hidden layers; therefore, it does not have a unique minimum. The number of minima increases with the complexity of a network architecture, or equivalently, the number of hidden layer units. Having the right network architecture which correctly represents the underlying model may alleviate this problem by smoothing the surface terrain of E_{LS}, therefore reducing the number of minima to those corresponding to solutions of smooth interpolation. For example, if we use a two-layer network when training with two data points, we will get a unique solution: the line that connects the two training points. However, designing the correct network architecture prior to learning is difficult. Most of the learning algorithms do not attempt to learn the network architectures. It is important to realize that the network learning can still be sensitive to outliers even with a correct architecture. The problem of erroneous training data cannot be solved by simply choosing the right network architecture. In our work, we investigate the possibility of dynamically altering the form of the objective function in order to achieve smoothing effects during approximation. The new algorithm approximates data instead of interpolating—an important feature in dealing with noisy data.

Noise sensitivity

The interpolating effects described above can have disastrous consequences when gross errors such as outliers are present in a training set. Recall that the input and output are related by $Y_p = f(X_p) + e_p$. The output deviation caused by the random noise term, e_p, can move training data away from their correct positions in an arbitrary manner. Consequently, the function approximator $\hat{f}$ obtained by the BP algorithm can oscillate wildly between interpolation points in the presence of gross errors. This catastrophic phenomenon is exemplified by two examples in Fig. 17.3. Figure 17.3(a) shows the result of approximating one segment of a sigmoid (logistic) curve. Twenty-five training samples containing three outliers are plotted in circular dots. The solid curve indicates the underlying true curve, and the dashed curve is the approximated curve plotted from 101 equally spaced testing samples which are not

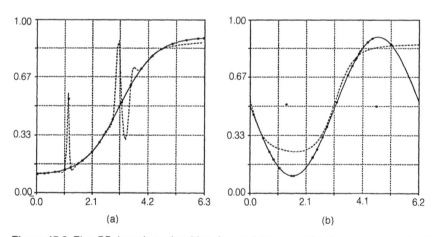

(a)   (b)

Figure 17.3 The BP learning algorithm is used to approximate one segment of sigmoid (logistic) function (a), and one segment of a sine function (b). The approximated function attempts to interpolate the training examples rather than approximate the underlying function.

used in the training set. Similarly, Fig. 17.3(b) shows the result of approximating one period of a sine curve. In the former example, the learning algorithm is terminated when the magnitude of the objective function falls below 10^{-6}. The learned curve interpolates almost all the training samples including the outliers. Poor approximation results occur in the neighbourhood of erroneous data points. In the latter example, the learning was terminated early before the objective function reaching the minimum. However, the learned curve is still incorrect near the outliers—the curve has been pulled away from the underlying model towards the outliers. This example demonstrates that inaccurate approximation *cannot* be avoided by terminating the BP algorithm earlier, since it is not a problem of overlearning. In both examples, the presence of gross errors leads to inaccurate approximation, especially when training data are sparsely distributed. The situation is not as bad for densely distributed training data. In that case, the worst that can happen to the learned curve is that a spurious sharp impulse may be generated where an outlier is present. These examples illustrate the noise sensitivity of the BP algorithm—the approximation process attempts to interpolate all the training data regardless of their quality. In the next section, we will show that the root of these problems comes from the use of the least squares objective function E_{LS}. We will develop a new learning algorithm to address these problems.

ROBUST BACK-PROPAGATION LEARNING ALGORITHM

In this section, we first discuss the inappropriateness of using the sum of squares of the residuals as an objective function for deriving the BP algorithm by analysing the optimality principle behind the least squares method. We then show how new objective functions can be obtained if the underlying noise model can be correctly identified. We devote the last part of this section to the development of a robust BP learning algorithm.

WHAT IS WRONG WITH THE LEAST SQUARES METHOD?

The method used to derive the BP algorithm by minimizing $E_{LS}(\mathcal{W})$ is known as the least squares (LS) method. The LS method, invented by Gauss and Legendre in the late eighteenth century, has been popular throughout history for two reasons. First, it produces the maximum likelihood (ML) estimator when the noise (residual) terms, e_p, are independent and identically distributed Gaussian random variables with zero mean. ML estimators are asymptotically efficient, that is to say, they achieve the maximum estimation accuracy with the minimum number of training samples. Second, the LS method produces the best linear unbiased estimator (BLUE) when the e_p are uncorrelated random variables with zero mean and finite variance, a result due to the Gauss–Markov theorem. The latter explains why the LS method is popular in linear regression analysis and Kalman filtering even when the noise distributions are not Gaussian. Ironically, it is exactly these two reasons that make the LS method inappropriate for deriving the BP learning algorithm. First, the distribution of e_p is never Gaussian. Most of the time it is unknown and cannot be reliably estimated in the presence of outliers. Second, as we discussed in the previous section, the parametric model represented by a network having more than two layers is not linear. Thus, the result of the Gauss–Markov theorem becomes irrelevant. With this analysis in mind, it is not too difficult to understand why the BP algorithm is extremely noise sensitive, and why it fails to approximate an underlying function in the vicinity of outliers.

When there are no outliers, it is sometimes possible to estimate the underlying noise probability distribution. In such cases, the LS method can be modified using the ML principle. More precisely, let $g_e(r)$ be the probability density function of e_p. The ML principle is equivalent to minimizing the objective function $\sum_{p=1}^{P} \sum_{j=1}^{m} \phi(r_{pj})$, where $\phi(r) = -\ln g_e(r)$. Clearly, $\phi(r) = r^2$ when $g_e(r) \propto e^{-r^2/2}$, and $\phi(r) = |r|$ when $g_e(r) \propto e^{-|r|}$. The former is the method of least squares (L_2 regression), the latter the method of least absolute values (L_1 regression). However, notice that the noise estimator is related to the function estimator by $\hat{e}_p = Y_p - \hat{f}(X_p)$, therefore in practice it is impossible to estimate the noise distribution reliably without a good estimation of the underlying function. This becomes a catch-22 situation—one needs a correct noise model to derive a function estimator, but one needs to know the underlying function to estimate the noise model. As will be seen later, an iterative strategy to circumvent this catch-22 situation is the centre of the robust BP learning algorithm.

The pioneering work in robust statistics by Huber (1981), Hampel (1974) and Hampel et al. (1986) gave rise to a set of new robust estimators called M-estimators. A class of M-estimators, which is capable of rejecting outliers, is obtained by modifying the shape of the objective function $\phi(r)$, or equivalently, its derivative $\psi(r) = d\phi(r)/dr$ to meet some robustification criteria. These criteria, in essence, require robust estimators to react smoothly to small perturbations such as rounding errors, to safeguard against large contamination (e.g. outliers), and to estimate the right quantity with the highest precision under ideal situations. In 1972, Hampel (1974) obtained the optimal $\psi(r)$ to be a three-part redescending odd function

$$\psi(r) = \begin{cases} r & |r| \leqslant a \\ c_1 \tanh(c_2(b - |r|)) \operatorname{sgn}(r) & a < |r| \leqslant b \\ 0 & |r| > b \end{cases}$$

under some minor assumptions about the noise model, where a and b are cut-off points. Notice that this $\phi(r)$ is not a convex function, so that $e^{-\ln\phi(r)}$ is no longer a probability density function. Thus, M-estimators are generally no longer the ML estimators. Figure 17.4 graphically shows the shape of $\psi(r)$ and its corresponding $\phi(r)$ functions. The shape of $\phi(r)$ plays an important role in restraining the effects of noise and rejecting gross errors. It behaves like r^2 for $|r| \leqslant a$. It responds to $|r|$ at a much slower rate than that of r^2 for $a < |r| \leqslant b$. It has constant response for $|r| > b$. Consequently, the effect of gross errors is reduced, and small perturbations are gracefully restrained, and the estimation process exercises the LS method whenever the error deviation is Gaussian-like. Following the spirit of robust statistics, we now

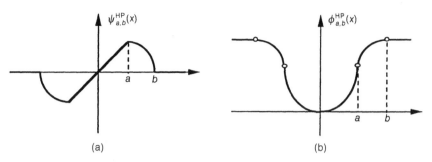

Figure 17.4 (a) The three-part redescending influence function $\psi(r)$. (b) $\phi(r)$ has different responses to residuals in different range.

present a robust BP learning algorithm that is noise-insensitive and capable of rejecting gross errors in function approximation.

ROBUST BP LEARNING ALGORITHM

In adapting a multi-layer neural network to learn an unknown function by minimizing the objective function E_{LS}, the BP learning process is coerced into accommodating all the training samples regardless of their quality. This leads to the undesirable interpolating behaviour of the BP learning algorithm. We now derive a robust BP learning algorithm by changing the form of the objective function so that it is similar to that of robust M-estimators, such that erroneous data do not increase the total energy in the order of squares of the residuals. In the following subsections, we first describe a time-varying energy function, which is the centre of our new algorithm. Then we derive the new algorithm analytically using the method of Lagrangian multipliers. Finally we explain the algorithmic details of the new learning method.

Time-varying objective function

We define a new objective function as

$$E_R(\mathcal{W}, \mathcal{T}, t) = \sum_{p=1}^{P} \sum_{j=1}^{m} \phi_t(r_{pj})$$

where $\psi_t(r) = d\phi_t(r)/dr$ has the shape of Hampel's redescending hyperbolic tangent estimator; however, it varies with learning time t. During the learning process, the shape of $\phi_t(r)$ reflects the knowledge about the underlying noise model at time t. In heuristic terms, the function $\phi_t(r)$ plays a role in filtering out noise effects in the data. The contribution of a single data point to the total energy E_R is no longer the square of the learning error (residual); instead, it is determined by the noise level of that point. More specifically, for data having minor noise contamination, the contribution to E_R is roughly of the order of r^2; for data having larger noise contamination, the contribution deteriorates from the order of r^2 in proportion to the level of contamination; for data having extremely large noise contamination, the contribution is limited to a constant. Since the noise statistics cannot be estimated reliably without the knowledge of the underlying function, they can only be iteratively estimated and refined as the estimator $\hat{f}$ approaches the true underlying function. In the current implementation, the shape of $\phi_t(r)$ is determined by the second-order statistics of the noise distribution at time t, which can be obtained using robust bootstrap methods (Efron and Tibshirani, 1984). The cut-off points a_t and b_t which determine the shape configuration of $\phi_t(r)$ can be computed in a number of ways. One method adopted in the current implementation assumes knowledge of the percentage of outliers in the training data, or of which the learning process wishes to tolerate. Let this percentage be q. The cut-off point a_t is computed as the qth largest magnitude of absolute residuals at time t, i.e. $a_t = |r_{(q)}(t)|$ where $|r_{(1)}(t)| \leqslant |r_{(2)}(t)| \leqslant \cdots \leqslant |r_{(q)}(t)|$. The cut-off point b_t is set to be $2a_t$. As we will show later, the amount of weight adjustment is proportional to the rate of change of $\phi_t(r)$ given by its derivative $\psi_t(r)$. At each time instance, about q per cent of the training data whose residual magnitude is smaller than $|r_{(q)}(t)|$ are used to their full content for learning. The rest of the training samples will have degraded influence on the learning process since they may be suspected outliers. Training data whose residual magnitude is larger than b_t will have no influence; they are rejected as outliers. In practice, $\phi_t(r)$ is changed at every Δt iteration to allow the network to settle down before the next update. At the beginning of the learning, $\phi_0(r)$ is close to the identity function ($a_0 \approx \infty$). The effect is that all training data are employed initially in order to get a crude initial estimation of the underlying function. One strategy for determining how many iterations are needed in order to obtain a good initial estimator is given on pages 230–231.

The robust BP learning algorithm is obtained by minimizing the new energy function $E_R(\mathscr{E}, \mathscr{T}, t)$, subject to the constraints imposed by the network dynamics by Equations (17.1). Next, we use the Lagrangian multiplier method to solve this constrained minimization problem analytically.

Learning algorithm derivation

In deriving the learning algorithm to determine the set of weight matrices such that the robust objective function $E_R(\mathscr{W}, \mathscr{T}, t)$ is minimized subject to the nonlinear constraints of network dynamics, we use a mathematical framework based on the Lagrangian multiplier method. This framework, inspired by optimal control theory, was introduced by Le Cun (1988) in deriving the BP learning algorithm.

The objective is to find the value of $\mathscr{W}^*$ that minimizes $E_R(\mathscr{W}, \mathscr{T}, t)$ subject to the nonlinear constraints of the network dynamics, i.e. $X_p^l - G_l(W^{l-1}X_p^{l-1}) = 0$, for $l = 1, \ldots, L-1, p = 1, \ldots, P$. Define the Lagrange function to be:

$$L(\Lambda, \mathscr{X}, \mathscr{W}) = \sum_{p=1}^{P} \sum_{j=1}^{m} \phi_t(r_{pj}) + \sum_{p=1}^{P} \sum_{l=1}^{L-1} \lambda_p^{lT}(X_p^l - G_l(W^{l-1}X_p^{l-1})) \qquad (17.8)$$

where $\Lambda = \{\lambda_p^l, p = 1, \ldots, P, l = 1, \ldots, L-1\}$ is the set of Lagrangian multiplier vectors, and $\mathscr{X} = \{X_p^l; p = 1, \ldots, P, l = 1, 2, \ldots, L-1\}$, with X_p^l the network output at layer l upon the presentation of training input X_p. The first term in Equation (17.8) is the time-variant objective function to be minimized and the second term encodes the constraints. By the principle of the Lagrangian multiplier method, solving $\mathcal{W}^*$ is equivalent to solving $\Lambda^*, \mathcal{X}^*, \mathcal{W}^*$ for $L(\Lambda^*, \mathcal{X}^*, \mathcal{W}^*) = \min_{\Lambda, \mathcal{X}, \mathcal{W}} L(\Lambda, \mathcal{X}, \mathcal{W})$. A necessary condition for the solutions is

$$\frac{\partial L(\Lambda, \mathcal{X}, \mathcal{W})}{\partial \Lambda} = 0 \tag{17.9a}$$

$$\frac{\partial L(\Lambda, \mathcal{X}, \mathcal{W})}{\partial \mathcal{X}} = 0 \tag{17.9b}$$

$$\frac{\partial L(\Lambda, \mathcal{X}, \mathcal{W})}{\partial \mathcal{W}} = 0 \tag{17.9c}$$

The computation for Equation (17.9a) is easy, since the first term of the Lagrange function (17.8) does not contain λs and the second term is linear in the λs. As a result, we obtain the forward propagating Equations (17.1) which guarantees the satisfaction of the constraints. From Equation (17.9b), we obtain two recursive formulae for the λ_p^ls, one for the output layer units, and the other for the hidden layer units:

$$\lambda_p^{L-1} = (\psi_t(r_{p1}), \dots, \psi(r_{pm}))^T \tag{17.10}$$

$$\lambda_p^l = W^{l^T} \nabla G_l(W^l X_p^l) \lambda_p^{l+1} \qquad l = 1, 2, \dots, L-2 \tag{17.11}$$

where T signifies matrix or vector transposition, the diagonal matrix ∇G_l is the Jacobian of G_l, and $\psi_t()$ is the derivative of $\phi_t()$. With the following variable substitution:

$$\delta_p^l = \nabla G_l(W^{l-1} X_p^{l-1}) \lambda_p^l \qquad l = 1, 2, \dots, L-1$$

Equations (17.10) and (17.11) are transformed into

$$\delta_p^{L-1} = \nabla G_l(W^{L-2} X_p^{L-2}) (\psi_t(r_{p1}), \dots, \psi(r_{pm}))^T \tag{17.12}$$

$$\delta_p^l = \nabla G_l(W^{l-1} X_p^{l-1}) W^{l^T} \delta_p^{l+1} \qquad l = 1, 2, \dots, L-2 \tag{17.13}$$

These are the back-propagating equations for computing the error signals.

We next solve Equation (17.9c) to obtain the weight adaptation equations. By differentiating L with respect to $\mathcal{W}^l$ for $l = 0, 1, \dots, L-2$, we obtain:

$$\frac{\partial L}{\partial W^l} = \sum_{p=1}^{P} \frac{\partial}{\partial W^l} [\lambda_p^{l+1^T} (X_p^{l+1} - G_l(W^l X_p^l))]$$

$$= \sum_{p=1}^{P} \nabla G_l(W^l X_p^l) \lambda_p^{l+1} X_p^{l^T} = - \sum_{p=1}^{P} \delta_p^{l+1} X_p^{l^T} = 0$$

Although this equation cannot be solved analytically, it can be solved iteratively using a gradient descent technique. Consider a dynamic system defined by a set of differential equations: $\partial W^l / \partial t = -\eta \partial L / \partial W^l$, $\eta > 0$. The equilibrium point of this dynamic system is the solution of $\partial L/\partial W^l = 0$. Assume L is a globally positive definite Lyapunov function (Vidyasagar, 1978) for the system. Because $\partial L/\partial t = [\partial L/\partial W^l][\partial W^l/\partial t] = -\eta[\partial L/\partial W^l]^2 \leqslant 0$, it guarantees that this equilibrium point is also a globally asymptotic stationary solution. Thus W^{l^*}, which minimizes $E_R(\mathcal{W}, \mathcal{T}, t)$ subject to the constraints, can be computed iteratively by the following difference equation:

$$W^l(t+1) = W^l(t) - \eta \frac{\partial L}{\partial W^l(t)} = W^l(t) + \eta \sum_{p=1}^{P} \delta_p^{l+1} X_p^{lT} \qquad (17.14)$$

The initial state $W^l(0)$ is randomly initialized. This is known as the **weight adaptation equation**. Notice that weights are adjusted after a complete cycle of all the training input presentations according to Equation (17.14). Sometimes in practice, weights are adjusted after a single input presentation. It is argued in Rummelhart *et al.* (1986) that when the learning rate η is small, the two adaptation rules are almost the same. As we see from the above derivation, the unique convergence of this solution depends upon whether the Lagrange function L is globally positive definite. In cases when L is locally positive definite, the solution converges to the local stationary state $\mathcal{W}^{l^*}$ only when the initial state $W^l(0)$ is within its domain of attraction.

The vector Equations (17.12)–(17.14) describe the robust BP learning algorithm, which can be written in terms of their scalar components as in Equations (17.5)–(17.7) for the BP algorithm. Comparing with the BP learning algorithm, the only change for the robust BP learning algorithm is the equation for computing error signals at output units. That is, in the robust BP learning algorithm, Equation (17.5) becomes:

$$\delta_{pj}^{L-1} = \dot{g}_{L-1}(a_{pj}^{L-1}) \, \psi_t(r_{pj}^{L-1}) \qquad (17.15)$$

The main difference between the robust BP algorithm and the BP algorithm is that the change in weights at the output layer during learning is proportional to $\psi_t(r)$ rather than the residual r. In this way the influence of noisy data on the learning process is gracefully suppressed, and the influence of gross errors is gradually eliminated.

Algorithmic details

We now describe the algorithmic details which are used in conducting all the experiments reported in this chapter. Currently, the algorithm needs three parameters: the percentage of outliers q; the number of iterations Δt between consecutive updates of $\phi_t(r)$; and the minimum energy level ε for the learning process to terminate. We now describe the new algorithm in the following detailed steps.

Step 1. Randomly initialize all the weights.

Step 2. Compute the estimated output using the forward-propagating equations for each training input. Compute the mean sum of squared residuals (MSSR).

Step 3. Compute the error signal for each output node using Equation (17.5), for any non-output node using the error propagating Equation (17.6).

Step 4. Update the network weights using the weight updating equation.

Step 5. If the difference between the current MSSR and the previous MSSR is small, then go to Step 2; otherwise go to Step 6.

Step 6. Set a counter, k, to zero, which is used for updating $\phi_t(r)$.

Step 7. Compute the estimated output using the forward-propagating equations for each training input. Compute the robust energy E_R. If $E_R < \varepsilon$, then terminate the learning process.

Step 8. If the counter k is a multiple of Δt, i.e. $k \bmod \Delta t = 0$, then sort the first q smallest absolute residuals: $|r_{(1)}| \leqslant \cdots \leqslant |r_{(q)}(t)|$. Update $\phi_t(r)$ by setting $a_t = |r_{(q)}(t)|$, and $b_t = 2a_t$.

Step 9. Compute the error signal for each output node using Equation (17.15), for any non-output node using the error propagating Equation (17.16).

Step 10. Update the network weights using the weight updating equation. Increment the counter k and go to Step 7.

Steps 1–5 compute an initial estimator for the underlying function. At this stage, the learning process uses the conventional back-propagation (BP) algorithm in which all the training data are employed indiscriminately. During this early stage, there is simply not enough information to judge the quality of the training data. It is critical to decide how many iterations are needed for this initial estimation. Too many iterations can lead to an initial estimator overfitting to the noise; this overfitting phenomenon can easily occur since the model is highly nonlinear. Too few iterations can lead to an initial estimator being incapable of representing the majority of the data; consequently, it yields a poor estimator for noise statistics which are vitally important for updating $\phi_t(r)$. A strategy employed here is to monitor the change in the MSSR for the training data. Usually, the MSSR remains large for a while initially. When the network under learning gets tuned into the training data, the MSSR often experiences a sharp decrease. This is an indication that the initially trained network has learned something from the data; then it is used to estimate the noise statistics, hence the quality of the training data. From this point on, learning relies crucially on the estimated knowledge about the quality of the data. More specifically, $\phi_t(r)$, updated at every Δt iteration, reflects the estimated knowledge of the data quality. The error signals at output nodes are computed in proportion to $d\phi_t(r)/dr$ instead of r^2. Therefore, the new algorithm adjusts the weights conservatively based upon the quality of the data. These concepts are reflected in Steps 6–10.

EXPERIMENTAL RESULTS

Our objective is to test the performance of the new learning algorithm in comparison with that of the conventional BP algorithm. In our experiment, we apply the new learning algorithm to approximate nonlinear univariate functions. Numerous such functions have been tested. While all the experiments conducted led to successful performance, only a few representative results are illustrated in this section to demonstrate the most important features of the new algorithm.

EXPERIMENT DESIGN

Training data and testing data

The learning algorithms attempt to approximate an underlying function from a set of training data $\mathcal{T}$. Training data are obtained by randomly sampling a chosen function experimentally. Specifically, let $f(x)$ be a selected nonlinear univariate function. Let (x_p, y_p) be a training input–output pair. The input x_p is generated by a uniform random variable whose range coincides with the domain of $f(x)$. The corresponding output y_p is the value of the function at x_p, i.e. $y_p = f(x_p)$. In order to simulate gross errors, a random deviation is added to the outputs of a fraction of the training pairs.

A testing set $\mathcal{S}$ is needed in order to evaluate the approximating capability of a trained network. It is obtained by sampling the chosen function at small and equally spaced intervals. Figure 17.5 shows training and testing data from a Gaussian function and a Gaussian derivative function. In both cases, 25 training data points plotted in circular dots were randomly generated, and 101 testing data points were generated at equally spaced intervals. The curves are plotted by joining the testing data points. This represents the actual underlying curve to be approximated from the training data provided.

Evaluation criterion

There are two quantitative measurements which reveal the status of a learning process: one is the mean sum of squares of the residuals of the training data, $MSSR_{\mathcal{T}}$; the other is the mean sum of squares of the residuals of the testing data, $MSSR_{\mathcal{S}}$. They are defined as follows:

$$MSSR_{\mathcal{T}} = \frac{1}{|\mathcal{T}|} \sum_{(x_p, y_p) \in \mathcal{T}} (y_p - \hat{y}_p)^2; \qquad MSSR_{\mathcal{S}} = \frac{1}{|\mathcal{S}|} \sum_{(x_p, y_p) \in \mathcal{S}} (y_p - \hat{y}_p)^2$$

where $\hat{y}_p$ is computed by forward-propagating input x_p through the network. We

(a)   (b)

Figure 17.5 Training and testing data for a Gaussian function (a) and a Gaussian derivative function (b). Circular dots represent 25 randomly sampled training data points. The curve is plotted by joining 101 equally spaced testing data points, and represents the original curve.

have already pointed out that the network tends to interpolate training data as $MSSR_{\mathscr{T}}$ approaches 0. Interpolating behaviour can be disastrous when a training set containing error. Therefore $MSSR_{\mathscr{T}}$ alone is not an appropriate figure of merit to reflect the status of the learning. In fact, we will show later in our experiment that smaller $MSSR_{\mathscr{T}}$ can lead to worse performance when erroneous training data are present. $MSSR_{\mathscr{S}}$, on the other hand, indicates how well a trained network behaves on the data not used during training. This behaviour is known as **generalization**. Assuming there are no errors in the testing set $\mathscr{S}$, $MSSR_{\mathscr{S}}$ correctly reflects how well a trained network has learned from the training data in approximating (or predicting) output values for new input data. For this reason, we use $MSSR_{\mathscr{S}}$ as an evaluation criterion in our experiment.

MAIN RESULTS

Figure 17.6 shows the results of training a feedforward network with the BP algorithm using the training set in Fig. 17.5(a). Figures 17.6(a)–(c) show the learning results at iterations 5000, 10 000, and 20 000, respectively. The learned curve is plotted from a point set $\{(x_p, \hat{y}_p) | p = 1, \ldots, |\mathscr{S}|\}$ where x_p is the input of a point from the testing

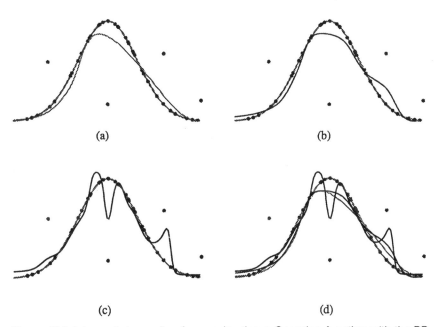

(a)   (b)

(c)   (d)

Figure 17.6 Intermediate results of approximating a Gaussian function with the BP algorithm using training data in Fig. 17.5(a). The thinner curves plotted in (a), (b) and (c) represent approximated curves at iteration 5000, 10 000 and 20 000, respectively. They are plotted against the training data (circular dots) and the original curve Approximated curves in (a), (b), and (c) are plotted together in (d) to illustrate the increasingly worsening effect of outliers.

set $\mathscr{S}$. In order to demonstrate the approximating behaviour of the network, the learned curve is plotted against the training data and the true underlying curve plotted by the testing data. As predicated, the network trained with the BP algorithm attempts to interpolate all the training data regardless of their quality. Intermediate results in Figs 17.6(a)–(c) are plotted together in Fig. 17.6(d) to illustrate the increasingly worsening effect of outliers.

Clearly, due to a fraction of the training data consisting of gross errors, the trained network fails to approximate the original curve. This phenomenon is further clarified by examining the mean sum of squares of the residuals for both the training data ($MSSR_{\mathscr{T}}$) and the testing data ($MSSR_{\mathscr{S}}$), which are plotted in Fig. 17.7(a). The   curve represents $MSSR_{\mathscr{T}}$, the solid curve represents $MSSR_{\mathscr{S}}$. The horizontal axis represents the number of iterations. At about iteration 1000, the MSSR for both the training and testing data experiences a sharp decrease—an indication that the network has made a qualitative leap in learning the underlying curve. However, from then on,

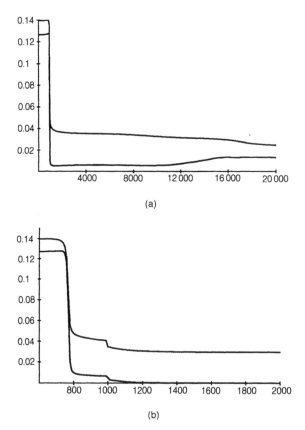

(a)

(b)

Figure 17.7 (a) The history of the mean sum of squares of the residuals for training curves, top and testing data (curves, lower·) using the BP algorithm, and (b) using the robust BP algorithm.

while $MSSR_{\mathscr{T}}$ decreases monotonically as iteration continues, $MSSR_{\mathscr{G}}$ increases almost monotonically. As $MSSR_{\mathscr{T}}$ approaches 0, the negative effect of the erroneous training data on the network becomes more and more severe. These erroneous points in the training data become increasingly influential. They literally pull the approximating curve away from the true underlying curve towards themselves. Such a disastrous effect may be alleviated by terminating the learning process at about iteration 1000 (see Fig. 17.7(a)); however, it cannot be eliminated, as illustrated in Fig. 17.6(a).

We then apply the robust BP algorithm to the same training set. At first all the training data were employed in an indiscriminate manner until the $MSSR_{\mathscr{T}}$ dropped sharply—in this case at around iteration 1000. Although the learned network is far from perfect at this point, it serves as a good initial estimator for the robust BP

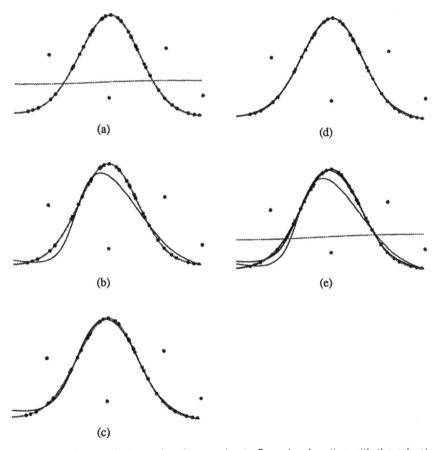

(a)

(d)

(b)

(e)

(c)

Figure 17.8 Intermediate results of approximate Gaussian function with the robust BP algorithm using training data in Fig. 17.5(a). The thinner curves plotted in (a), (b), (c) and (d) represent approximated curves at iteration 700, 1000, 1200 and 6000, respectively. They are plotted together in (e) to demonstrate the evolving learning process.

algorithm. From then on, $\phi_i(r)$ is updated at every $\Delta t = 100$ iterations, and the learning process attempts to minimize the new time-varying energy function $E_R = \Sigma_p \phi_i(r_p)$. Figure 17.8 shows a sequence of intermediate results for this experiment.

It is interesting to note that the network finds a line, significantly biased towards the outliers, going through the set of training points at iteration 700. All the training data were employed with the BP algorithm prior to iteration 1000. The effect of outliers can be clearly observed from Fig. 17.8(b). From Fig. 17.7(a), we know that the effect of outliers will increasingly worsen if no error-resistant measure is taken. The results in Fig. 17.8(b) nevertheless serve as an initial estimator. Figures 17.8(c) and 17.8(d) show the learned curves at iterations 1200 and 6000. Figure 17.8(e) shows the intermediate results in Figs 17.8(a)–(d) together to demonstrate the evolving learning process. Clearly, the influence of outliers is successfully eliminated. Figure 17.7(b) shows the MSSR history for the training data and testing data. After

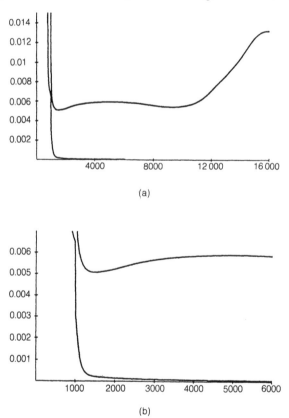

(a)

(b)

Figure 17.9 Comparison of the history of the mean sum of squares of the residuals for the testing data using the BP and robust BP algorithm. (a) and (b) are plotted from the same data with different scales. The higher curves in (a) and (b) are the $MSSR_{\mathscr{y}}$ history for the BP algorithm, the lower curves are for the robust BP algorithm.

the first update of $\phi_i(r)$ at the 1000th iteration, the network undergoes another fall in MSSR. From then on $MSSR_{\mathscr{S}}$ monotonically decreases along the iteration time. It reaches about 2×10^{-4} at iteration 1200. However, $MSSR_{\mathscr{T}}$ remains almost constant after 1200 iterations. This is mainly due to the errors in the training data.

To compare the performance of the BP and robust BP algorithms, we plot, together, the history of the $MSSR_{\mathscr{S}}$ for both algorithms in Fig. 17.9. The data used are from the same set as used for plotting $MSSR_{\mathscr{S}}$ in Figs 17.7(a) and 17.7(b), but they are plotted together with a different scale. Figure 17.9(a) plots the $MSSR_{\mathscr{S}}$ history up to iteration 16 000. In order to show fine changes in $MSSR_{\mathscr{S}}$, Fig. 17.9(b) is plotted up to iteration 6000. While the $MSSR_{\mathscr{S}}$ for the BP algorithm (the higher curve) tends to increase continuously from iteration 1000, the $MSSR_{\mathscr{S}}$ for the robust BP algorithm (the lower curve) decreases monotonically. The latter reaches 10^{-4} at iteration 6000.

We have repeated this experiment on a wide class of nonlinear univariate functions. Similar qualitative performances have been observed. Figures 17.10(a)–(c) illustrate three additional examples of approximating a Gaussian derivative function, a logistic function, and a segment of sine function using the new learning algorithm. The corresponding results using the conventional BP algorithm are shown in Figs 17.10(d)–(f). These experiments support and validate the claims made in the previous theoretical analysis. That is, the robust BP algorithm is capable of learning in the presence of erroneous training data; the learning process approximates rather than interpolates training data; and the new algorithm converges at a much faster rate than that of the BP algorithm.

APPROXIMATE PIECEWISE SMOOTH FUNCTIONS

It has been proved by Hornik *et al.* (1989) that multi-layer feedforward neural networks are capable of approximating arbitrary real-valued continuous functions. This powerful feature allows neural networks to approximate piecewise smooth functions. There are many applications for fitting piecewise curves or surfaces in object recognition. Many computer vision techniques for fitting piecewise functions require pregrouping of the data for each smooth function prior to fitting (Besl and Jain, 1988; Sinha and Schunck, 1992). This task of pregrouping, also known as **segmentation**, has been proved to be extremely difficult in computer vision research. One obvious advantage of neural networks for fitting piecewise smooth functions is that the pregrouping is no longer needed. Figure 17.11(a) shows the results of approximating two sigmoid (logistic) curves using a single network. We observe that the network is able to approximate both curves while preserving the discontinuities.

However, in reality, a fraction of the data is always contaminated due to sensory noise. Because of sensitivity to gross errors in the training data, conventional learning methods will inevitably have difficulty in preserving discontinuities where piecewise functions meet as well as removing gross errors such as outliers. Will the new robust method developed in this chapter be able to perform discontinuity-preserving and outlier-removing function fitting when erroneous data are present? To address this

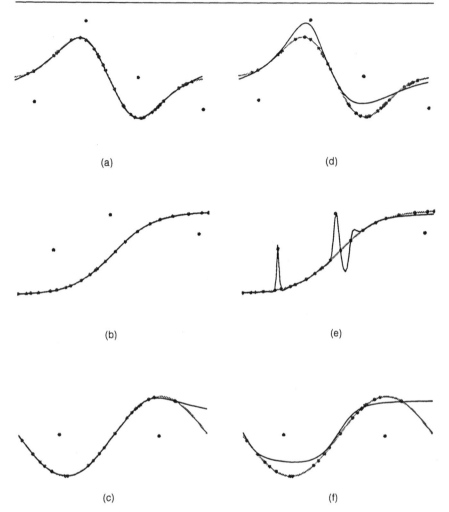

(a)   (d)

(b)   (e)

(c)   (f)

Figure 17.10 Learning results using the robust BP algorithm (a, b, c) and the conventional BP algorithm (d, e, f) to approximate a Gaussian derivative function, a logistic function, and a segment of sine function.

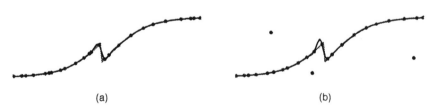

(a)   (b)

Figure 17.11 Learning results of approximating two sigmoid (logistic) curves without outliers (a), and with outliers (b).

question, we deliberately misplace several data points and apply the robust method to the same set of data as in Fig. 17.11(a). As we can see in Fig. 17.11(b), the new method demonstrates its potential in fitting piecewise smooth functions using noise data; and the fitting is capable of preserving discontinuities as well as removing outliers. Currently, we are conducting more experiments in this direction. The research results focusing on discontinuity-preserving noise-filtering operations in signal and image reconstruction applications will be reported in a future paper.

CONCLUSIONS

The main advantages of the robust BP learning algorithm can be explained in the context of Hebbian learning. The Hebbian learning rule states that weights change in proportion to the product of the residual error and input, i.e.

$$\Delta w_{ji} \propto r_{pj} o_{pi} \qquad (17.16)$$

The BP learning algorithm using this learning rule is shown to be extremely sensitive to gross errors in the training data. The reason for this noise sensitivity is simply that the residual errors from erroneous training data cause unrealistic and unnecessary adjustments of network weights. This leads to slow, sometimes incorrect, learning results when there are noise and outliers present in the training data. The robust BP learning algorithm proposed in our work states that weights change in proportion to the product of the residual error under the redescending transformation $\psi_t(r)$ and input, i.e.

$$\Delta w_{ji} \propto \psi_t(r_{pj}) o_{pi} \qquad (17.17)$$

This time-varying transformation has led to three major advantages. First, the algorithm is robust against gross errors in the training data. The redescending influence function varies with learning time and has decreasing tolerance for data having larger residuals. Thus, the influence of noise in the training data is gracefully suppressed and the influence of outliers is gradually eliminated. This phenomenon resembles human learning behaviour in that erroneous and problematic environmental inputs have decreasing effects on human beings growing towards maturity. The learning pattern of elder humans tends to be more and more conservative. This form of conservatism is quantitatively captured in the time-varying shape of the three-part redescending function. Second, the minimization of the robust objective function E_R does not lead to the interpolation of all the training data. Thus, the network approximates rather than interpolates the training data. This property often yields smooth approximating results. Last, in addition to the ability to reject outliers during learning, the new algorithm significantly improves the rate of convergence compared to that of the BP learning algorithm. The reason behind this improvement is that training data that produce large error signals do not perturb the network weights as much as they do in the BP algorithm. Thus, the algorithm will not undo what it has learned upon input presentations having large error signals. This desirable property is in principle similar to the minimum disturbance principle (Widrow *et al.*, 1988) and the least commitment heuristics used in machine learning in artificial intelligence.

REFERENCES

Besl, P. and Jain, R. (1988) Segmentation through variable-order surface fitting. *IEEE Transactions on Pattern Analysis and Machine Intelligence*, **10**, 167–192.

Diaconis, P. and Shahshahani, M. (1984) On linear functions of linear combinations. *Journal of the American Statistical Association*, **5**, 175–191.

Efron, B. and Tibshirani, R. (1984) Bootstrap methods for standard errors, confidence intervals, and other measures of statistical accuracy. *Statistical Science*, **1**, 54–77.

Friedman, J.H. (1985) Classification and multiple regression through projection pursuit. LCS 12 Technical Report, Department of Statistics, Stanford University, and Stanford Linear Accelerator Center.

Friedman, J.H. and Stuetzle, W. (1981) Projection pursuit regression. *Journal of the American Statistical Association*, **76**, 817–823.

Hampel, F.R. (1974) The influence curve and its role in robust estimation. *Journal of the American Statistical Association*, **69**, 383–393.

Hampel, F.R., Rousseeuw, P.J. and Ronchetti, E. (1981) The change-of-variance curve and optimal redescending *M*-estimators. *Journal of the American Statistical Association*, **76**, 643–648.

Hampel, F.R., Rousseeuw, P.J., Ronchetti, E.M. and Stahel, W.A. (1986) *Robust Statistics—The Approach Based on Influence Function*, New York: John Wiley.

Hartman, E.J., Keeler, J.E. and Kowalski, J.M. (1990) Layered neural networks with Gaussian hidden units as universal approximations. *Neural Computation*, **2**, 210–215.

Hecht-Nielsen, R. (1989) Theory of the backpropagation neural network. *Proceedings of the IEEE 3rd Conference on Neural Networks*.

Hornik, K., Stinchcombe, M. and White, H. (1989) Multilayer feedforward networks are universal approximators. *Neural Networks*, **2**, 359–366.

Huber, P.J. (1981) *Robust Statistics*. New York: John Wiley.

Ji, C., Snapp, R.R. and Psaltis, D. (1990) Generalizing smoothness constraints from discrete samples, *Neural Computation*, **2**, 188–197.

Le Cun, Y. (1988) A theoretical framework for back propagation. In D. Touretzky, G. Hinton, and T. Sejnowski (eds), *Proceedings of 1988 Connectionist Models Summer School*, San Mateo, CA: Morgan Kaufmann, pp. 21–28.

McKay, N. (1989) Some mathematical properties of three-layer perceptron-like neural nets. Research Memo 50-510, General Motors Research Labs., Warren Michigan.

Poggio, T. and Girosi, F. (1990) Networks for approximation and learning, *Proceedings of the IEEE*, **78**, 1481–1497.

Rummelhart, D.E., Hinton, G.E. and Williams, R.J. (1986) *Parallel Distributed Processing: Explorations in the Microstructure of Cognition*, Vol. 1. Cambridge, MA: MIT Press, Chapter 8.

Sinha, S., and Schunck, B. (1992) A two stage spline-based surface reconstruction algorithm. *IEEE Pattern Analysis and Machine Intelligence*, **14**, 36–55.

Vidyasagar, M. (1978) *Nonlinear Systems Analysis*, Englewood Cliffs, NJ: Prentice Hall.

Widrow, B. and Hoff, M.E. (1960) Adaptive switching circuits. *1960 IRE WESCON Convention Record*, Part 4, pp. 96–104.

Widrow, B., Winton, R.G. and Baxter, R.A. (1988) Layered neural nets for pattern recognition. *IEEE Transactions on Acoustic Signals and Signal Processing*, **36**, 1109–1118.

Maximum likelihood training of neural networks

18

H. Gish

INTRODUCTION

Gish (1990) took a probabilistic view of neural networks and showed the equivalence between neural networks[1] trained with maximum likelihood (ML), maximum mutual information (MMI), and the Kullback–Leibler criteria. This chapter extends that initial work by first further exploring the meaning of the ML criterion in the context of neural networks. It is assumed from the outset that the network is a model for a posterior probability of the class membership of the data. This leads to a partially specified probability model for the observations which, however, still provides us with an ML estimate of our posterior probability model.

The chapter then looks at the design of the network from the point of view of mutual information, and notes that the basis for the equivalence between the methods relies on the use of the ML criterion to estimate a posterior probability. The MMI approach also leads to an interesting 'distance' measure interpretation for conditional entropy. An important aspect of the equivalence between criteria is that the network that is designed with one criterion becomes imbued with the properties of the other. The equivalence enables us to have the satisfaction of designing a network that maximizes information about unknown classes and still be able to use the estimation methods and asymptotic properties associated with ML estimation. These asymptotic properties include the various methods of model selection that are based on the likelihood function.

An algorithm is also developed for finding the ML estimate of a network with multiple layers. This is a procedure wherein the parameters of the individual nodes are estimated sequentially, using a Newton–Raphson algorithm at each step in the sequence. It is applied to several examples and the use of ML-based criteria for comparing models is illustrated.

In this chapter we have only considered the case where observed feature vectors come from one of two classes. The results can be extended without much difficulty to the more general case.

Artificial Intelligence Frontiers in Statistics: AI and statistics III. Edited by D.J. Hand. Published in 1993 by Chapman & Hall, London. ISBN 0 412 40710 8

THE MAXIMUM LIKELIHOOD CRITERION

Consider that x_i, $i = 1, \ldots, N$, denotes a set of feature vectors, and that each x_i belongs to either class C_1 or class C_2. We consider the classifier network as being described by the mapping, $f(x, \theta)$, which is parametrized by θ and maps each x_i onto the interval [0, 1]. Our goal is the selection of the parameters, θ, such that $f(x, \theta)$ is an estimate of $P_\theta(C_1|x)$, the probability that class C_1 has occurred given that we have observed x.

The criterion we employ to accomplish this is the maximum likelihood criterion. The likelihood of an observation, u, is $p_\theta(u)$, the value of the probability density function of u. In the classification problem an observation is taken to be the pair, (x_i, C^i), consisting of the feature vector, x_i, and its class membership, C^i, which will be equal to C_1 or C_2.

We can therefore express the log-likelihood of our training vectors $u_i = (x_i, C^i)$, $i = 1, \ldots, N$, as

$$L = \prod_{i=1}^{N} p_\theta(u_i) \tag{18.1}$$

Using Bayes's formula we write

$$p_\theta(u_i) = P_\theta(C^i|x_i)p(x_i) \tag{18.2}$$

and obtain

$$L = \prod_{i=1}^{N} P_\theta(C^i|x_i)p(x_i) \tag{18.3}$$

$$= \prod_{i=1}^{N} P_\theta(C^i|x_i) \prod_{i=1}^{N} p(x_i) \tag{18.4}$$

If we now collect together those terms for which $x_i \in C_1$ and $x_i \in C_2$, we can rewrite the above equation as

$$L = \prod_{x \in C_1} f(x, \theta) \prod_{x \in C_2} [1 - f(x, \theta)] \prod_{i=1}^{N} p(x_i) \tag{18.5}$$

and correspondingly, the log-likelihood

$$\log L = \sum_{x \in C_1} \log f(x, \theta) + \sum_{x \in C_2} \log [1 - f(x, \theta)] + \sum_{i=1}^{N} \log p(x_i) \tag{18.6}$$

where we have substituted $f(x, \theta)$ for $P_\theta(C_1|x)$ and have used $P_\theta(C_2|x) = 1 - P_\theta(C_1|x)$.

Let us observe that the log-likelihood, given above, has only been partially specified. That is, we have not modelled the marginal probability density, $p(x)$, which can be expressed in terms of the models for each of the classes, i.e.

$$p(x) = p(x|C_1)P(C_1) + p(x|C_2)P(C_2) \tag{18.7}$$

where $p(x|C_j)$, $j = 1, 2$, are the probability density functions for each of the classes and $P(C_j)$, $j = 1, 2$, are the prior probabilities of the class occurrences.

Although we have not modelled $p(\mathbf{x})$, we do not have to. Modelling $f(\mathbf{x}, \boldsymbol{\theta})$ and obtaining its ML estimate can be done independently of the marginal probability component, $p(\mathbf{x})$. This independence allows us to maximize the log-likelihood given by Equation (18.6) by maximizing the first two terms and ignoring the third.

The output of our network, $f(\mathbf{x}, \boldsymbol{\theta})$, will be a sigmoidal transformation of a function that we will call $z(\mathbf{x}, \boldsymbol{\theta})$. This transformation, which does not introduce any loss of generality as to the modelling of the posterior probability, is expressed as

$$f(\mathbf{x}, \boldsymbol{\theta}) = \frac{1}{1 + e^{-z(\mathbf{x}, \boldsymbol{\theta})}} \tag{18.8}$$

If we invert the sigmoid we obtain

$$z(\mathbf{x}, \boldsymbol{\theta}) = \log \frac{f(\mathbf{x}, \boldsymbol{\theta})}{1 - f(\mathbf{x}, \boldsymbol{\theta})} \tag{18.9}$$

Now let us write

$$f(\mathbf{x}, \boldsymbol{\theta}) = P_\theta(C_1 | \mathbf{x}) \tag{18.10}$$

in order to emphasize that $f(\mathbf{x}, \boldsymbol{\theta})$ is a probability.

From Equations (18.9) and (18.10) and, as before, using $P_\theta(C_2 | \mathbf{x}) = 1 - P_\theta(C_1 | \mathbf{x})$, we can write,

$$z(\mathbf{x}, \boldsymbol{\theta}) = \log \frac{P_\theta(C_1 | \mathbf{x})}{P_\theta(C_2 | \mathbf{x})} \tag{18.11}$$

$$= \log \frac{p_\theta(\mathbf{x} | C_1)}{p_\theta(\mathbf{x} | C_2)} + \log \frac{P(C_1)}{P(C_2)} \tag{18.12}$$

The above equations show that the neural network is in effect modelling the log-likelihood ratio of the two classes. The direct modelling of the ratio allows the network to be efficient in its use of parameters since each of the probability density functions is not modelled separately as is done with other types of classifiers. We will give an example of this efficiency in a later section.

The heart of the model we are considering is the function $z(\mathbf{x}, \boldsymbol{\theta})$. If we restrict $z(\mathbf{x}, \boldsymbol{\theta})$ to be a linear function of the features, i.e.

$$z(\mathbf{x}, \boldsymbol{\theta}) = \theta_0 + \boldsymbol{\theta}'\mathbf{x} \tag{18.13}$$

we have a model that has been studied extensively in the statistical literature under the name of logistic discrimination. Some early works include Cox (1970), and Day and Kerridge (1967), and more recently Anderson (1982). When viewing this type of model as a neural network, it constitutes a network with no hidden layers. In a neural network with a single hidden layer we have (with $\boldsymbol{\theta} = (\alpha, \beta)$)

$$z(\mathbf{x}_i, \boldsymbol{\theta}) = \beta_0 + \sum_{j=1}^{m} \beta_j \Lambda \left(\alpha_{0,j} + \sum_{k=1}^{p} \alpha_{k,j} x_{i,k} \right) \tag{18.14}$$

where $\Lambda(v) = [1 + \exp(-v)]^{-1}$, $x_{i,k}$ is the kth of the p components in input vector $\mathbf{x}_i$. Also each of the Λ-components of the m terms of the sum represent the output of a node of the hidden layer, with the input weights (parameters) to a hidden node being the α-terms and the weights of the output node being the β-terms. Thus neural networks with layers comprise significantly more complex models than are typically dealt with in logistic discrimination. The significant consequence of this additional complexity is that of having more complex partitions of feature space. Having $z(\mathbf{x}, \boldsymbol{\theta})$ linear restricts the partitioning of the feature space into only two regions which are split by the hyperplane defined by $z(\mathbf{x}, \boldsymbol{\theta}) = 0$. This restriction is overcome by having the hidden layer, where each node in the hidden layer effectively defines a partition of the feature space, with the output node combining the information from the hidden layer nodes.

Figure 18.1(b) is an illustration of a neural network that has a single hidden layer, i.e. two layers including the input as a layer. The values at the input nodes are multiplied by the weights on the links from the input nodes to the nodes of the hidden layer. At each node in the hidden layer the weighted inputs are summed, a bias term is added, and the resulting quantity is logistically transformed. The network output node operates on the outputs from the hidden layer in the same way as the network inputs were processed by the hidden layer. Figure 18.1(a) illustrates the input–output relations for a generic node.

Although the hidden layer does provide the capability of having more complex partitions, there are other ways of approaching the problem. An important alternative is the inclusion of somewhat more complex features into the feature set which will allow a single-layer network to construct more complex boundaries. A prime example of this approach is the inclusion of product terms, $x_{i,k}x_{i,l}$, which will allow for the creation of quadratic boundaries. These boundaries are again defined by the set of those $\mathbf{x}$ for which $z(\mathbf{x}, \boldsymbol{\theta}) = 0$. As we will see in an example, this approach can generate multiple partitions with a single layer.

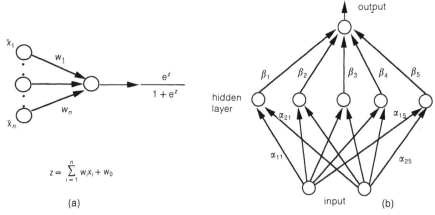

Figure 18.1 (a) Model of a generic node. (b) Diagram of a neural network: two input nodes, five nodes in the hidden layer.

MAXIMUM MUTUAL INFORMATION AND MAXIMUM LIKELIHOOD

An alternative to maximum likelihood for specifying a model is an approach which is based on Shannon's mutual information. The mutual information gives the amount of information that one random variable (or vector) provides about another. The application of this information-theoretic approach to modelling requires the expression of the mutual information between the random quantities in terms of the underlying parametric models. By adjusting the model parameters to maximize the mutual information when evaluated over the training set, we obtain the maximum mutual information (MMI) estimate of the parameters. Below we will derive an expression for the MMI estimate and relate it to the ML estimate.

The definition of the mutual information for our application is given by

$$I(C, \mathbf{X}) = \int \sum_{i=1}^{2} p(\mathbf{x}, C_i) \log \frac{p(\mathbf{x}, C_i)}{P(C_i)p(\mathbf{x})} d\mathbf{x} \qquad (18.15)$$

which can be written in a more informative way as

$$I(C, \mathbf{X}) = H(C) - H(C|\mathbf{X}) \qquad (18.16)$$

where $H(C)$ is the entropy of the class occurrences based on the prior probabilities of the class occurrences and is given by

$$H(C) = - \sum_{i=1}^{2} P(C_i) \log P(C_i) \qquad (18.17)$$

It is a measure of the uncertainty as to which class will occur before any observation is made.

This uncertainty as to which class will occur is reduced after we have observed a feature vector, $\mathbf{x}$. The uncertainty about C after making the observation is given by the conditional entropy $H(C|\mathbf{X})$. The reduction in uncertainty about the occurrence of C by having observed $\mathbf{x}$ is the information that $\mathbf{x}$ provides about C. We have

$$H(C|\mathbf{X}) = - \int p(\mathbf{x}) \left[\sum_{i=1}^{2} P(C_i|\mathbf{x}) \log P(C_i|\mathbf{x}) \right] d\mathbf{x} \qquad (18.18)$$

After using $P(C_i|\mathbf{x}) = p(\mathbf{x}|C_i)P(C_i)/p(\mathbf{x})$ in the above expression we obtain

$$H(C|\mathbf{X}) = - \int \sum_{i=1}^{2} p(\mathbf{x}|C_i)P(C_i) \log P(C_i|\mathbf{x}) d\mathbf{x} \qquad (18.19)$$

Maximizing the mutual information is, of course, equivalent to minimizing the conditional entropy $H(C|\mathbf{X})$. We note that in the above expression we are taking the expectation of $-\log P(C|\mathbf{x})$. If we replace $P(C_1|\mathbf{x})$ with its parametric representation, $f(\mathbf{x}, \boldsymbol{\theta})$, and $P(C_2|\mathbf{x})$ with $1 - f(\mathbf{x}, \boldsymbol{\theta})$, and perform the expectation over the training data, we find that minimizing conditional entropy is equivalent to $\min_{\boldsymbol{\theta}} \tilde{H}_{\boldsymbol{\theta}}(C|\mathbf{X})$, where

$$\tilde{H}_{\boldsymbol{\theta}}(C|\mathbf{X}) = \frac{-1}{N} \left[\sum_{\mathbf{x} \in C_1} \log f(\mathbf{x}, \boldsymbol{\theta}) + \sum_{\mathbf{x} \in C_2} \log [1 - f(\mathbf{x}, \boldsymbol{\theta})] \right] \qquad (18.20)$$

is the empirical conditional entropy.

Minimizing this conditional entropy is of course equivalent to maximizing the log-likelihood which is given by Equation (18.6). This demonstrates the equivalence of the two criteria for the neural network application. This equality of criteria is a result of our using the ML criterion for the estimation of the parameters of a posterior probability as opposed to the parameters of a class-conditional probability density function. Also, the formulation of the likelihood function and the equivalence of the criteria will apply to functions, $f(\mathbf{x}, \boldsymbol{\theta})$, that do not rely upon the functions having the structure of a neural network.

Let us consider minimizing the conditional entropy for a linear $z(\mathbf{x}, \boldsymbol{\theta})$. This implies a linear partitioning of feature space into regions for the two classes, i.e. the hyperplane defined by $z(\mathbf{x}, \boldsymbol{\theta}) = 0$, for which $f(\mathbf{x}, \boldsymbol{\theta}) = 1/2$. This partition is not the same as would be obtained by minimizing the conditional entropy based on the fraction of the class members found on each side of the boundary. The entropy we are computing depends on how far the points are from the boundary, and on which side of the boundary the points fall. The boundary between the classes in our case is a consequence of modelling the posterior probability. The hyperplane that minimizes the conditional entropy based on counts can be very useful but is typically very difficult to obtain.

The conditional entropy also has an interpretation as a distance measure. Following the logistic discrimination study of Day and Kerridge (1967), who considered the negative of the log-likelihood (which is proportional to our empirical conditional entropy) as a distance measure, we can write our conditional entropy as

$$\tilde{H}(C|\mathbf{X}) = \frac{1}{N}\left[\sum_{\mathbf{x} \in C_1} d_1(\mathbf{x}) + \sum_{\mathbf{x} \in C_2} d_2(\mathbf{x}) \right] \tag{18.21}$$

where

$$d_1(\mathbf{x}) = \log\left(1 + e^{-z(\mathbf{x}, \boldsymbol{\theta})}\right) \tag{18.22}$$

is the 'distance' measure for members of C_1 and

$$d_2(\mathbf{x}) = \log\left(1 + e^{z(\mathbf{x}, \boldsymbol{\theta})}\right) \tag{18.23}$$

for C_2. These 'distance' measures are asymmetric with respect to the partition and will emphasize those vectors that are near the boundary or are misclassified.

The distance-measure interpretation serves to emphasize the general nature of the modelling since it can be given an interpretation that is not dependent on probability models. It can also lead to the tailoring of the distance measure to specific circumstances, such as robustness to outliers.

AN ALGORITHM FOR MAXIMUM LIKELIHOOD
TRAINING OF NETWORKS

For a large variety of ML estimation problems and for the logistic discrimination problem in particular, a useful iterative algorithm is a variant of the Newton–Raphson algorithm. This variant is known as **Fisher's method of scoring** (Seber and Wild,

1989). In the following we will briefly describe this method and our application of it to neural networks with at least one hidden layer.

In general, if $\log L$ denotes the log-likelihood depending on the parameters $\boldsymbol{\beta}$, and with $\boldsymbol{\beta}_n$ the estimate of the parameters at iteration stage n, the parameter estimate at stage $n + 1$ given by the method of scoring is

$$\boldsymbol{\beta}_{n+1} = \boldsymbol{\beta}_n + I(\boldsymbol{\beta}_n)^{-1} \frac{\partial \log L}{\partial \boldsymbol{\beta}} \bigg|_{\boldsymbol{\beta}_n} \tag{18.24}$$

where $I(\boldsymbol{\beta})$ is Fisher's information matrix and is given by

$$I(\boldsymbol{\beta}) = E\left(-\frac{\partial^2 \log L}{\partial \boldsymbol{\beta} \partial \boldsymbol{\beta}'} \right) \tag{18.25}$$

where E denotes expected value. It is convenient that under rather general conditions the information matrix is also given by

$$I(\boldsymbol{\beta}) = E\left(\frac{\partial \log L}{\partial \boldsymbol{\beta}} \frac{\partial \log L}{\partial \boldsymbol{\beta}'} \right) \tag{18.26}$$

avoiding the need to compute the second-order derivatives.

The strategy we have employed in applying the scoring method to neural networks with a single layer is to apply it to different subsets of the parameters (weights) in the network while holding the remaining weights fixed. That is, the log-likelihood will be maximized in stages, a procedure generally referred to as a **Gauss–Seidel** iterative method.

With the network in its initial state we select a node from the hidden layer and adjust its weights so as to maximize the likelihood. This change in weights alters the output that will be generated by this node. We therefore follow the changes in the weights of the node in the hidden layer by updating the weights of the output node. We then proceed to the next node in the hidden layer and follow the adjustment of its weights by the adjustment of the output weights. We continue in this manner until all the nodes in the hidden layer have been updated and then the procedure is repeated. Thus the algorithm has two iterations going on: one is the iteration of the scoring algorithm, i.e. the Newton–Raphson iterations as the node is updated; and the other is the iterative procedure of cycling through the nodes of the network.

Let $\mathbf{x}_i$, as before, denote the input features to the neural network, and let $\mathbf{u}_i$ denote the vector of outputs of the nodes in the hidden layer. If we let $\boldsymbol{\beta}$ denote the weights on $\mathbf{u}_i$, then maximizing the log-likelihood with respect to $\boldsymbol{\beta}$, while keeping the other weights fixed, is the typical procedure used in logistic discrimination. Writing $\log L$ as

$$\log L = \sum_{i=1}^{N} (y_i \boldsymbol{\beta}' \mathbf{u}_i - \log(1 + e^{\boldsymbol{\beta}' \mathbf{u}_i})) \tag{18.27}$$

where $y_i = 1$ for $\mathbf{x}_i \in C_1$ and $y_i = 0$ for $\mathbf{x}_i \in C_2$, and $\mathbf{u}_i$ is the hidden layer output corresponding to input $\mathbf{x}_i$. This gives

$$\frac{\partial \log L}{\partial \boldsymbol{\beta}} = \sum_{i=1}^{N} (y_i - P_i) \mathbf{u}_i \tag{18.28}$$

where

$$P_i = \frac{e^{\boldsymbol{\beta}' \mathbf{u}_i}}{1 + e^{\boldsymbol{\beta}' \mathbf{u}_i}} \tag{18.29}$$

which is the probability that $\mathbf{u}_i$, and equivalently $\mathbf{x}_i$, according to the model, is a member of C_1. Also,

$$\boldsymbol{I}(\boldsymbol{\beta}) = \sum_{i=1}^{N} P_i(1 - P_i)\mathbf{u}_i\mathbf{u}_i' \tag{18.30}$$

With the gradient from Equation (18.28) and information matrix given above, the weights from the hidden layer to the output layer can be optimized.

Let $u_{i,j}$ denote the output of the jth node in the hidden layer and let $\alpha_{k,j}$ be the weight from $x_{i,k}$, the kth component of feature vector $\mathbf{x}_i$, to node j of the hidden layer. Then

$$\frac{\partial \log L}{\partial \alpha_{k,j}} = \sum_{i=1}^{N} \frac{\partial \log L}{\partial u_{i,j}} \frac{\partial u_{i,j}}{\partial \alpha_{k,j}} \tag{18.31}$$

$$\frac{\partial \log L}{\partial u_{i,j}} = \beta_j(y_i - P_i) \tag{18.32}$$

and

$$\frac{\partial u_{i,j}}{\partial \alpha_{k,j}} = P_{i,k}(1 - P_{i,k})x_{i,k} \tag{18.33}$$

where β_j is the kth component of $\boldsymbol{\beta}$ and

$$P_{i,j} = \frac{e^{\alpha_j' x_i}}{1 + e^{\alpha_j' x_i}} \tag{18.34}$$

The above equation can be viewed as a measure of the probability that x_i is a member of a hidden state that is represented by node j of the hidden layer.

From the above we obtain the desired gradient

$$\frac{\partial \log L}{\partial \alpha_{k,j}} = \sum_{i=1}^{N} \beta_j(y_i - P_i) P_{i,j}(1 - P_{i,j})x_{i,k} \tag{18.35}$$

Using the above equation for the gradient of the log-likelihood and Equation (18.26) we have for the Fisher information matrix for the parameters at the jth node of the hidden layer

$$\boldsymbol{I}(\alpha_j) = \sum_{i=1}^{N} \beta_j^2 P_i(1 - P_i) P_{i,j}^2(1 - P_{i,j})^2 \mathbf{x}_i\mathbf{x}_i' \tag{18.36}$$

The preceding two equations now permit us to apply the scoring method to any node in the hidden layer.

EXAMPLES

In the following examples we will be dealing with a single data set. We will use it to design three different networks. These examples will illustrate the behaviour of the training algorithm and also the methods for comparing the trained models.

The examples are based on the traditional XOR problem in which there are two classes, with each class comprised of disjoint components. In particular the pairs $(0, 1)$ and $(1, 0)$ constitute one class and the values $(0, 0)$ and $(1, 1)$ constitute the other. This is a traditional example for showing that a neural network with no hidden layer will not be able to separate the classes and that a network with a hidden layer is required. The problem is discussed in some detail in Rumelhart *et al.* (1986).

Our examples will be based on a noisy version of the XOR problem. We have two classes, each being a Gaussian mixture with two components; for class C_1 one component is a bivariate Gaussian centred on $(0, 5)$, with an identity covariance matrix. The other component is similar but located at $(5, 0)$. Class C_2 has its components centred on $(0, 0)$ and $(5, 5)$. They also have identity covariance matrices. Each feature vector is a pair, (x_1, x_2), selected from either of the two mixtures.

From these models for the classes we generated 40 exemplars from each of the classes which formed our training set, giving us a total of $N = 80$ samples.

Example 18.1: Two networks with a single hidden layer

The first model that we will consider will have five nodes in the hidden layer with the inputs being the (x_1, x_2) pairs from the data set. The structure for this model is the same as illustrated in Fig. 18.1(b). We applied our algorithm to this model in the following manner: each time the algorithm was applied to a node, five iterations of the scoring algorithm were performed on the first visit and ten iterations on subsequent visits; each node in the hidden layer was visited five times. These choices were arbitrary and represent our initial explorations with the algorithm. The initial conditions consisted of setting the β to a vector of 1s, and the weights on the second layer were initialized by randomly selecting each weight from a uniform distribution.

In Fig. 18.2 we have plotted the conditional entropy after each visit to a node. The iteration number indicates the visits to the nodes in the hidden layer. The nodes were cycled through a total of five times giving a total of 25 iterations. The figure also shows the number of errors made by the model after a visit to a node. We find a reasonable correspondence between reduction in entropy and errors except when the entropy is small. In Fig. 18.3 we have plotted the decision boundary obtained from the estimated model. Some of the jagged nature of the boundary is due to the sparsity of points in the data set. The resulting decision boundary and the fairly rapid convergence indicate that the proposed algorithm may be quite useful. The conditional entropy, $\tilde{H}(C|X)$, at the termination of the algorithm, was 0.22 bits.

In a similar manner the experiment was performed for a model that had six nodes in the hidden layer. The resulting conditional entropy was 0.07 bits. For the five-node model there are a total of 21 free parameters and for the six-node model there are 25 free parameters.

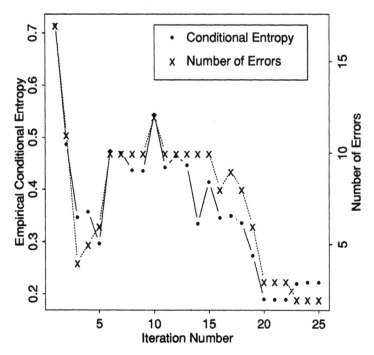

Figure 18.2 Conditional entropy and errors plotted against iteration number.

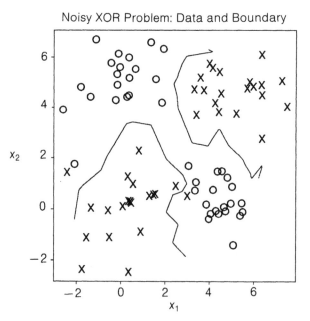

Figure 18.3 Two inputs; one hidden layer with five nodes.

Several methods for selecting between models based on the likelihood function have been developed that take into account the number of parameters in the models and some incorporating the amount of data used. Our purpose is to illustrate the applicability of these methods to the selection of neural networks.

The approach developed by Akaike (1973) considers the expected log-likelihood as the measure of goodness of fit of a probability model. The likelihood of observations obtained via the ML estimate of parameters represents a biased estimate of the expected log-likelihood and must be corrected. This bias is a function of the number of parameters in the model and the process of correction results in the criterion

$$\text{AIC}(k) = - \max_\theta \log L + k \tag{18.37}$$

where k is the number of parameters and the model for which the criterion is minimized is the model to be chosen.

An important alternative to the AIC criterion was developed by Rissanen (1978), based on the notion of minimum description length (MDL), the minimum amount of information needed to describe the observations. This approach yields the criterion

$$\text{MDL}(k, N) = - \max_\theta \log L + \frac{k}{2} \log N \tag{18.38}$$

where N is the total number of observations and, again, the model is chosen for which the criterion is minimum. For $N > 8$ the MDL criterion puts a heavier weight on k, the number of parameters, than AIC and will therefore typically result in models with fewer parameters than selected via AIC.

In the problem we are considering, since one network actually resides in a subspace of the parameter space of the other, it is possible to approach this model selection problem by using the asymptotic properties of the generalized likelihood ratio. Basically, if

$$\lambda = \frac{\sup_{\theta \in \Omega_0} L_0}{\sup_{\theta \in \Omega_1} L_1} \tag{18.39}$$

is the ratio of two likelihoods of two models whose parameters have been estimated by the ML criterion, and $\Omega_0 \subset \Omega_1$ and the dimensionality difference between Ω_1 and Ω_0 is r, then under the assumption that the model represented by L_0 is true,

$$- 2 \log \lambda \to \chi_r \tag{18.40}$$

i.e. is asymptotically distributed as a chi-squared distribution with r degrees of freedom (see Rao, 1965). Using this distribution we can test for the significance of the differences in the log-likelihoods for the two models.

In Table 18.1 we have shown the model selection criteria for the networks with five and six nodes in the hidden layer. (Note that $- \max_\theta \log L$ is equal to $N\tilde{H}(C|\mathbf{X})/\log_2 e$). From the table we see that the AIC criterion prefers the six-node network whereas the MDL criterion gives a slight preference to the five-node network.

Table 18.1 Model selection criteria for networks with 5 and 6 nodes in the hidden layer

Criterion	5-node	6-node
AIC	33.2	28.88
MDL	58.2	58.6

Application of the likelihood ratio method to our example shows that $-2\log\lambda$ has a value of approximately 16.66. The probability that a value this large or larger would be observed assuming that the five-node model is true is about 0.0022, a value sufficiently small to make the six-node model a better choice. Thus, for this example we have the AIC criterion selecting the six-node model with MDL slightly in favour of the five-node model.

Example 18.2: A network with no hidden layer (single layer)

Another experiment that was performed with the same data set was to consider a single-layer network, i.e. with no hidden layer, with the inclusion of the product of the two features as a third feature. The network is shown in Fig. 18.4. Five iterations were performed. The decision boundary is shown in Fig. 18.5. The resulting conditional entropy was equal to 0.135 bits. Thus, by inclusion of the product of features as a feature, we have accomplished with a single-layer network with four parameters the modelling capabilities of a two-layer network with 21 parameters. Although this example has benefited from symmetries existing in the data set it does point up the need to consider issues of feature generation versus model structure. It is also of interest to note from Fig. 18.5 that the single layer generated two boundaries.

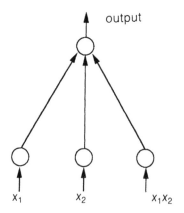

Figure 18.4 No hidden layer: three input nodes.

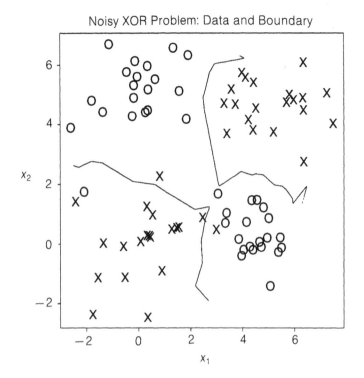

Figure 18.5 Three input nodes: x_1, x_2, x_1, x_2; no hidden layer.

Needless to say that small number of parameters in this model will make it the choice, by either the AIC or MDL criteria, over the five-node or six-node models. In particular, we obtained for this model

$$AIC = 11.48$$

$$MDL = 16.24$$

The generalized likelihood ratio approach is not applicable in comparing this model to the other two.

For the data set that we have generated, the optimal decision boundary is given by the horizontal and vertical lines passing through $x_2 = 2.5$ and $x_1 = 2.5$, respectively. In Fig. 18.6 we have plotted the boundaries from the three networks that we have considered along with the optimal boundary. The figure shows that the boundaries generated by the networks differ most markedly in the lower left-hand quadrant, with the five-node boundary being significantly different than the other two and seemingly differing most from the optimal. It seems that the five-node boundary, narrowly selected by the MDL criterion, is the least desirable, and the MDL criterion's predilection for model parsimony is apparently unwarranted in this example.

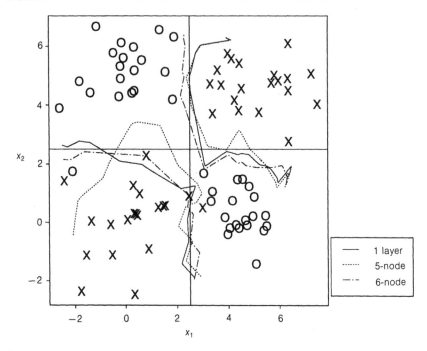

Figure 18.6 The different boundaries and truth.

Examination of additional examples would be useful in understanding the various criteria.

DISCUSSION

We have taken a probabilistic view of neural networks. This has led us to relationships of networks with statistical models and information theory. We have illustrated some benefits of these relationships by using an ML algorithm to optimize an information-theoretic criterion for the network. The information-theoretic criterion, in turn, had a 'distance' interpretation.

We also believe that the algorithm for finding the ML estimate shows much promise. The incorporation of variable step sizes, line searches, and better selection of initial conditions are possible areas for improvement. The algorithm can be extended to handle additional layers and more than two classes.

An important question is how networks developed with the ML criterion compare with networks developed with the traditional mean square error (MSE) criterion. Gish (1990) showed that networks trained using MSE are actually developing posterior probability estimators that are minimizing the average MSE to the true probabilities. The relationship of the data to the decision boundary between the classes, with the MSE criterion, is through the mechanism of the MSE in the probabilities. We have seen that when using the ML criterion the relationship of the data to the decision boundary is rather direct, with the distance from the boundary of those points that are misclassified having the greatest impact on the model. Many questions in comparing the criteria remain to be answered.

Our purpose in the section on model selection. is to illustrate how the usual methods, based on ML estimation, are applicable to neural networks. The issue of which selection criterion is appropriate is an issue we have not addressed and is, in general, a topic of much importance beyond the application to neural networks.

NOTE

1. In this chapter neural network will mean a feedforward network or multi-layer perceptron.

REFERENCES

Akaike, H. (1973) Information theory and an extension of the maximum likelihood principle, in *Proceedings of the Second International Symposium on Information Theory*, Petrov, B.N. and Csaki, F. (eds), Budapest: Akademia Kiado, pp. 267–281.

Anderson, J.A. (1982) Logistic discrimination, in *Handbook of Statistics*, Krishnaiah, P.R. (ed.), vol. 2, New York: North-Holland, pp. 169–191.

Cox, D.R. (1970) *The Analysis of Binary Data*, London: Methuen.

Day, N.E. and Kerridge, D.F. (1967) A general maximum likelihood discriminant. *Biometrics*, **23**, 313–323.

Gish, H. (1990) A probabilistic approach to the understanding and training of neural network classifiers, in *IEEE International Conference on Acoustics Speech and Signal Processing*, Albuquerque, NM, pp. 1361–1364.

Rao, C.R. (1965) *Linear Statistical Inference and Its Applications*, New York: Wiley, pp. 350–352.

Rissanen, J. (1978) Modeling by shortest data description. *Automatica*, **14**, 465–471.

Rumelhart, D.E., Hinton, G.E. and Williams, R.J. (1986) Learning internal representations by error propagation, *Parallel Distributed Processing*, vol. 1. Cambridge, MA, MIT Press, Chapter 8.

Seber, G.A.F. and Wild, C.J. (1989) *Nonlinear Regression*, New York: Wiley, pp. 685–686.

A connectionist knowledge-acquisition tool: CONKAT

19

A. Ultsch, R. Mantyk and G. Halmans

INTRODUCTION

Knowledge acquisition for expert systems poses many problems. Expert systems depend on a human expert to formulate knowledge in symbolic rules. It is almost impossible for an expert to describe knowledge entirely in the form of rules. In particular, it is very difficult to describe knowledge acquired by experience. An expert system may, therefore, not be able to diagnose a case whereas the expert is able to do so. The question is how to extract experience from a set of examples for the use of expert systems.

Machine learning algorithms such as 'learning from example' claim to be able to extract knowledge from experience. Symbolic systems such as ID3 (Quinlan, 1984) and VERSION-SPACE (Mitchell, 1982) are capable of learning from examples. Connectionist systems claim to be superior to these systems in generalization and handling noisy and incomplete data. Queries to expert systems often contain inconsistent data. For every data set the rule-based systems have to find a definite diagnosis. Inconsistent data can force symbolic systems into an indefinite state. In connectionist networks a distributed representation of concepts is used. The interference of different concepts allows networks to generalize (Hinton *et al.*, 1986). A network computes for every input the best output. Because of this, connectionist networks perform well in handling noisy and incomplete data. They are also able to make a plausible statement about missing components. The CONKAT (CONnectionist Knowledge Acquisition Tool) system presented in this chapter, uses a rule-based expert system with an integrated connectionist network to benefit from the described advantages of connectionist systems. The main purpose of CONKAT is to show that the linkage between classical expert systems and connectionist networks is possible and to point out one way in which this could be done.

In the following sections CONKAT will be presented in detail. In the next section we will describe the learning phase. In this phase a neural network is trained and the

Artificial Intelligence Frontiers in Statistics: AI and statistics III. Edited by D.J. Hand. Published in 1993 by Chapman & Hall, London. ISBN 0 412 40710 8

symbolic rules are generated. Then we describe CONKAT's diagnosis phase. CONKAT is a rule-based expert system with a connectionist reasoning component, that is to say, it is able to use the knowledge of the trained networks and is also able to query these networks. The fourth section presents first applications and their results with CONKAT.

THE LEARNING PHASE

During CONKAT's learning phase, regularities in the data are discovered and formulated as symbolic rules. Starting from a set of exemplary cases (n-dimensional real-valued vectors), a neural network is trained. The connectionist networks used are unsupervised learning models which are able to detect irregularities in data; that is, the neural network, eventually followed by a clustering algorithm, provides a classification of the cases.

One of the ideas behind CONKAT is to express such a neural classification by IF–THEN rules (PANDA, 1990). This is the task of the rule extractor module (Fig. 19.1). Based on the classification provided by the neural networks, the rule extractor builds a minimal decision tree. In order to do this, the rule extractor needs disjoint and named intervals of the data components. The decision tree is translated into PROLOG rules. With these rules it is possible to classify the existing cases using only a few selected components of the data vectors. With his domain-specific knowledge, an expert relates the provided classification to a diagnosis. Therefore, another aim of CONKAT is to assist the expert in acquiring rules from examples and unclassified data.

Figure 19.1 shows the structure of the neural knowledge acquisition module. Cases from the case database are encoded by the data transformation machine (DTM). They are then used to train the neural networks. For CONKAT we use a Kohonen network and a competitive learning network.

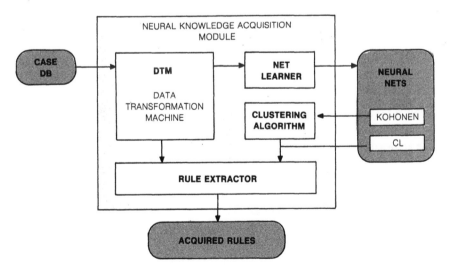

Figure 19.1 The neural knowledge acquisition module.

The Kohonen network used consists of an input layer and an output layer (Kohonen, 1984). The input layer has n units representing the n components of a data vector. The output layer is a two-dimensional array of units arranged on a grid. The number of output units is determined experimentally. Each unit in the input layer is connected to each unit in the output layer with an associated weight. The weights are initialized randomly, taking the smallest and the greatest value of each component (of all vectors) as boundaries. They are adjusted according to Kohonen's (1984) learning rule. The rule applied uses the Euclidean distance and a simulated Mexican hat function to realize lateral inhibition. In the output layer neighbouring units form regions, which correspond to similar input vectors. These neighbourhoods form disjoint regions, thus classifying the input vectors. The automatic detection of this classification is realized by a clustering algorithm. A suitable clustering algorithm is proposed, for example, by Kodratoff (1988).

Our competitive learning network is a three-layered network with input, hidden and output layers. The number of input units depends on the size of the binary-encoded patterns of the input vectors. The number and grouping of the units in the hidden layer are determined experimentally. Each output unit represents a diagnosis. Between every two neighbouring layers all connections exist. The weights on these connections are initialized with random values and normalized (Rumelhart and Zipser, 1985). To train the network all binary input patterns are randomly presented to the units in the input layer until the network has reached a stable state, i.e. no further significant weight changes take place. The 'winner-take-all' principle assures that each input pattern activates exactly one output unit, i.e. exactly one diagnosis is found.

With the help of the trained networks CONKAT is able to classify an arbitrary case. It is classified either by its location in a specific area in the output layer of the Kohonen network or by the activation of one output unit in the competitive learning network. The rule extractor builds up a decision tree with the cases classified by the neural networks. The leaves of the decision tree correspond to the possible diagnoses. A description of a path from the root to a leaf yields a PROLOG rule (ACQUIRED RULES in Fig. 19.1).

THE DIAGNOSIS PHASE

CONKAT is a diagnosis system with a connectionist reasoning component (PANDA, 1990). The system contains three knowledge bases, as shown at the bottom of Fig. 19.2. One knowledge base consists of the rules formulated by an expert (EXPERT RULES). These are of the same type (mainly IF—THEN rules) as the rules generated by the rule extractor (ACQUIRED RULES). The third knowledge base is the neural network trained in CONKAT's learning phase (NEURAL NETS).

If CONKAT is asked for a diagnosis of a new case, it first tries to apply the expert rules. If this query fails, it is possible to consult the rules that CONKAT has acquired during the learning phase. If the query fails again, CONKAT passes the case to the trained network. Appropriately encoded, the case is presented to the input units of the network. Since neural nets are able to generalize, they always deliver a diagnosis. The

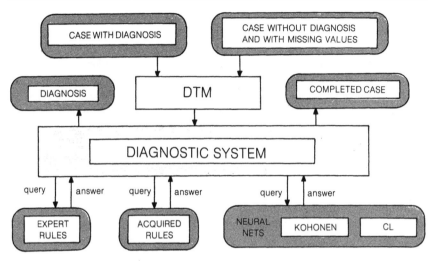

Figure 19.2 The diagnostic expert system.

Kohonen network maps the case to a region in the output layer, which corresponds to a certain diagnosis. In the competitive learning network exactly one output unit relating to a specific diagnosis is activated (Fig. 19.2).

Because of the use of neural networks and their property of generalization it is even possible to diagnose incomplete cases (vectors with missing components). The completion of such data is achieved through the ability of the network to find the most similar complete vector (Fig. 19.2). The incomplete vector is arranged in Kohonen's feature map after learning and the missing components can be reconstructed using the corresponding weight vectors.

With the translation of the neural network's classification to symbolic rules, CONKAT avoids one disadvantage of a purely subsymbolic system. The translated rules can be used for an explanation component. If the only way to diagnose a case is to present it to the neural network, however, there is no possibility of explaining the diagnosis.

FIRST RESULTS

We realized our system on a Sun workstation in C and PROLOG. The connectionist network was implemented using C. The expert-system shell and the rule extractor were implemented in PROLOG.

We tested CONKAT with a data set taken from a medical domain. This contained blood analysis values from 20 patients (20 vectors with 11 real-valued components), together with diagnoses of diabetes. These were selected from a set of 1500 patients (Deichsel and Trampisch, 1985). We selected this data set because it has been extensively analysed using classical cluster techniques. With this data set we were able to check the results of CONKAT.

Each component was divided into disjoint intervals depending on the distribution of its values (DTM in Fig. 19.1). This interval description was binary coded and used to train the competitive learning network. The competitive learning network used in this example was a three-layered network consisting of 58 input, 15 hidden and five output units. The hidden units were organized into three groups of five units each. It was found that the competitive learning algorithm is in principle able to classify the data set correctly. One problem, however, lies in a possible lack of stability of the classification. Different sequences of data presentation lead to different classifications. We found that the classifications depend particularly on the initial values of the weights. The analysis of 400 experiments showed that, in about 60% of the classifications, the competitive learning model corresponds to the correct classification of the data (Deichsel and Trampisch, 1985).

For the Kohonen network, the data set was transformed to give the components equal scope. The transformation also depends on the distribution of the values of the components. We found that the z-transformation (Deichsel and Trampisch, 1985) performs well. For a data set of 20 vectors with 11 components, we chose a Kohonen network with 400 output units arranged in a 20 × 20 grid (Fanihagh *et al.*, 1990). Normally Kohonen's learning algorithm tends to an equal distribution of the units representing the data set. One possibility for avoiding this kind of distribution is the addition of a suitable set of randomly generated data vectors in order to improve the separation of the learned classifications. Another way to achieve this is provided by the *U*-matrix method (Ultsch and Siemon, 1989). The Kohonen network was trained by presenting the input vectors and the additional vectors to the input layer about 200 times. After the learning process, the automatic detection of the regions could be realized using a clustering algorithm. It turned out that Kohonen's network and the clustering algorithm of Kodratoff (1988) are able to elicit the correct classifications from the test data in all cases.

The classifications provided by the network, together with the interval description, were then processed by the rule extractor. Classifying the 20 cases, CONKAT generated four classification rules which can be transformed into diagnosis rules. Figure 19.3 shows a portion of one of the classification trees based on the medical data.

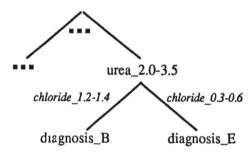

Figure 19.3 Portion of a classification tree for the diagnosis of diabetes.

For the generation of the rules one node of this tree and its daughters are translated into a rule like the following:

patient_data(Chloride,_)
   with urea_between_2.0_and_3.5
   is_a diagnosis_E:-
         chloride(patient_data(Chloride,_),*chlorid_between_0.3_and_0.6*)

This rule describes the edge named *chlorid_between_0.3_and_0.6* beginning at the node urea_between_2.0_and_3.5 and which ends at the leaf 'diagnosis_E'. In this case the urea value of the patient is in the range 2.0–3.5. The next decision is represented by the edge *chlorid_between_0.3_and_0.6* and therefore a connection is established between the patient with urea_between_2.0_and_3.5 and *chlorid_between_0.3_and_0.6* and the node 'diagnosis_E'.

With the rules generated it is possible to classify the existing cases using only a few selected components. With his domain-specific knowledge, an expert relates the provided classification to a diagnosis. In the example above, the classification 'diagnosis_E' may then be changed into the diagnosis 'metabolic_acidosis'. Therefore, CONKAT assists the expert in acquiring rules from examples.

With the data set described we also tested whether CONKAT is able to make a plausible statement concerning missing components. A trained Kohonen network (300 learning steps), consisting of 400 output units, is able to diagnose and to complete vectors with up to five missing components correctly. The accuracy of the reconstruction and the correctness of the classification depend on the missing component. If there were enough components left to diagnose the vector correctly, the obtained accuracy of the completion was around 95% of the vectors of the test data set.

DISCUSSION

The implementation of CONKAT shows the usefulness of the combination of a rule-based expert system with a neural network. Unsupervised neural networks are capable of extracting the regularities from the data. The application of such networks allows noisy and incomplete data to be handled. This is of particular importance because the system is able to make a plausible statement concerning the missing components. For example, it is possible to compute the range for these components. So CONKAT realizes an ability that expert systems often lack.

Due to the distributed subsymbolic representation, neural networks are typically not able to explain the inferences of the system. CONKAT avoids this disadvantage by extracting symbolic rules from the network. The rules acquired can be used like the rules of the expert. In particular, it is therefore possible to explain the inferences of the connectionist system.

Gallant (1988) also uses a neural network as a knowledge base in his connectionist expert system. Furthermore, he extracts rules from the trained network. One important difference from CONKAT is that Gallant constructs rules depending on the weights. Therefore, Gallant uses locally represented knowledge and a supervised learning model. His major goal was 'to make a connectionist model behave as much as possible like an expert system' (Gallant, 1988). In contrast to Gallant, we wish to improve an expert system by integrating a neural network. Moreover, Gallant's 'connectionist expert system' cannot handle data for which no diagnoses are provided, because it uses a supervised learning algorithm (Gallant, 1988). The main difference between CONKAT and other work in the field of classification and rule extraction (Weiss and Kapouleas, 1989; Draper et al., 1987; Gorman and Sejnowski, 1988; Denker et al., 1987; McMillan et al., 1991) is that CONKAT is a *complete* system which combines the tasks of classification and rule extraction and, in addition, uses the knowledge directly from the learned neural networks.

CONKAT is useful in two aspects. First, the system is able to learn from examples with a known diagnosis. With this extracted knowledge it is possible to diagnose new unknown examples. Another ability of CONKAT is to handle a (large) data set for which a classification or diagnosis is unknown. For such a data set CONKAT proposes classification rules to an expert. The expert is then able to refine the classification rule to a diagnosis rule.

A possible extension of our system is the accumulation of knowledge from examples for which no rules exist. After the set of examples has reached a certain size, the network can be trained with this new data set. Thus the system gains the experience from the new examples without using more memory space, because the neural networks retain their size.

CONCLUSION

The CONKAT system presented in this chapter is a rule-based expert system with a connectionist module. The system is characterized by two phases. In the learning phase the neural networks are trained and PROLOG rules are extracted from the trained networks. In the diagnosis phase it is possible to query the expert rules, the acquired rules or the trained network. The integration of a connectionist module realizes 'learning from examples'. Furthermore, CONKAT is able to handle noisy and incomplete data. First results show that the combination of a rule-based expert system with a connectionist module is not only feasible but also useful.

CONKAT is one of the possible ways of combining the advantages of the symbolic and subsymbolic paradigms. It is an example of how a rule-based expert system can be equipped with the ability to learn from experience using a neural network.

ACKNOWLEDGEMENTS

We would like to thank the following members of the student research group PANDA (PROLOG And Neural Distributed Architectures)—F. Fanihagh, A. Lütgendorf, M. Mempel, P. Rossbach, B. Schneider and F. Wegmann—who did major parts of the design and implementation of CONKAT. Special thanks to Frank Wegmann and Volker Weber for their help in preparing this chapter. This work has been supported in part by the Forschungspreis Nordrhein-Westfalen.

REFERENCES

Denker et al. (1987) Large automatic learning, rule extraction, and generalization. *Complex Systems*, 877–922.

Deichsel, G. and Trampisch, H.J. (1985) *Clusteranalyse und Diskriminanzanalyse*, Stuttgart: Gustav Fisher Verlag.

Draper, J., Frankel, D., Hancock, H. and Mize, A. (1987) A microcomputer neural net benchmarked against standard classification techniques, in *IEEE First International Conference on Neural Networks*, San Diego, Vol. 4, pp. 651–658.

Gallant, S.L. (1988) Connectionist expert systems. *Communications of the ACM*, **31**, 152–169.

Gorman and Sejnowski (1988) Analysis of hidden units in a layered network trained to classify sonar targets. *Neural Networks*, **1** (2), 57–89.

Hinton, G.E., McClelland, J.L. and Rumelhart, D.E. (1986) Distributed representations, in *Parallel Distributed Processing: Explorations in the Microstructure of Foundations*, Vol. 1, Rumelhart, D.E. and McClelland, J.L. (eds), Cambridge, MA: MIT Press.

Kodratoff, Y. (1988) *Introduction to Machine Learning*, London: Pitman.

Kohonen, T. (1984) *Self-Organization and Associative Memory*, Berlin: Springer-Verlag.

McMillan, C., Mozer, M.C. and Smolensky, P. (1991) The connectionist scientist game: rule extraction and refinement in a neural network. Paper presented to the *13th Annual Conference of the Cognitive Science Society*, Chicago.

Mitchell, T.M. (1982) Generalization as search. *Artificial Intelligence*, **18**, 203–226.

PANDA Group (1990) PROLOG and neural distributed architectures. Project Report, University of Dortmund, Germany.

Quinlan, J.R. (1984) Learning efficient classification procedures and their application to chess end games, in *Machine Learning: An Artificial Intelligence Approach*, Michalski, R., Carbonell, J.G. and Mitchell, T.M. (eds), Berlin: Springer-Verlag.

Rumelhart, D.E. and Zipser, D. (1985) Feature discovery by competitive learning. *Cognitive Science*, **9**, 75–112.

Ultsch, A. and Siemon, H.P. (1989) Exploratory data analysis: using Kohonen networks on transputers. Department of Computer Science, Technical Report no. 932, Research Report in Computer Science, University of Dortmund.

Weiss, S.M. and Kapouleas, I. (1989) An empirical comparison of pattern recognition, neural nets, and machine learning classification methods, in *Proceedings of the 11th International Joint Conference on Artificial Intelligence*, Detroit, Vol. 1, pp. 781–787.

Connectionist, rule-based and Bayesian decision aids: an empirical comparison

20

S. Schwartz, J. Wiles, I. Gough and S. Phillips

INTRODUCTION

Diagnosis is one of the doctor's central tasks. As an intellectual activity, it is a variant of the more general skill of classification—assigning entities to different classes or categories. Classification is easy when each category (or, in the present case, each disease) has a specific, reliably detected, sign. Unfortunately, such 'pathognomonic' signs are rare. The common signs of illness (such as fever or pain) are shared by many different diseases and most laboratory test results have more than one possible cause. This non-specific relationship between signs and diseases ensures that there will always be an element of uncertainty in medical diagnosis. This uncertainty can be reduced by the discovery of more sensitive clinical signs, but it can never be eliminated. No test is perfectly accurate, signs can be misleading, and even the best treatments do not always succeed (see Schwartz and Griffin, 1986, for more on the probabilistic nature of medical decision-making).

Complicating matters even further is the frequent lack of any specific causal theory relating diagnostic signs to underlying pathophysiology. Consider, for example, patients who present at hospital casualty rooms complaining of acute abdominal pain. There are many possible causes: appendicitis, perforated ulcer, urinary tract infection, and so on. Doctors use a pattern of signs to discriminate among these conditions. For example, it is most common for appendicitis to occur in males, to begin with a central pain that moves to the right lower abdominal quadrant and to be accompanied by vomiting, loss of appetite, and so on. But this is not always the case. Females also get appendicitis and sometimes it begins with pain in the lower right quadrant. Some patients mimic the complete appendicitis pattern but turn out to have some other illness. Because there is no clear physiological theory to explain why appendicitis should produce a particular pattern of signs, we are left with purely empirical correlations—correlations with values distinctly less than 1.0 (de Dombal, 1984).

Given the uncertain relationship between signs and illnesses, it is often difficult for

Artificial Intelligence Frontiers in Statistics: AI and statistics III. Edited by D.J. Hand. Published in 1993 by Chapman & Hall, London. ISBN 0 412 40710 8

doctors to decide whether a patient's abdominal pain requires an immediate operation or whether it is safe to merely 'watch and wait'. There are risks either way. An infected appendix allowed to fester may perforate, creating a potentially lethal peritonitis. This situation can be avoided by an early operation. On the other hand, removing a healthy appendix needlessly exposes the patient to the risks of surgery. Because the risks of perforation are greater than those of an operation, all surgeons support the philosophy of 'when it doubt, cut it out'. An unavoidable result of this decision rule is that 25% or more of all appendix operations result in the removal of perfectly healthy organs (Adams *et al.*, 1986). Clearly, there would be considerable savings in money, time and surgical morbidity, if these unnecessary operations could be avoided (and if diseased appendixes could be removed as quickly as possible). It has often been suggested that computerized decision aids might be able to help doctors make better diagnoses. However, obtaining the necessary expertise, in a form suitable for computer coding, is often a major problem. System designers consult medical experts but, in many cases, these experts have considerable difficulty explaining how they go about diagnosing patients. In recent years, it has been suggested that this 'knowledge elicitation bottleneck' can be broken by using the machine learning techniques developed by cognitive scientists (Gallant, 1988; Schwartz, 1989; Schwartz *et al.*, 1989).

CONNECTIONIST NETWORKS

One increasingly popular approach uses connectionist networks to produce diagnostic advisors (Bounds *et al.*, 1988; Hart and Wyatt, 1989; Gallant, 1988). The argument in favour of connectionist networks derives from extensive research showing that much human expertise resides in complex pattern recognition. Chess grand masters, for example, excel because they have a large memory store of game patterns (de Groot, 1965). Similarly, in the medical domain, expert radiologists appear to differ from newly trained doctors mainly in their ability to recognize abnormal patterns quickly (Hillard *et al.*, 1985). Because connectionist networks can learn to recognize ill-defined patterns, they should—in principle at least—be able to learn to make difficult diagnoses even in the absence of a causal theory relating signs to diseases.

It should be kept in mind that the term 'connectionist' is a generic one that refers to many different types of network. Thus, although several attempts to use connectionist networks for diagnostic tasks have been published (Bounds *et al.*, 1988; Hart and Wyatt, 1989; Gallant, 1988), no two have used exactly the same structure or learning rules. Not surprisingly, therefore, the results have been equivocal. In general, studies using simple networks and artificial laboratory data find connectionist networks to work well (Gallant, 1988; Hunt, 1989), while those using actual patient data have found them to perform rather poorly (Hart and Wyatt, 1989). The reason for this difference is not entirely clear. It may result from differences in the specification of the various networks, or, possibly, in the relative completeness of the respective training sets. (Real patient data are never as complete, or 'clean', as artificial data.) In the present research, we used a feedforward network with one hidden layer which learned to classify patients using back-propagation (Rumelhart and McClelland, 1986).

RULE INDUCTION

Connectionist networks are not the only way that cognitive scientists have modelled learning from experience. Quinlan's (1983; 1986; 1988) ID3 algorithm tackles classification by breaking down the overall problem into a series of subclassifications. Specifically, ID3 constructs a decision tree. Each tree node represents the influence of the most diagnostic sign at that point in the sequence. At each node, the value of the sign is used to partition the cases into separate groups. The algorithm is then invoked recursively on the separate groups. The process continues until all the cases at a node fall into the same partition. When that happens a 'leaf node' is created and given a unique label.

ID3 is a particularly attractive alternative to a connectionist network because it is 'rule-based'. To convert the decision tree into a collection of conjunctive rules one simply traces each path from the root to a leaf (Quinlan, 1987). The rules generated by ID3 may produce new insights into the relationship between signs and diseases. In contrast, connectionist networks represent knowledge mathematically and are difficult to interpret in terms specific to a problem domain. ID3's 'divide and conquer' strategy is also consistent with the way doctors are actually taught to make diagnoses (Schwartz and Griffin, 1986). Indeed, branching algorithms, in the form of flow charts, are common in medical textbooks (Komaroff, 1982). In the present research, we directly compared ID3 and connectionist networks by applying both to the same set of patient data.

BAYESIAN PROBABILITY REVISION

A third approach to diagnosis, one that has been used extensively in designing decision aids in the domain of acute abdominal pain, is based on the probabilistic relationship between signs and diseases. Specifically, the conditional probabilities of the signs given the various diseases are combined using Bayes's formula to yield the posterior probability of the disease given a specific pattern of signs (see Sox *et al.*, 1988). This approach does not claim to model human learning or cognition. Nevertheless, it was included in our research because of the normative status of Bayesian probability revision and its widespread use in medicine. We applied Bayesian probability revision in a simplistic manner, making the unrealistic, but exceedingly common, assumption that the various signs were independent.

RESEARCH AIMS

This research project was conceived as an attempt to apply cognitive science techniques to a domain in which categorization carries 'life-or-death' implications. Specifically, the aims of the present research were: to compare the effectiveness of a back-propagation network, ID3 and Bayesian probability revision in classifying acute abdominal pain patients using a set of real patient data; to examine the 'practical value' of the three techniques by comparing their performance with the performance of trained doctors; and to ask whether the various techniques produce new insights concerning the relationships between signs and diseases.

METHOD

PROBLEM DOMAIN: DIFFERENTIAL DIAGNOSIS OF ACUTE ABDOMINAL PAIN

The data used in this study were collected prospectively from 276 patients over the age of 12 who presented to the casualty room at the Royal Brisbane Hospital complaining of acute abdominal pain (see Gough, 1988, for details). For our present purpose, the most important aspects of the data were the signs gathered for each patient, the doctor's initial diagnostic impression and the final diagnosis which served as the definitive criterion of accuracy or the 'gold standard'.

Although there were 41 diagnostic signs, each could take on at least two values. For example, there were six values for the sign 'aggravating factors' ('movement', 'coughing', 'food', and so on). We coded each value of a sign as either present or absent. Thus, the total number of inputs available for classification purposes was not 41 but 159. As is often the case with real patient data, some signs were not available for some patients. We assumed that such absent signs were distributed randomly across patients and diagnostic groups. For analytical purposes, we made the arbitrary decision to treat such signs as 'absent'.

As noted earlier, there are many possible causes of acute abdominal pain, but from the point of view of the casualty room doctor there are really only two important categories: either the patient needs an operation or the patient does not need an operation. Because most patients who need an operation are suffering from acute appendicitis, the differential diagnosis often boils down to whether the patient has appendicitis, some other serious illness or, for want of a better term, 'non-specific abdominal pain'. These were the three diagnostic categories used in the present research.

DESIGNATION OF TRAINING AND TESTING SETS

The **holdout** method (Weiss and Kapouleas, 1989) was used to partition the cases into training sets (which included approximately 90% of the cases) and testing sets of about 10% of the cases. To ensure that the training and testing sets were equally difficult, we compared doctors' performances on the two sets. Only those sets which were equally difficult for doctors (approximately the same percentage correct) were retained for further use. The holdout method was used in preference to other methods (bootstrapping, for example) because of the enormous length of time it takes to train a network on a new training set (several days of continuous computing, in some cases). To ensure any differences obtained were reliable, we ran both ID3 and the network many times, altering various parameters (for details, see below).

BACK-PROPAGATION PROCEDURES

The network consisted of three layers (input, hidden, output). Each input unit was completely connected to each hidden unit and each hidden unit was completely connected to each output unit. There were no lateral connections among the units. The input layer consisted of one unit for each value of a sign. For example, sex had two input units: 1 0 encoding a male, and 0 1 encoding female. At first glance it might seem odd to code sex separately for male and female; after all, the two units would be perfectly correlated (-1.00). The reason for adopting the present approach is that the back-propagation algorithm can only learn on the presence of information and not on its absence. The amount of weight adjustment is proportional to a unit's activation. Encoding sex as a single unit (with, say, 0 standing for male and 1 for female) would mean that, for males, the activation would be 0, no error term could be calculated, and there would be no weight adjustment.[1] To enable fair comparisons, we coded the data in a similar manner for ID3 but not for Bayesian probability revision which, as already stated, was conducted under the assumption that the signs were independent. (Coding for Bayesian probability revision is described below.)

On the output side, we had a similar coding for appendicitis, non-specific abdominal pain and other illness. This method minimized the amount of decoding required and maximized the separation (or dissimilarity) between different inputs. It did mean, however, that there were a great number of weights to be updated. The approach taken was to use only a few diagnostic signs (a small, fast, network) as a starting point. We then added inputs checking performance at each stage. This allowed us to determine whether some subset of inputs produces optimum classification (or whether all diagnostic signs are necessary).

Using the implementation provided by McClelland and Rumelhart (1988), training involved repeated epochs (one forward and one backward pass through the network for all cases). For the purpose of calculating errors, the output with the highest weight was selected as the system's 'conclusion'. Processing continued until the network error (sum of the squares of the difference between the desired output and the actual output for each output unit) was minimized.

Back-propagation is not deterministic. The success of a network depends to a large extent on the starting weights and the number of hidden units. To optimize the network's classification performance, the number of hidden units was progressively varied. Also varied were the network parameters: *lgrain*, *lrate*, and *weights*. *lgrain* was set to either *pattern* (weights were adjusted after each pattern was presented), or *epoch* (errors were accumulated and the weights adjusted after all patterns were presented). *lrate* is the fraction of the error used when updating the weights. Nets with different numbers of hidden units were each run three times, each time with different random starting weights. After finding the best combination of factors and parameters, the data set was repartitioned and back-propagation rerun over the new training set using the new parameters.

ID3 PROCEDURES

Unlike back-propagation, ID3 is a deterministic algorithm. ID3 (C4.5) has various parameters with which to 'fine-tune' performance. Over 80 parameter combinations were tried altogether. The parameters used in the present research included:

1. Confidence Factor (CF). Under certain circumstances, ID3 may 'overfit' the training data. That is, the trees generated are specific to the particular training set and may not generalize well to the test set. Overfitting is usually caused by rules formulated to explain 'noise' in the data. The CF attempts to suppress noise by pruning the tree. To ensure that only noise, not relevant information, was eliminated, ID3 was run over a range of CFs.
2. Gain Criterion versus Gain Ratio Criterion. These are two methods of choosing which feature to place at the root of each decision tree (and subtree). The gain criterion tends to favour signs with large numbers of values, while the gain ratio is either neutral or slightly biased toward signs with few values.
3. Windowing versus No Windowing. If windowing is specified, then a tree is constructed from a subset of the training data called a 'window'. The process, called a **cycle**, of generating a tree from a window is repeated until all items not in the current window have been classified correctly. The size of the initial window and the window increment rate—the maximum number of items that can be added to a window at each cycle—may be specified by the researcher. If windowing is not specified then a single tree is constructed from the entire training set.
4. Subsetting versus No Subsetting. Subsetting partitions the training set into subsets which are examined separately. Subsetting, like the CF, is an attempt to improve generalization by restricting the tree's sensitivity to a specific training set.

BAYESIAN PROCEDURES

As noted earlier, the conditional probability of the signs given each of the three classifications was calculated for the training set cases, making the unrealistic, but nevertheless common, assumption that the signs were conditionally independent—the probability of a sign given a classification is not affected by the presence or absence of other signs. (To make this assumption as reasonable as possible, the data were recoded so that each sign could take on $n - 1$ values.) The test cases were classified into diagnostic groups by assigning each case to the group for which its signs produce the highest posterior probability. We also had available a more reliable set of conditional probabilities collected from more than 6000 cases by the Organisation Mondiale de Gastro-Enterologie (OMGE) and coded in a commercial computer program called MEDICL. For comparison purposes, we also used MEDICL to classify our patients.

RESULTS

BACK-PROPAGATION

Varying the network parameters produced dramatic effects. We also found considerable differences in performance depending on the criterion used to define a 'correct' classification. Table 20.1 summarizes system performance under a number of different conditions, using a relatively lenient criterion of 'correct'. A classification was considered to be correct if the output unit with the maximum activation corresponded

Table 20.1 Effects of varying the number of hidden units, *lgrain* and *lrate* on classification performance. The net had 159 input units

Hidden units	*lgrain*	*lrate*	Epochs	Error (sums of squares)	No. of correct test case classifications (out of 30)
30	epoch	0.5	100	246	0
30	pattern	0.5	100	246	0
30	epoch	0.1	100	194	16
30	pattern	0.02	100	13	15
30	pattern	0.02	400	10	15
30	pattern	0.02	1000	8	15
125	epoch	0.01	300	38	19
125	epoch	0.5	100	194	16
125	pattern	0.5	100	194	16
125	pattern	0.01	100	14	17
125	pattern	0.01	500	7	18
220	epoch	0.5	100	194	16
220	pattern	0.5	100	296	0
220	epoch	0.02	100	194	16
220	pattern	0.02	100	194	16

to the patient's final diagnosis (the present 'gold standard'). Stricter criteria, using various thresholds for counting a classification as correct, significantly reduced performance.

All other things being equal, setting *lgrain* equal to *epoch* produced better performance than *lgrain* set to *pattern* and slow learning rates were better than high rates (although these required many epochs to reach convergence). The optimal number of hidden units was somewhat less than the number of input units, but not much less (125). To a great extent, learning depended on the initial starting weights. 'Good' initial weights allowed for faster and more accurate learning because the network's starting position in the error-weight space was in the neighbourhood of a global optimum (that is, the network was not prevented from eventually moving into a global optimum by an intervening local optimum). Nevertheless, after 2000 epochs, all networks managed to classify 99% of the training set and at least 16 test cases correctly. A conservative conclusion, therefore, is that back-propagation networks will always get at least 16 test cases correct.

Although it might be expected that at least some of the signs were redundant, reducing the number of inputs always resulted in degraded performance. From a practical viewpoint, this meant that back-propagation was not able to suggest any way in which the amount of data collected for each case could be reduced. Table 20.2 summarizes the performance of the most successful network (125 hidden units, *lgrain* = *epoch* and *lrate* = 0.01). As may be seen, after 300 epochs, this network correctly classified 232 of the 246 training cases and 19 of the 30 test cases. After 3000 epochs, the network had reduced its error considerably and was able to classify correctly more than 99% of the training cases. So, there is no doubt about the network's ability to learn. Nevertheless, its best performance on the test cases was 63%. This suggests that there were insufficient training cases for the network to learn all the possible variations. Note, also, that test case accuracy was not well correlated with training case accuracy; the network that performed best on the test cases missed 14 of the training cases. Thus, it appears that there was some tradeoff between learning the specifics of the training set and generalization to the testing set. Given the small numbers of cases in the training sets, the network may have overfitted the training data, thus reducing its ability to generalize to the test set.

It should be noted that the optimal network learned fairly slowly, taking 72.5 hours (real time) to converge, running continuously on the Sun 3/50.

Table 20.2 Performance on the training and test sets at selected stages of learning. (The net had 159 input units, 125 hidden units, *lgrain* = *epoch* and *lrate* = 0.01)

Epochs	Error (sums of squares)	Number of correctly classified cases	
		Training set ($N = 246$)	Testing set ($N = 30$)
100	71.3	204	16
300	37.8	232	19
500	22.0	237	18
700	15.7	240	17
1000	9.8	242	17
1500	5.9	244	17
2000	4.8	244	18
3000	4.4	244	18

RULE INDUCTION

Table 20.3 summarizes the effects of changing various parameters on ID3's classific-ation performance. The values in the table are the number of correctly classified test cases out of 30 averaged over 10 trees and rounded to the nearest whole number. Each cell in Table 20.3 represents ID3's performance using a specific combination of parameters. For example, the first row of the table shows that with a CF of 10% and windowing, ID3 was correct on 16 cases using the gain criterion but only classified 15 cases correctly using the gain ratio criterion. As noted earlier, 80 trees were constructed using different combinations of parameters. Using the gain criterion, and subsetting, most of these trees were able to get 16 of 30 test cases correct.

Looking at Table 20.3 as a whole, it is apparent that, unlike back-propagation, ID3's performance was not greatly affected by changes to its parameters. The only exception was the gain criterion, which was consistently better than the gain ratio criterion—a rather unusual finding (see Quinlan, 1986, for example).

Why should the gain criterion outperform the gain ratio criterion? The answer undoubtedly lies in the specific characteristics of the abdominal pain domain. The gain criterion resulted in smaller and 'shallower' trees than the gain ratio criterion. In addition, signs with many values, such as 'pain onset site' (13 values), and age (8 values),

Table 20.3 Classification performance of ID3 under varying conditions (using all inputs)

Pruning confidence factor (%)	Gain criterion			Gain ratio criterion		
		No window (single tree)			No window (single tree)	
	Windows	No subset	Subset	Windows	No subset	Subset
10	16	16	16	15	15	15
20	15	15	15	14	14	11
30	16	16	15	13	13	11
40	16	16	16	13	11	11
50	15	15	16	12	11	11
60	15	15	16	12	11	11
70	14	15	16	12	11	11
80	14	15	16	13	11	11
90	14	15	16	12	11	11
100	14	15	16	11	11	11

tended to be used first (they were closer to the tree root) when trees were constructed using the gain criterion, whereas signs with relatively few values were closer to the root in trees constructed using the gain ratio criterion. In the present domain, signs with only a few values tend to be less diagnostic than those with many values. For example, 'severity of pain' could take on only two values: 'moderate' or 'severe'. This highly subjective distinction does not differentiate well among classifications. Because the gain ratio criterion favours such signs it may have focused excessively on less diagnostic information.

An important exception to this argument is the two-valued clinical sign, 'rebound tenderness' versus 'no rebound tenderness', which was regarded as the most important sign and chosen as the root of the tree under both the gain and the gain ratio criteria. This ability to provide new insight into the data—in this case, the identification of the most important sign—is one of the benefits ID3 has over back-propagation.

Neither the gain nor the gain ratio are inherently superior (both need to be tested to determine which is best for a particular domain). The same is true of the remaining parameters. For example, windowing (using various initial window sizes) made little difference because the number of training cases from which trees were constructed with windowing (an average of 220) was almost the same as without windowing (246). Also, because tree generation required up to ten cycles, it was actually slower than building a single tree from the total number of items. The finding that windowing is slower than no windowing is not typical (see Quinlan, 1986); it emphasizes the differences among problem domains.

Although the data used here were undoubtedly noisy, there is little evidence that ID3 was seriously overfitting the data. Pruning the tree using the CF and subsetting produced only small improvements in performance. Nevertheless, it is possible that some of the signs were redundant to ID3's diagnostic process. To find out, we examined ID'3 performance with reduced numbers of inputs. In contrast to back-propagation, ID3's performance actually improved when the number of inputs was reduced. ID3's best performance—18 cases correctly classified out of 30—was recorded using only 11 inputs, the gain criterion, subsetting and a CF of 20%. Cross-validation, conducted by repartitioning the set into a new training and a new test set, produced the same results. It would appear that ID3 was able to extract general rules from the training cases, thereby eliminating the need for the specifics of each case.

ID3, at its best, was slightly less accurate than back-propagation (18 versus 19 correct), but the two procedures performed similarly, correctly classifying 16 cases or so. ID3 was rather more robust than back-propagation, getting 16 of 30 correct regardless of the parameters used. It was also of more practical value because it was able to identify the most important signs. (Reducing the amount of data necessary for a diagnosis can save time, money and sometimes lives.) Finally, ID3 had a great advantage in speed. In single-window mode, a complete run required only a few minutes.

CONDITIONAL PROBABILITIES

Conditional probabilities were calculated from the training set for each sign given each classification. Using Bayes's formula, these probabilities were used to calculate the posterior probability of each of the three classifications given the particular pattern of signs presented by a case. Using a lenient criterion, each case was assigned to the category with the highest posterior probability. This method classified correctly 17 of the 30 test cases. The MEDICL program (which uses the conditional probabilities derived from the OMGE survey of 6000 cases) classified 19 of the 30 cases correctly but its results are not strictly comparable to those obtained in the present study because it uses a slightly different set of categories.

The sign with the highest conditional probability for appendicitis was 'pain migrating to the right lower quadrant'. This is related to but not the same as the 'rebound tenderness' sign designated most important by ID3. There was only modest overlap between ID3's most important 11 signs and those signs with high conditional probabilities.

PHYSICIANS

The doctors' initial diagnoses correctly classified 21 of the 30 cases in the test set. Thus, their performance was better than any of the other procedures. This is to be expected given their far greater experience with the domain and the strong possibility that some of the information they gained from their examination of the patients was not coded in the 159 signs.

The doctors' 'hit' rate of 70% of the test set compares well with their average hit rate of 76% of training cases. As noted earlier, this outcome was achieved by design to ensure the test set cases were no more difficult than the training set.

SENSITIVITY AND SPECIFICITY

The raw performance of the various classification algorithms is summarized in Table 20.4. As may be seen, no technique reached the level of the doctors. But, given their greater experience and their probable access to additional information, this is not surprising. What is remarkable is how well the various techniques performed given the relatively small training set and their remarkable similarity; the differences among the various classification procedures were too small to justify statistical analysis. This does not mean, however, that the techniques are interchangeable. We used the contingency coefficient (Siegel, 1956) to determine whether the various techniques were all getting the same test cases right (or wrong). These coefficients were remarkably low. For example, the average correlation between the classifications reached by the Bayesian and ID3 procedures was only 0.35, and no correlation exceeded 0.55. It seems safe to conclude that the three procedures are attacking the classification problem in rather different ways.

Of course, accuracy is only part of the story. As noted earlier, different types of error have different costs. For example, it is less costly to misdiagnose a non-specific pain patient as having appendicitis than to misdiagnose an appendicitis patient as non-specific pain. In the first instance, the patient will have an unnecessary operation. In the second case, the patient may die. Looked at this way, pure accuracy becomes less important than **sensitivity** (true-positive rate) and **specificity** (true-negative rate).

Table 20.4 Summary of the best performances

Diagnostician	Appendicitis	Other serious illness	Non-specific pain	Total correct cases*
Back-propagation	14	2	3	19
ID3	15	0	3	18
Bayesian	12	1	4	17
Doctors[†]	14	2	5	21
Maximum possible cases	16	5	9	30

*Best performance
[†]Initial diagnosis only

Table 20.5 Sensitivity, specificity and predictive value of diagnosticians for diagnosing appendicitis

Diagnostician	Sensitivity (true positive rate or $1 - \alpha$)	Specificity (true negative rate or $1 - \beta$)
Back-propagation	88	36
Bayesian	69	36
ID3	94	21
Doctors	88	50

These are summarized in Table 20.5 for appendicitis versus the other two categories.

As may be seen, all three techniques have a relatively high sensitivity for appendicitis, with ID3 performing best. However, ID3's specificity is low. This means that its success in diagnosing appendicitis is achieved by over-using the diagnosis. Because of their large number of false positives, the predictive values (true positives/all positives) of ID3, back-propagation and Bayesian probability revision are all relatively low. Overall, the doctors probably perform best. They miss only two cases of appendicitis and have the lowest number of false positives. However, this is not a fair comparison because only doctors explicitly take into account the cost of mistakes.

DISCUSSION

The present results are limited by the small number of training and test cases. It has been suggested that a training set should contain at least five cases per item of input data per classification (Wasson *et al.*, 1985). In the present case, this would mean a set of 2400 cases. Few medical data bases are anywhere near this size. We should also note that the Bayesian analysis might have performed better if signs were not considered conditionally independent (Seroussi *et al.*, 1986).

Given the relatively small training set, all of the techniques—but especially back-propagation and ID3—performed remarkably well. Still, there were various differences among the techniques worth noting. First, ID3 was generally more robust than back-propagation. For most runs, it achieved 15 or 16 correct classifications. Its worst performance was 11 out of 30. Back-propagation, on the other hand, was highly susceptible to the learning rate parameter. If not set low enough, the algorithm would not converge at all. Back-propagation was also a much slower process than either ID3 or Bayesian probability revision (it took days as compared with minutes running on the same machine). Finally, the workings of ID3 are more accessible than back-propagation; its trees can easily be stated as rules. It was able to give new insights into the data, emphasizing the importance of a particular sign and identifying the 11 most important signs. Back-propagation, in contrast, appeared to be learning patterns specific to the training set and provided no new insights into the data.

Because the techniques made errors on different cases, they cannot be considered mere substitutes for one another. In the clinic, the best technique would be the one that minimized costly errors. In this regard, none of the techniques was as good as the doctors. But this comparison is unfair because none of the techniques was designed to attend to errors. It is possible that they would have performed as well as the doctors had they been biased away from certain categories. Such biases could easily be added to the network and to the Bayesian analysis (by requiring higher output thresholds for certain diagnoses). ID3 could also be trained to be cautious about certain diagnoses. If they had acceptable accuracy rates, techniques which take into consideration the costs and benefits of various outcomes have the potential to become realistic clinical decision aids. The design of such techniques constitutes an important path for future research.

ACKNOWLEDGEMENTS

Preparation of this chapter was assisted by grants from the Australian Research Council and the NHMRC Public Health and Development Committee to the first named author. The authors are also grateful to ICL for access to the MEDICL program.

NOTE

1. An alternative technique is to use the symmetrical sigmoid function as an activation function. It has a range from -1 to $+1$, so males could be coded -1 and females $+1$.

REFERENCES

Adams, I.D., Chan, M., Clifford, P.C., Cooke, W.M., Dallos, V., de Dombal, F.T., Edwards, M.H., Hancock, D.M., Hewett, D.J., McIntyre, N., Somerville, P.G., Spiegelhalter, D.J., Wellwood,

J. and Wilson, D.H. (1986) Computer aided diagnosis of acute abdominal pain: a multicentre study. *British Medical Journal*, **298**, 800–804.

Bounds, D.G., *et. al.* (1988) A multilayer perceptron network for the diagnosis of low back pain, in *Proceedings of the San Diego Conference on Neural Networks*, Vol. 2, pp. 481–489.

de Dombal, F.T. (1984) Computer based assistance for medical decision making. *Gastroenterology and Clinical Biology*, **8**, 135–137.

de Groot, A.D. (1965) *Thought and Choice in Chess*, The Hague: Mouton.

Dietterich, T.G., Hild, H. and Bakiri, G. (1990) A comparative study of ID3 and backpropagation for English text-to-speech mapping, in *Proceedings of the International Workshop on Machine Learning*, Austin, TX.

Gallant, S.I. (1988) Connectionist expert systems. Communications of the *ACM*, **31**, 152–169.

Gough, I. (1988) A study of diagnostic accuracy in suspected acute appendicitis. *Australia and New Zealand Journal of Surgery*, **58**, 555–589.

Hart, A. and Wyatt, J. (1989) Connectionist models in medicine: an investigation of their potential, in *AIME89: Proceedings of the Second European Conference on Artificial Intelligence in Medicine*, Hunter, J., Cookson, J. and Wyatt, J. (eds), Lecture Notes in Medical Informatics, Heidelberg: Springer-Verlag, 115–124.

Hillard, A., Myles-Worsley, M., Johnston, W. and Baxter, B. (1985) The development of radiological schemata through training and experience: a preliminary communication. *Investigative Radiology*, **18**, 422–425.

Hunt, E.B. (1989) Connectionist and rule-based representations of expert knowledge. *Behavior Research Methods, Instruments and Computers*, **21**, 88–95.

Komaroff, A.L. (1982) Algorithms and the 'art' of medicine. *American Journal of Public Health*, **72**, 10–12.

McLelland, J.L. and Rumelhart, D. (1988) *Explorations in Parallel Distributed Processing*, Cambridge, MA: MIT Press.

Quinlan, J.R. (1983) Learning efficient classification procedures and their application to chess endgames, in *Machine Learning: An Artificial Intelligence Approach*, Vol. 1, Michalski, R.S., Carbonell, J.G. and Mitchell, T.M. (eds), Palo Alto, CA: Morgan Kaufmann, pp. 149–166.

Quinlan, J.R. (1986) Induction of decision trees. *Machine Learning*, **1**, 81–106.

Quinlan, J.R. (1987) Simplifying decision trees. *International Journal of Man–Machine Studies*, **27**, 221–234.

Quinlan, J.R. (1988) An empirical comparison of genetic and decision-tree classifiers, in *Proceedings of the Fifth International Conference on Machine Learning*.

Rumelhart, D.E. and McClelland, J.L. (1986) *Parallel Distributed Processing: Explorations in the microstructure of cognition: Vol. 1: Foundations*, Cambridge, MA: MIT Press.

Schwartz, S. (1989) Computer consultants in the clinic, in *Proceedings of the XXIV International Congress of Psychology: Vol. 9. Clinical and Abnormal Psychology*, Lovibond, P. and Wilson, P. (eds), Amsterdam: North-Holland.

Schwartz, S. and Griffin, T. (1986) *Medical Thinking: The psychology of medical judgment and decision making*, New York: Springer-Verlag.

Schwartz, S., *et al.* (1989) Clinical expert systems versus linear models: Do we really have to choose? *Behavioral Science*, 34, 305–311.

Seroussi, B. and the ARC & AURC Cooperative Group (1986) Computer-aided diagnosis of acute abdominal pain when taking into account interactions. *Methods of Information in Medicine*, **25**, 194–198.

Siegel, S. (1956) *Non-Parametric Statistics*, New York: McGraw-Hill.

Sox, H.C., Blatt, M.A., Higgins, M.C. and Marton, K.I. (1988) *Medical Decision Making*, Boston: Butterworths.

Wasson, J.H., Sox, H.C., Neff, R.K. and Goldman, L. (1985) Clinical prediction rules: applications and methodological standards. *New England Journal of Medicine*, **313**, 793–799.

Weiss, S.M. and Kapouleas, I. (1989) An empirical comparison of pattern recognition, neural nets and machine learning classification methods, in *Proceedings of the 11th International Joint Conference on Artificial Intelligence*, Detroit, Vol. 1, pp. 781–787.

PART FIVE
Text manipulation

A statistical approach to aligning sentences in bilingual corpora

21

W.A. Gale and K.W. Church

INTRODUCTION

The extent of linguistic data has traditionally both stimulated a statistical response, such as work by Zipf (1932) and Yule (1944), and at the same time frustrated such approaches for inability to manage more than a small corpus by hand. Cheap computers have encouraged the development of large machine-readable corpora, such as the Associated Press newswire, which provides 40 million words per year. They have also made management of such large corpora possible. This leaves the stimulus and reduces the frustration, which has led to an increased demand for new statistical approaches to linguistics, as exemplified in Church and Gale (1991), Gale and Church (1990), and Church *et al.* (1990).

Just in the last few years, large *bilingual* corpora have become available for the first time. These are bodies of text such as the Canadian Hansards (parliamentary debates) which are available in multiple languages (such as French and English). Bilingual corpora have supported research in both machine translation (see, for example, Brown *et al.*, 1990) and bilingual lexicography (Klavans and Tzoukermann, 1990). A useful first step in the study of bilingual corpora is the sentence alignment task, to identify correspondences between sentences in one language and sentences in the other language. This task is a first step towards the more ambitious task of finding words which correspond to each other.

This chapter describes a statistical approach to sentence alignment. The input is a pair of texts such as:

English	*French*
According to our survey, 1988 sales of mineral water and soft drinks were much higher than in 1987, reflecting the growing popularity of these products.	Quant aux eaux minérales et aux limonades, elles rencontrent toujours plus d'adeptes. En effet, notre sondage fait ressortir des ventes nettement supérieures

Artificial Intelligence Frontiers in Statistics: AI and statistics III. Edited by D.J. Hand. Published in 1993 by Chapman & Hall, London. ISBN 0 412 40710 8

English	French
Cola drink manufacturers in particular achieved above-average growth rates. The higher turnover was largely due to an increase in the sales volume. Employment and investment levels also climbed. Following a two-year transitional period, the new Foodstuffs Ordinance for Mineral Water came into effect on April 1, 1988. Specifically, it contains more stringent requirements regarding quality consistency and purity guarantees.	à celles de 1987, pour les boissons à base de cola notamment. La progression des chiffres d'affaires résulte en grande partie de l'accroissement du volume des ventes. L'emploi et les investissements ont également augmenté. La nouvelle ordonnance fédérale sur les denrées alimentaires concernant entre autres les eaux minérales, entrée en vigueur le 1er avril 1988 après une période transitoire de deux ans, exige surtout une plus grande constance dans la qualité et une garantie de la pureté.

The output identifies the correspondence between sentences. Most English sentences correspond to exactly one French sentence, but it is possible for an English sentence to correspond to two or more French sentences. The first two English sentences below illustrate a particularly hard case where two English sentences correspond to two French sentences. No smaller alignments are possible because the clause 'sales...were higher...' in the first English sentence aligns with (part of) the second French sentence. The next two alignments below illustrate the more typical case where one English sentence aligns with exactly one French sentence. The final alignment matches two English sentences to one French sentence. These alignments agreed with the results produced by a human judge.

English	French
According to our survey, 1988 sales of mineral water and soft drinks were much higher than in 1987, reflecting the growing popularity of these products. Cola drink manufacturers in particular achieved above-average growth rates.	Quant aux eaux minérales et aux limonades, elles rencontrent toujours plus d'adeptes. En effet, notre sondage fait ressortir des ventes nettement supérieures à celles de 1987, pour les boissons à base de cola notamment.
The higher turnover was largely due to an increase in the sales volume.	La progression des chiffres d'affaires résulte en grande partie de l'accroissement du volume des ventes.
Employment and investment levels also climbed.	L'emploi et les investissements ont également augmenté.
Following a two-year transitional period, the new Foodstuffs Ordinance for Mineral Water came into effect on April	La nouvelle ordonnance fédérale sur les denrées alimentaires concernant entre autres les eaux minérales, entrée en

English	French
1, 1988. Specifically, it contains more stringent requirements regarding quality consistency and purity guarantees.	vigueur le 1er avril 1988 après une période transitoire de deux ans, exige surtout une plus grande constance dans la qualité et une garantie de la pureté.

Aligning sentences is just a first step towards constructing a probabilistic dictionary for use in aligning words in machine translation, or for constructing a bilingual concordance for use in lexicography. Although there has been some previous work on the sentence alignment (see Brown *et al.*, 1990; Kay, personal communication; Warwick, personal communication), the alignment task remains a significant obstacle preventing many potential users from reaping many of the benefits of bilingual corpora, because the proposed solutions are often unavailable, unreliable, and/or inefficient (Zampolli, personal communication; Warwick, personal communication).

Almost all of the previous work on sentence alignment has yet to be published. Kay's approach (Kay, personal communication) is said to be very 'heavy'; it ought to be possible to achieve fairly reasonable results with much less computation (Warwick, personal communication). In Brown *et al.* (1990) there is a very brief discussion of sentence alignment. In its entirety, it reads:

> We have been able to extract about 3 million pairs of sentences by using a statistical algorithm based on length. Approximately 99% of these pairs are made up of sentences that are actually translations of one another.

Brown *et al.*'s discussion was justifiably brief because sentence alignment was not the main topic of their paper; they were trying to argue for a revival of statistical methods for machine translation.[1] Nevertheless, it would be helpful to the research community to provide a more detailed discussion of a practical sentence alignment algorithm and its evaluation.

This chapter will describe a program for aligning sentences based on a very simple statistical model of character lengths. The model makes use of the fact that longer sentences in one language tend to be translated into longer sentences in the other language, and that shorter sentences tend to be translated into shorter sentences. A probabilistic score is assigned to each pair of proposed sentence pairs, based on the ratio of lengths of the two sentences (in characters) and the variance of this ratio. This probabilistic score is used in a dynamic programming framework in order to find the maximum likelihood alignment of sentences.

It is remarkable that such a simple approach can work as well as it does. An evaluation was performed based on a trilingual corpus of 15 economic reports issued by the Union Bank of Switzerland (UBS) in English, French and German ($N = 14\,680$ words, 725 sentences, and 188 paragraphs in English and corresponding numbers in the other two languages). The method correctly aligned all but 4% of the sentences. Moreover, it is possible to extract a large subcorpus which has a much smaller error rate. By selecting the best-scoring 80% of the alignments, the error rate is reduced from 4% to 0.7%. There were roughly the same number of errors in each of the English–French and English–German alignments, suggesting that the method may be fairly language-independent.

The methods described here were developed and tested on the UBS corpus. They have been applied to Hansards dating from 1986 to 1988. The Hansards that we aligned contained about 1 million sentences, and are available for linguistic research through the Association for Computational Linguistics DCI project.

PARAGRAPH ALIGNMENT

The method described here relies heavily on the lengths of sentences and of paragraphs. To know these lengths, one must know where the sentence and paragraph boundaries lie. On the UBS corpus we used very simple heuristics and hand checking to locate the sentence boundaries. On the Hansards we used a sophisticated set of heuristics from Church's (1988) part-of-speech tagger. From the viewpoint of the work described here, these sentence markers should be regarded as part of the input.

The sentence alignment program is a two-step process. First, paragraphs are aligned, and then sentences within a paragraph are aligned. It is fairly easy to align paragraphs in our trilingual corpus of Swiss banking reports since the boundaries are usually clearly marked. However, there are some short headings and signatures that can be confused with paragraphs. Moreover, these short 'pseudo-paragraphs' are not always translated into all languages. Fortunately, it is very easy to distinguish the short 'pseudo-paragraphs' from real ones by a simple threshold on length. Figure 21.1 shows very clearly a bimodal distribution. It turns out that 'pseudo-paragraphs' have less than 50 characters and that real paragraphs have more in this small corpus.

Since pseudo-paragraphs can be distinguished on the basis of length, the paragraph

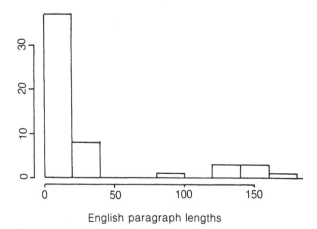

English paragraph lengths

Figure 21.1 This histogram shows the frequency of English paragraphs whose lengths were less than 200 characters. There is a clear gap around 50, with headings and signatures in the short group and contents paragraphs in the long group.

aligning algorithm is very simple. It begins by considering the first two paragraphs in a pair of documents. When considering two paragraphs, if both have length over 100 characters, or both have lengths under 50 characters, then align them, and continue. However if one of the paragraphs is short, while the other is long, then align the short paragraph with null, and continue.

For English and German, this simple algorithm made no mistakes. The algorithm was unable to align one of the French documents with either English or German. It turned out that the French translation of the document had omitted one long paragraph and had duplicated a heading and another long paragraph. (This document was excluded for the purposes of the remainder of this experiment.) If this frequency of paragraph errors were found in a larger corpus, it might be worthwhile building a more elaborate paragraph alignment mechanism.

We will show below that paragraph alignment is an important step, so it is fortunate that it is extremely easy. In aligning the Hansards we found that many days' proceedings already had a perfect alignment of paragraphs. For robustness, we did align the paragraphs, using the same method as that described below for alignment of sentences within a paragraph. However, in this application we used some invariant formatting commands as the fixed reference points (instead of paragraph boundaries) and the paragraph boundaries as the movable partitions (instead of sentence boundaries).

A DYNAMIC PROGRAMMING FRAMEWORK

Now, let us consider how sentences can be aligned within a paragraph. The program makes use of the fact that longer sentences in one language tend to be translated into longer sentences in the other language, and that shorter sentences tend to be translated into shorter sentences. A probabilistic score is assigned to each proposed pair of sentences, based on the ratio of lengths of the two sentences (in characters) and the variance of this ratio. This probabilistic score is used in a dynamic programming framework in order to find the maximum likelihood alignment of sentences. We were led to this approach after noting that the lengths (in characters) of English and German paragraphs are highly correlated (0.991), as illustrated in Fig. 21.2.

Dynamic programming is often used to align two sequences of symbols in a variety of settings, such as genetic code sequences from different species, speech sequences from different speakers, gas chromatograph sequences from different compounds, and geologic sequences from different locations (Kruskal, 1983). We could expect these matching techniques to be useful, as long as the order of the sentences does not differ too radically between the two languages. Details of the alignment techniques differ considerably from one application to another, but all use a distance measure to compare two individual elements within the sequences, and a dynamic programming algorithm to minimize the total distances between aligned elements within two sequences. Placing the sentence alignment problem within this setting, we find we need to develop a new distance measure, but that standard dynamic programming techniques suffice.

Kruskal and Liberman (1983) describe distance measures as belonging to one of two classes: **trace** and **time-warp**. The difference becomes important when a single element

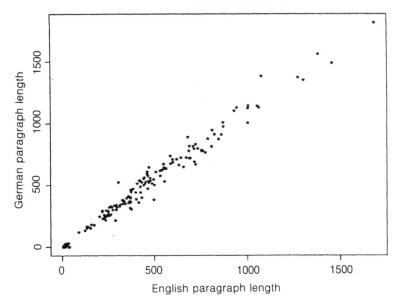

Figure 21.2 The horizontal axis shows the length of English paragraphs, while the vertical scale shows the lengths of the corresponding German paragraphs. The correlation between their lengths is 0.991.

of one sequence is being matched with multiple elements from the other. In trace applications, such as genetic code matching, the single element is matched with just one of the multiple elements, and all of the others are ignored. In contrast, in time-warp applications such as speech template matching, the single element is matched with each of the multiple elements, and the single element will be used in multiple matches. Interestingly enough, our application does not fit into either of Kruskal and Liberman's classes because our distance measure needs to compare the single element with the aggregate of the multiple elements.

The ideal distance measure is based on probability, because new information can then be combined with old information in a consistent way. Since a distance measure must be additive to use the dynamic programming techniques, the logarithm of the probability is used. Thus we seek an estimate of the probability that two sentences or groups of sentences match. This will be the basis of the distance function $d(\cdot;\cdot)$ used later for the dynamic programming.

We use the following simple model. Each character in one language gives rise to a random number of characters in the other language. These random variables are independent and identically distributed with a normal distribution. The model is then specified by the mean and standard deviation of the distribution. Let c be the expected number of characters in language 2 per character in language 1, and s^2 be the variance of the number of characters in language 2 per character in language 1. Then the expected number of characters in the sentences translating a group of sentences of length l_1 in language 1 is $l_1 c$ with variance $l_1 s^2$. Let $\delta = (l_2 - l_1 c)/\sqrt{l_1 s^2}$. Then if sentences of length l_1 and l_2 in the two languages do match, δ has a normal distribution with mean 0 and variance 1.

To estimate $P(\text{match}|\delta)$, we appeal to Bayes's theorem, which says that this conditional probability is proportional to $P(\delta|\text{match}) \, P(\text{match})$. Both of these factors are fitted from the UBS data, and applied to both UBS and Hansards. We take $P(\delta|\text{match})$ to be the probability that a random variable, z, with a standardized (mean 0, variance 1) normal distribution, has magnitude at least as large as $|\delta|$. That is,

$$P(\delta|\text{match}) = 2(1 - P(|\delta|))$$

where

$$P(\delta) = \frac{1}{\sqrt{2\pi}} \int_{-\infty}^{\delta} e^{-t^2/2} dt$$

We take the prior probability of a match, $P(\text{match})$, to depend only on the number of sentences being matched in the two languages, as given in Table 21.1. The distance measure used thus depends on c, s, and the probabilities of matching given numbers of sentences.

We can estimate c, the ratio of lengths, from the aggregate lengths of the matched paragraphs. This gives German characters per English character $= 81\,105/73\,481 = 1.1$, and French characters per English character $= 72\,302/68\,450 = 1.06$. As explained later, using these precise and language-dependent quantities does not improve the performance over $c = 1$, which gives a symmetric and simpler model. If *align* is used with languages for which the character lengths differ by more than 10%, this parameter should be reviewed.

Our model assumes that variance s^2 is proportional to length. Figure 21.3 does not contradict this assumption, and allows us to estimate the constant of proportionality. The line shown in the figure is the result of a robust regression. The result for English–German is $s^2 = 7.3$, and for English–French $s^2 = 5.6$. Again, the sensitivity study described later showed that the differences between these two slopes were not important, so we combined the data across the languages, and used $s^2 = 6.8$ in *align*.

A sentence in one language normally matches exactly one sentence in the other language (1–1), but we allow three other possibilities as well: 1–0, 2–1, and 2–2. That is, there are four categories for matching groups of sentences: one to zero (1–0), one to one (1–1), two to one (2–1), and two to two (2–2). The 1–0 category includes all sentences in either language aligned with nothing in the other. Likewise, the 2–1 category includes cases with a pair of sentences in either language aligned with one in the other. The probabilities of the categories were determined from the data as shown in Table 21.1. The fractions shown in the right-hand column times the probability $P(\delta|\text{match})$, or actually the logarithm of the product, gives the distance measure for a proposed alignment. The distance measure thus has six parameters: $c, s, f_{01}, f_{11}, f_{21}$ and f_{22}. We also refer to this distance measure as the **probability score**. The distance measure was then used in a dynamic programming subroutine written by Mike Riley to align the sentences.

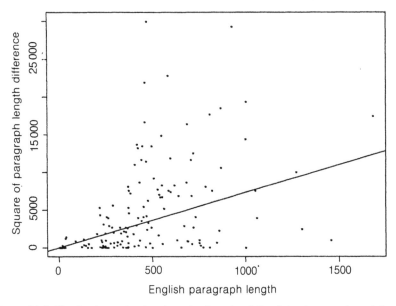

Figure 21.3 The horizontal axis plots the length of English paragraphs, while the vertical axis shows the square of the difference of English and German lengths, an estimate of variance. The variance increases with length. We model the increase as proportional to length, as shown by the line. Five extreme points lying above the top of this figure have been suppressed since they contributed nothing to the robust regression.

Table 21.1 Categories of matching

Category	Frequency	Proportion
1–0	13	0.0099
1–1	1167	0.89
2–1	117	0.089
2–2	15	0.011

The algorithm is summarized in the following equations. Let $s_i, i = 1, \ldots, I$, be the sentences of one language, and $t_j, j = 1, \ldots, J$, be the translations of those sentences in the other language. Let $D(i, j)$ be the minimum distance between sentences s_i and t_j, under the maximum likelihood alignment. $D(i, j)$ is computed by minimizing over six cases which impose a set of slope constraints, where d denotes the distance for an insertion, deletion, substitution, etc. d is a function of four arguments:

1. let $d(x_1, y_1; 0, 0)$ be the cost of substituting x_1 with y_1;
2. $d(x_1, 0; 0, 0)$ be the cost of deleting x_1;
3. $d(0, y_1; 0, 0)$ be the cost of inserting y_1;
4. $d(x_1, y_1; x_2, 0)$ be the cost of contracting x_1 and x_2 to y_1;

5. $d(x_1, y_1; 0, y_2)$ be the cost of expanding x_1 to y_1 and y_2; and

6. $d(x_1, y_1; x_2, y_2)$ be the cost of merging x_1 and x_2 and matching with y_1 and y_2.

$D(i, j)$ is defined by the following recursion with the initial condition $D(i, j) = 0$:

$$D(i, j) = \min \begin{cases} D(i, j-1) + d(0, t_j; 0, 0) \\ D(i-1, j) + d(s_i, 0; 0, 0) \\ D(i-1, j-1) + d(s_i, t_j; 0, 0) \\ D(i-1, j-2) + d(s_i, t_j; 0, t_{j-1}) \\ D(i-2, j-1) + d(s_i, t_j; s_{i-1}, 0) \\ D(i-2, j-2) + d(s_i, t_j; s_{i-1}, t_{j-1}) \end{cases}$$

The transition distances d are given by

$$d(s_1, t_1; s_2, t_2) = -\log \left(2(1 - P(|\delta|)) f \left(\sum_{i=1}^{2} \chi(s_i), \sum_{j=1}^{2} \chi(t_j) \right) \right)$$

This equation breaks the transition distances into two parts, one based on the aggregate lengths of the sentences being aligned, and the other based on the number of sentences from each language being matched. The factor $2(1 - P)$ gives the area under the tails of a normal distribution with mean 0 and variance 1, measured from δ out, where

$$\delta = \frac{c(\operatorname{len}(t_1) + \operatorname{len}(t_2)) - (\operatorname{len}(s_1) + \operatorname{len}(s_2))}{\sqrt{s^2(\operatorname{len}(s_1) + \operatorname{len}(s_2))}}$$

This factor δ measures how far the aggregate lengths of the sentences aligned differ, compared to the usual differences of matching regions. The function χ just counts sentences:

$$\chi(s) = \begin{cases} 0 & \operatorname{len}(s) = 0 \\ 1 & \operatorname{len}(s) > 0 \end{cases}$$

and the factors $f(m, n)$ are based on Table 21.1. For instance, $f(1, 1) = 0.89$, while $f(1, 2) = f(2, 1) = 0.089$.

EVALUATION

A primary judge did all of the alignments, and two corroborative judges independently aligned selected hard paragraphs. In doing so, they used materials prepared with the three languages in columns, each paragraph begun on a new page, and having numerically corresponding sentences beginning at the same level. They did the alignment by drawing lines between sentences that shared a clause. If a sentence was not translated, they wrote in a zero and drew a line to that.

All of the sentences were aligned by the primary judge. The primary judge is a native speaker of English with a reading knowledge of French and of German. Each of the other two judges aligned one of the languages in 43 paragraphs with 230 sentences, chosen as particularly difficult. One of the other judges is a native speaker of French and fluent in English; the other is a native speaker of German and fluent in English. The primary judge was found to be wrong on one alignment in each of French and German in the hard group of paragraphs.

The 43 hard paragraphs were selected by making all three pairs of alignments, then attempting to trace each sentence from English to German, from there to French, and from there back to English. The 43 paragraphs included all sentences in which this process could not be completed around the loop. This group of paragraphs, selected through a trilingual criterion, had 82% of the errors made by the program, another justification for calling them 'hard'. Since the primary judge had an error rate of only 0.5% on these hard paragraphs, we believe that one judge is sufficient for defining the correct alignment of the sentences.

In evaluating the program, we considered only the English–German and English–French alignments, because the independence of the French–German alignments given the other two is questionable. Errors are reported with respect to the alignments judged correct. That is, each correct alignment that is not called for by the program is scored as an error. This convention allows us to compare the performances of different algorithms. There were 36 errors out of 621 total alignments for English–French and 19 errors out of 695 alignments for English–German. Overall, there were 55 errors on 1316 alignments, or 4.2%. Table 21.2 shows the errors by match category.

The table shows that 1–1 alignments are by far the easiest. The 2–1 alignments, which come next, have four times the error rate for 1–1. The 2–2 alignments are harder still, but a majority of the alignments are found. The 3–1 and 3–2 alignments were not possible for the algorithm, so naturally it missed all three of these. The most surprising category is 1–0, in which all cases were missed. Furthermore, of the sentences that the algorithm is most inclined to assign to the 1–0 category, none belong there, so adding inducements of various kinds to make 1–0 assignments introduces new errors without reducing the errors reported on here. Apparently when the assumption that the two documents are translations of each other fails, which is the case for the 1–0 category, then we have considerable trouble finding the correct alignment without knowledge of the languages.

We investigated the possible dependence of the error rate on four variables. First was length of the alignment; the average of the total lengths of the sentences grouped

Table 21.2 Complex matches are more difficult

Category	English–French			English–German			Total		
	N	Error	%	N	Error	%	N	Error	%
1–0	8	8	100	5	5	100	13	13	100
1–1	542	14	2.6	625	9	1.4	1167	23	2.0
2–1	59	8	14	58	2	3.4	117	10	9
2–2	9	3	33	6	2	33	15	5	33
3–1	1	1	100	1	1	100	2	2	100
3–2	1	1	100	0	0	0	1	1	100

Table 21.3 Probability score predicts errors

Variable	Coefficient	Std. dev.
score	0.071	0.011
complexity	0.52	0.47
sentence length	0.0013	0.0029
paragraph length	0.0003	0.0005

for each language. Second was the paragraph length: the average length of the paragraphs in which the alignment occurred. Third was the complexity: whether the alignment was of the 2–1 or 2–2 category. Fourth was the probability score: the score assigned to the alignment if the program was correct, or the maximum score for any sentence involved in the true alignment if it was not correct.

Since each alignment was judged either right or wrong, we used logistic regression of the judgement on the four variables. The coefficients and their standard deviations are shown in Table 21.3. The table shows that the probability score assigned to an alignment is a good predictor of whether it is in error. No other factor was significant after considering the probability score. Thus we could not improve the probability score by including any more information from these sources. The fact that the probability score predicts the error rate may be useful, as suggested by Fig. 21.4.

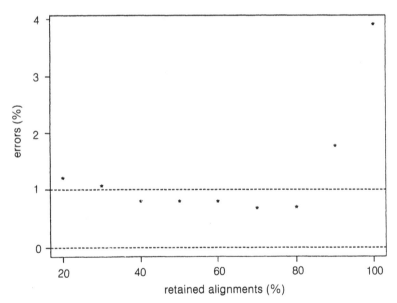

Figure 21.4 By retaining the alignments with the lowest p% of the probability scores, as shown on the horizontal axis, the error rate among the retained alignments is as shown on the vertical axis. An error rate of about 0.7% can be obtained with 80% of the alignments retained.

Less formal tests of the error rate in the Hansards suggests that the overall error rate is about 2%, while the error rate for the easy 80% of the sentences is about 0.4%. Apparently the Hansards have better translations than these UBS reports.

EVALUATION OF ALGORITHM VARIATIONS

We considered several variations of the algorithm used in *align*. There are also several possible extensions of the algorithm.

VARIATIONS

One variation considered was the use of words instead of characters to measure the lengths of the sentences. Words might be expected to be worse, since fewer words per sentence suggests more uncertainty. In fact, doing so raised the English–French errors from 36 to 50, and the English–German errors from 19 to 35. The total errors were thereby increased from 55 to 85, or from 4.2% to 6.5%.

Lengths measured in characters are better because they have a lower ratio of standard deviation to length at the mean sentence length. We have modelled variance as proportional to sentence length, $V = s^2 l$. Thus the standard deviation is $\sigma(l) = s\sqrt{l}$. At the mean sentence length, m, the ratio of standard deviation to mean is thus $\sigma(m)/m = s\sqrt{m}/m = s/\sqrt{m}$. Using the true alignments, we find for character lengths $s^2 = 6.5$, and for word lengths $s^2 = 1.9$, while the mean character length is 117 and the mean word length is 17. Thus the $\sigma(m)/m$ ratios are 0.22 for characters and 0.33 for words. This means that the character lengths are less noisy than are the word lengths, and thus better for comparison.

We tried the algorithm ignoring paragraph boundaries. The English–French errors were increased from 36 to 84, and the English–German errors from 19 to 86. The overall errors were increased from 55 to 170. Thus a top-down approach reduces errors by a factor of 3. This suggests that alignments by clause or phrase within a sentence would be valuable, if parsers were available for each language.

Originally we did not allow for 2–2 alignments. Table 21.2 shows that the program was right on 10 out of 15 actual 2–2 alignments. This was achieved at the cost of introducing two spurious 2–2 alignments. Thus in 12 tries, the program was right 10 times and wrong twice. This is significantly different from random, since there is less than 0.02 chance of getting 10 or more heads out of 12 flips of a fair coin. Thus it is worthwhile including the 2–2 alignment possibility.

We showed that the best estimates of the model parameters, c and s^2, are somewhat different from the values $c = 1$ and $s^2 = 6.8$ that we used. When we used the exact estimates for each individual language pair, then we found exactly the same total number of errors. The models were not exactly the same, however, as there were four changes (two right and two wrong) for French and two changes (one right, one wrong) for German. Since the parameters we use are somewhat less dependent on a particular language, and since their use makes little difference for the two language pairs we have studied, their use seems desirable.

EXTENSIONS

We rejected one document in which one paragraph was omitted and two paragraphs were duplicated. A more powerful paragraph alignment could be built that might handle this case. The distance measure for paragraphs would be the average distance between their sentences under the best sentence alignment. We expect this measure would discriminate strongly between paragraphs that should be aligned and others, so that a paragraph aligning very poorly with any available paragraph could be called **inserted**. This measure essentially pushes the paragraph measure down a level to sentences.

For sentence alignment, it would undoubtedly be useful to use a probabilistic dictionary to augment the probability based on lengths. That is, we could push the sentence match down to the word level. This would be possible because our distance measure is a probability; it is thus possible to combine other sources of information with it in a principled fashion. We expect that this would give as good a sentence matching job as could be done by hand. However, it is not clear if it is necessary, since alignments with less than 1% errors are likely to suffice for many purposes, such as building a probabilistic dictionary.

CONCLUSIONS

This chapter has proposed a method for aligning sentences in a bilingual corpus. We want to emphasize that the method is based on a probabilistic model, which was described in a previous section. The model is based on the observation that the lengths of aligned paragraphs have a 0.991 correlation. This suggests that aligned sentences should also have nearly the same lengths. The model quantifies this observation.

The method is also accurate. Overall, there was a 4.2% error rate on 1316 alignments. The alignments included both English–French and English–German corpora. Since the probability score assigned to each alignment is predictive of errors, it is also possible to select out 80% of the alignments that have an overall error rate of 0.7%.

The method is not strongly dependent on the language pair. Both English–French and English–German corpora were processed using the same parameters. The model does have six parameters that could vary with the language pair. They each have operational interpretations, and can be estimated for a particular language pair if necessary.

It is better to count the sentence lengths in characters than in words, because there is less variability in the ratios of sentence lengths so measured. Using words as units increases the error rate by half, from 4.2% to 6.5%. The algorithm works by aligning paragraphs first. Omitting this step increases errors by a factor of 3.

Using a probability-based model allows extensions that include other sources of information. After the preliminary alignments are made, a probabilistic dictionary can be built. The information from this dictionary could be fed back to the sentence alignment task to improve it. Also, if a more powerful paragraph alignment is needed, the total scores for sentence alignment within a pair of paragraphs would be a useful distance measure for the paragraphs.

ACKNOWLEDGEMENTS

We thank Susanne Wolff and Evelyne Tzoukermann for their pains in aligning sentences. Susan Warwick provided us with the UBS trilingual corpus and posed the problem addressed here. Code from Michael D. Riley was used for dynamic programming.

NOTE

1. Statistical methods were popular in the 1950s when machine translation·was first unsuccessfully tried, but they are being revived with the availability of extensive machine-readable bilingual corpora.

REFERENCES

Brown, P., Cocke, J., Della Pietra, S., Della Pietra, V., Jelinek, F., Lafferty, J., Mercer, R. and Roossin, P. (1990) A statistical approach to machine translation. *Computational Linguistics*, **16**, 79–85.

Church, K. (1988) A stochastic parts program and noun phrase parser for unrestricted text. Second Conference on Applied Natural Language Processing, Austin, TX.

Church, K. and Gale, W. (1991) A comparison of the enhanced Good–Turing and deleted estimation methods for estimating probabilities of English bigrams. *Computer Speech and Language*, **5**(1).

Church, K., Gale, W., Hanks, P. and Hindle, D. (1991) Using statistics in lexical analysis. In *Lexical Acquisition: Using On-line Resources to Build a Lexicon*, Zernik, U. (ed.), Hillsdale, NJ: Lawrence Erlbaum.

Gale, W.A. and Church, K.W. (1990) Estimation procedures for language context: poor estimates are worse than none. *Proceedings in Computational Statistics*, Heidelberg: Physica-Verlag, 69–74.

Klavans, J. and Tzoukermann, E. (1990) The BICORD system, *COLING-90*, 174–179.

Kruskal, J. (1983) *Time Warps, String Edits, and Macromolecules: The Theory and Practice of Sequence Comparison*, Reading, MA: Addison-Wesley.

Kruskal, J. and Liberman, M. (1983) The symmetric time-warping problem: from continuous to discrete, in *Time Warps, String Edits and Micromolecules,* Kruskal, J., Reading, MA: Addison-Wesley.

Yule, G.U. (1944) *Statistical Studies of Literary Vocabulary*, Cambridge: Cambridge University Press.

Zipf, G.K. (1932) *Selected Studies of the Principle of Relative Frequency in Language*. Cambridge, MA: Harvard University Press.

Probabilistic text understanding

22

R.P. Goldman and E. Charniak

INTRODUCTION

For some time we have been interested in the problems posed by uncertainty in text understanding, in particular the problem of representing and reasoning about altern- ative interpretations of natural language texts. We are concerned with problems like pronoun reference, prepositional phrase attachment, and lexical ambiguity. We are particularly interested in plan recognition as it is needed in text understanding: understanding the meanings of stories by understanding the way the actions of characters in the story serve purposes in their plans. Our work builds upon earlier work in script- and plan-based understanding of stories like that of Cullingford (1978), Wilensky (1983), Wong (1981), Charniak (1986) and Norvig (1987).

There are three features of our approach which distinguish it from other approaches to natural language processing. First, we treat the problem of natural language under- standing as an abduction problem. Second, we use probability theory as our theoretical framework. Finally, we tightly integrate all aspects of natural language processing, syntax, semantics and pragmatics (largely plan recognition). Our research is also set apart from earlier efforts by the use of a single-blind testing methodology.

We see text understanding as a particular case of the problem of abduction (reasoning from effects to causes), or diagnosis. This approach to natural language understanding is outlined in Charniak and McDermott (1985) and Hobbs *et al.* (1988). In particular, for the case of simple, declarative text, we simplify by viewing the language user as a transducer. The language user observes some thing (event or object) in the 'real world', and translates this thing into language. Our task is to reason from the language to the intentions of the language user and thence to the events described.

We have adopted probability theory as our framework because it is the most thoroughly understood way of reasoning about uncertainty. Unlike other ways of treating uncertainty (e.g., default logic), probability provides a unified 'currency' in which we can weigh evidence from different sources. Furthermore, it is at least in some

Artificial Intelligence Frontiers in Statistics: AI and statistics III. Edited by D.J. Hand. Published in 1993 by Chapman & Hall, London. ISBN 0 412 40710 8

sense normative (see, for example, the proof given in Cox, 1946). Finally, techniques for graphically representing probability distributions have made the task of drawing up and reasoning with probabilistic models much less onerous than they were once thought to be.

The third distinguishing feature of our approach is the extent to which it integrates all levels of processing. Previous work in text understanding has almost universally assumed that the text will be preprocessed by an intelligent parser (e.g., Cullingford (1978), Hobbs *et al.* (1988), Norvig (1987), Wilensky (1983)). We, on the other hand, read the text word-by-word through a parser that simply returns all possible parses of the input. Because the structure of the sentence is described in the same framework as its semantics and pragmatics, constraints from different levels can be assembled into a global interpretation. For example, in deciding the proper attachment of the prepositional phrase 'with some poison' in 'Janet killed the boy with some poison' we can make use of our knowledge about murder, or about a particular poison-wielding boy that Janet dislikes.

We have developed a 'single-blind' technique for testing our story–understanding program, WIMP3. One of us has developed a set of pairs of synonymous stories whose syntax and vocabulary were restricted so that they could be handled by WIMP3's parser. These pairs were randomly assigned to either test or training set. The other author developed the program until it was able to handle the training set, and then used it to align the training set stories with their counterparts in the test set (which were kept hidden from him). We believe that this testing methodology allows us to eliminate some of the problems of 'wishful programming', while still being within the capabilities of small research groups.

Our previous work in this area was codified in the program WIMP2, described in Charniak and Goldman (1988). WIMP2 used a multiple-context deductive database, based on the Assumption-based Truth Maintenance System (ATMS) presented in deKleer (1986). The contexts of WIMP2's database were tagged with probabilities.[1] Having a multiple-context database to cache inferences made it possible to consider simultaneously different interpretations, such as alternate meanings of words, or different parses, and their consequences. The probabilities made it possible to allocate WIMP2's computational resources to considering the most likely candidate explanations.

We were unhappy with WIMP2 for three reasons. First of all, it was difficult to reconcile the logical approach which motivated the use of the ATMS with the use of probability theory. This problem manifested itself particularly acutely in difficulties we had assigning a clear semantics to the probabilities we used. This difficulty arises because of the difference between the logical operation of material implication and the probabilistic operation of conditioning. Nilsson (1986) discusses this issue. On a more concrete level, the system was making every inference twice over: the system reasoned deductively using the ATMS, and in an *ad hoc* probabilistic way when managing the probabilities of the ATMS's contexts. Finally, the deductive framework is unable to distinguish evidential and causal support for propositions. When a new justification is added to a proposition, one cannot tell whether it represents evidence for that proposition, in which case it would confirm hypotheses which explain the proposition; or whether it represents an alternative explanation, in which case it should compete with other explanations. That is to say, when writing rules of the form $A \rightarrow B$, there is no way to distinguish between the interpretations 'A is evidence for possible explanation B' and 'A can cause B to be true'. For these reasons, we decided to try a wholly probabilistic approach to the problem.

Our current program, WIMP3, operates by turning problems of text interpretation into probability questions of a form such as 'What is the probability that this word has this sense, given the evidence?' (where the evidence is the text of the story so far). WIMP3 does this on a word-by-word basis. The probability questions are formulated in a language we have developed for this purpose. This formulation is given additional structure by a graphical representation, belief networks. In the next section of this chapter, we will outline the language we have developed to express linguistic problems in terms of random variables. Following that, we will show how these sets of random variables can be structured as belief networks, and how this structure helps us assess the quantities we need for our probabilistic models. We will then show how WIMP3 actually constructs and solves these problems. We conclude with a discussion of the single-blind testing of WIMP3, and brief summary. Sample test data for the single-blind tests are given in an appendix.

THE FORMALISM

In adopting a probabilistic approach, our first problem was to come up with a clear understanding of what probabilities we were to manipulate. We have created a simplified formal language suited to expressing facts about a text's structure and meaning. We have devised a semantics for this language which allows us to treat its formulae as random variables. We show that this semantics gives us the guidance we need to devise a consistent set of probabilities for use by a story-understanding program. A slightly earlier version of this language is described in Charniak and Goldman (1989; 1990); we will only have space for a brief discussion here.

We will start by describing the notation of our language and showing how it is used to formulate the text-understanding problem. For convenience sake, we will proceed in a way which follows the traditional syntax–semantics–pragmatics distinction. First, we will show how the language can be used to describe the input text itself. Then, we will go on to discuss how it can express semantic propositions. Finally, we will show the part of the language used for plan recognition, and how this provides contextual information.

We use four kinds of proposition in formulating text-understanding problems:

(1) (word-inst x *root*) The token x is an instance of the root form *root*. We use the term 'token' to emphasize the distinction between a *word* and a particular *use* of that word in a text.

(2) (syn-rel *type* x y) The tokens x and y are in a syntactic relation of *type*.

(3) (inst x *type*) The thing x is of *type*. x is some thing in the real world, typically one

named by a token in the text. We use the word 'thing' here because x may be an object, an event, or a state. We will sometimes use the word 'entity' for the same purpose.

(4) ($= = x\,y$) The objects x and y are the same. We use the symbol $= =$ rather than $=$ because, for the sake of efficiency, we use an antisymmetric 'better-name' relation, rather than true equality. This distinction need not trouble the reader.

This notation is similar to the first-order predicate calculus, but all terms are ground, i.e. no quantifiers are used. These propositions (we will also refer to them as 'statements') will be treated as random variables over the sample space {*true, false*}. In the following paragraphs we will give a more concrete description of the use of this formalism.

The parser and morphological analyser component of WIMP3 generate the statements that describe the natural language input. These statements are of two types:

- specifying the words used (word-inst), and
- specifying relations between words used (syn-rel).

For example, the sentence

Jack gave Mary the ball.   (22.1)

would be translated into:

(word-inst word1 Jack)
(word-inst word2 give)
(syn-rel subject word1 word2)
(word-inst word3 Mary)
(syn-rel indirect-object word3 word2)
(word-inst word4 ball)
(syn-rel object word4 word2)
(syn-rel det word4 the)

The morphological analyser translates inflected forms like 'gave' into root forms. We simplify by ignoring issues of tense and time. In the case of syntactic ambiguity, the parser will generate and describe all possible parses of the input; semantic and pragmatic information will be used to choose between them. If the sentence were

Mary killed the boy with the poison.   (22.2)

The parser would give both (syn-rel with poison12 boy11) and (syn-rel with poison12 kill10). From now on, for ease of understanding, we will use more descriptive names, like poison12 and boy11, rather than word1, word2, etc.

For semantic processing, we need to be able to express hypotheses about the denotations of the various words. Associated with each word instance is a constant representing the denotation of that word. We will write this as the same constant, but in bold face. We might more clearly write (denotation word n), but this is too bulky. For example, if **ball27** was the constant representing the denotation of 'ball' in Sentence

22.1, two possible denotations might be expressed as:

(inst **ball27** bouncing-ball)
(inst **ball27** cotillion)

In order to express case relations, we have functions for the various cases. To return to Sentence 22.2 the case relations corresponding to the different prepositional phrase attachments would be:

(= =(accompaniment **boy11**) **poison12**)
(= =(instrument **kill10**) **poison12**)

Moving into the grey area between semantics and pragmatics that Hobbs and Martin (1987) call 'local pragmatics', equality statements are used to express coreference. For example, part of the task of understanding

Jack gave Mary the ball. She liked it. (22.3)

is to reason from

(inst **mary24** girl-)
(inst **she28** girl-)

to (= =**she28 mary24**).

Finally, we need to be able to express a plan-based understanding of our texts, such that when reading a story like

Jack went to the liquor-store. He paid for some bourbon. (22.4)

our program can understand the entire story as a description of a plan to buy liquor.

Making such inferences requires generic information about plans and their parts. To do so, we have a frame-based database of events and objects. Events and objects are represented by instances of frames, which are organized into an *isa hierarchy*. In such a system, frames representing different types are arranged in a taxonomy according to the *isa* relation. For example, mammal isa animal, elephant isa mammal, etc. A particular elephant is an instance of the frame elephant. A frame which isa another frame inherits the features of its supertype. So elephants are warm-blooded, bear live young, have mammary glands, etc.

Components, or *slots*, of frames are represented by functions. For example, the trunk of an elephant called Clyde, might be represented by the function (trunk-of Clyde). The actions which make up a particular plan are slot-fillers of that plan instance. For example, in order to shop at a store, one must first go to the store. If we have a particular instance of shopping, say shop1, we would refer to the corresponding event of going to the store as (go-step shop1). Finally, we need to be able to represent constraints between the different slots of a plan. For example, we know that when one goes to a store as part of a plan to shop, one goes to the particular store at which one intends to buy something. The destination of the go-step of the shopping is the same as the store-of the shopping. In our notation: (= =(destination (go-step shop1)) (store-of shop1)).

Here is an example of the kind of generic knowledge we have given WIMP3:

(inst ?shop shopping-) → (and (inst (go-stp ?shop) go-)
                   (= = (agent (go-stp ?shop)) (agent ?shop))
                   (= = (destination (go-stp ?shop))
                    (store-of ?shop)))

Currently, WIMP3's database contains 148 frames, which are translated into 513 constraints in its internal database format. The internal format is resembles the example above, but 'syntactic sugar' is removed, and typed variables are used.

For plan recognition in this framework, we need another kind of proposition: the proposition that a given frame-instance (say, **went2**) fills a slot in a given kind of plan (say, robbery). This would be written

(and (inst *robbery27* robbery)
    (= = (go-step *robbery* 27) **went2**))   (22.5)

The atom *robbery27* is written in *italic* to indicate that it represents a different kind of constant. *robbery27* is a function of **went2**. Analogously to the denotation constants, these constants might more clearly be written (robbery-explanation-of **went2**). A detailed discussion of plan recognition in WIMP3 is given in Charniak and Goldman (1991).

These propositions are virtually identical to those in our earlier logic for semantic interpretation (presented in Charniak and Goldman, 1988), but their semantics is different. Following probability theory, the expressions of our language get their semantics from a sample space. This sample space is the set of all possible stories from some restricted universe. The word constants denote random variables whose domain is particular words. Likewise, the denotations of words are random variables whose domain is the set of entities (including events). Propositions are random variables whose domain is the set {*true, false*}. A more extensive discussion of the semantics of our language can be found in Charniak and Goldman (1989).

NETWORK REPRESENTATIONS

In the past, it has been difficult to apply probability theory to reasoning problems because such problems did not have sufficient structure. It appeared that in order to apply probability theory, it would be necessary to have a gigantic joint probability table, giving the probability of all possible combinations of the propositions of interest. Recently, however, there has been renewed interest in graphical representations for probability distributions, such as Markov fields, belief networks and influence diagrams. Belief networks have provided us with a way of structuring the problem of text understanding as a problem of probabilistic inference, and have simplified the problem of acquiring the requisite numbers.

Belief networks are a way of representing probabilistic dependency information in directed acyclic graphs. Such graphical representations have been used in various fields for some time (Lauritzen and Spiegelhalter, 1988, cite uses in path analysis as early as (1921). These graphs make possible qualitative reasoning about influence, and, for distributions with desirable properties of conditional independence, make it possible to calculate posterior probabilities efficiently. Pearl (1988) gives a thorough account of the properties of such networks.

The nodes in a belief network represent random variables, and the edges represent direct dependence. We follow Pearl's suggestion that edge direction be assigned causally: nodes at the tails of edges should be (direct) causes of the nodes at their heads. There are three advantages to belief networks as representations for probability distributions. First, properties of conditional independence can be read off a belief network. Second, the probability distribution corresponding to a belief network may be represented locally. For each node, it suffices to provide a conditional probability distribution for each combination of values of its parent nodes. Finally, while in general the problem of determining the posterior distribution of a partially instantiated belief network is NP-hard (proof given in Cooper, 1990), considerable attention has been devoted to finding efficient approaches to evaluating such networks.

We give a partial belief network representation of the analysis of the sentence

Jack went to the liquor-store (22.6)

in Fig. 22.1. This fraction of the network is intended to capture the following relations: A liquor-shopping plan can cause there to be a go action. It will also cause there to be a liquor-store, the store-of the liquor-shopping. The liquor-store mentioned in the story might be this liquor-store. The store-of the liquor-shopping will be the destination of the go-step of the liquor-shopping. These relations might be realized in the text sequence we have observed.

These are two things to be noted about this diagram. First of all, the 'explanatory' variables (e.g., liquor-shop4) are tied to the actions of characters, rather than to all entities mentioned. The liquor-store simply supports the liquor-shop hypothesis, and that only in the presence of evidence for its being the destination of Jack's going. This avoids having objects call into consideration every plan in which they could play a role.

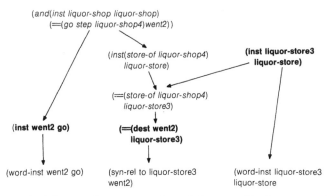

Figure 22.1 A belief network representation of the story 'Jack went to the liquor-store...'

For example, we do not consider a shopping explanation when we read 'Jack bombed the supermarket.'

Note also that this is an immensely simplified version of the networks which are, in fact, used in our program. First of all, we have suppressed many of the nodes that would appear in the network for this interpretation (e.g., the nodes specifying Jack as the agent of the go). Second, the networks we construct are not restricted to contain only nodes which are part of *correct* interpretations of the text!

QUANTIFYING THE NETWORKS

The belief network representation makes it easier for us to specify the probability distributions we need. It permits us to express the distribution as a collection of conditional probabilities involving only a few nodes (no more than three in our current program). This representation is compact, and these low-order probabilities are easier to assess subjectively. While in principle we could collect statistics to find these distributions, this is not practical. In this section we will give a brief discussion of the kinds of conditional probability we need. For the sake of clarity, we will proceed from the leaves of the belief networks to their roots, in parallel with the syntax–semantics–pragmatics distinction and our treatment in an earlier section.

The leaves of the network are the word-inst and syn-rel nodes. For this portion of the graph, the belief network formalism requires us to provide $P(\text{denotation})[P(d)]$ and $P(\text{word}|\text{denotation})[P(w|d)]$. In principle, these could be computed from a labelled corpus. In practice, this is not necessary. Because the denotation variable is a function of the word variable, its probability is zero in the absence of the word. Accordingly, we are only concerned with $P(d|w)$. We can specify these by ranking the various different denotation hypotheses relative to each other. Precise values of $P(d)$ and $P(w|d)$ are not important: it is enough to know the values of the products $P(d)P(w|d)[= P(w, d)]$ *relative to other possible denotations for the same word*. The exact values of the products are not necessary since they will be normalized. This also enables us to do without $P(w)$.

We do not have a good theory of the conditional probabilities for the syn-rel nodes, which are of the form $P(\text{syn-rel}|\text{relation in the real world})$. For example, $P((\text{syn-rel to store2 go1})|(= =(\text{destination go1}) \text{store2}))$. We use a rough subjective estimate of how often a given construction will be used to convey some relation. For example, the instrument relation is more often expressed by 'with' phrases than is the accompaniment relation (which is often expressed by conjunction).

Deeper in the graph, we need probabilities for propositions which represent explanatory hypotheses. For example, the hypothesis that Jack has a shopping plan which explains his going to the liquor-store. This is the node labelled (and (inst liquor-shop4 liquor-shop) ($= =$ (go-step liquor-shop4) went2)) in Fig. 22.1.

The task of quantifying this part of the diagrams may seem wellnigh impossible, because it appears to require that we specify a prior probability for this node. In fact, however, this is not the case. This part of the network is subject to the same analysis as that of the words and denotations. Again, this is because the explanatory nodes have meaning only in the presence of their explanands. So we can specify this part of our network by ranking the different explanations for each action relative to each other.

In our program, we have tried to tie these probabilities to the frequencies of events in the real world. We want to do this so that, for example, when our program reads a sentence like 'Mary went to the airport' it will favour explanations like 'Mary is going to meet someone on the plane', or 'Mary will fly somewhere', over 'Mary is going to hijack the plane.' This is justified because we are assuming the speaker is a transducer. For example, in 'Tom Clancy world' the probabilities would be radically different.[2] In theory, one could collect statistics for these networks by labelling a large corpus of stories, and counting nodes in the instantiated networks.

Finally, we need probabilities of the form $P((==\text{thing1 thing2})|(\text{inst thing1 type1})$, $(\text{inst thing2 type1}))$. These are needed for coreference and for linking together explanations for different objects (we must recognize that the shopping plan that explains the presence of a liquor-store in a story is the same shopping plan that explains the presence of a bottle of bourbon). These are difficult to assess. We do not have time to discuss the problem fully here; we do so in an earlier paper (Charniak and Goldman, 1990). In brief, these probabilities must be sensitive to the number of different objects of each type which are likely to appear in a given story. For example, in simple stories like those WIMP3 reads,

$$P((==\text{person1 person2})|(\text{inst person1 person}), (\text{inst person2 person}))$$
$$< P((==\text{airport1 airport2})|(\text{inst airport1 airport}), (\text{inst airport2 airport}))$$

because it is very likely that there will be more than one person discussed in a single story, but unlikely to be more than one airport. We hope that discourse theory (e.g., Grosz and Sidner, 1986; Webber, 1987) will shed some light on these quantities, and that notions like focus can be used as conditioning events.

THE PROGRAM

In the preceding, we have developed our probabilistic model of stories. Now we will show how we use this model to do story understanding. Because the network models we have outlined here are impractically large, we proceed by incrementally constructing and evaluating relevant portions of the full network. These partial networks are constructed by network-building rules which are similar to forward-chaining rules. We have experimented with evaluating them using a number of different belief network algorithms.

WIMP3 works as follows: an all-paths parser reads the English-language input, one word at a time, and yields statements about the input. These statements are given to the network-construction component. This component extends the belief network corresponding to the input. Then the network evaluation component computes the posterior distribution of the network. If any hypotheses seem particularly good or bad, they may be accepted or rejected categorically. After this, if necessary, the network may be further extended, and re-evaluated. At this point, control is returned to the parser,

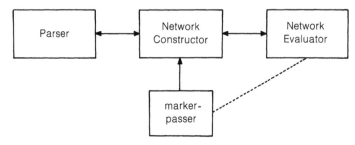

Figure 22.2 The architecture of WIMP3.

which reads the next word. A picture of the architecture of this system is given in Fig. 22.2.

We have developed network-building rules for constructing belief networks which correspond to story-understanding problems. These rules operate on WIMP's deductive database. Statements in this database are also random variables in a belief network. The network construction rules are similar to conventional forward-chaining or production rules. They are augmented in two ways.

1. It is possible to add hyperedges between statements mentioned in the rules. These hyperedges are similar to the justifications in a data-dependency system (for an introduction to truth maintenance systems and data dependencies, see Charniak and McDermott, 1985, Chapter 7). They differ in that rules are not restricted to adding a single hyperedge from the antecedent nodes to the consequent. An arbitrary number of edges may be added, and the edges may point in any direction (as long as the resulting network is still acyclic).
2. Network-building rules contain information used to set the conditional probability matrices of the nodes in the network.

A sample network-building rule is given in Fig. 22.3. This rule handles part of the problem of finding an appropriate referent for a word (the rules for plan recognition are similar, but more complex). The meaning of the rule is:

> If a node is added to the network describing a new word-token (line 1 of the rule), and if there is a statement in the database specifying that one of the senses of this word is ?frame (line2), add a node for an instance of the type ?frame (line 3). Draw an arc between this newly-added node (?C), and the word-inst node (?A) (line 4). Construct the conditional probability matrix for the word-inst node using information from the word-sense statement (?PROB) (line 4).

$$(\rightarrow (\text{word-inst } ?i \text{ } ?word) \text{ :label } ?A$$
$$(\rightarrow \leftarrow (\text{word-sense } ?word \text{ } ?frame \text{ } ?PROB)$$
$$((\text{inst } ?i \text{ } ?frame) \text{ :label } ?C)$$
$$:\text{prob}((?C \Rightarrow ?A) ((t|t = ?PROB) (t|f = prior))))$$

Figure 22.3 One of the rules used to handle word-sense relationships.

For example, if we were to learn of a use of the word 'bank' (say, bank1) and knew of two meanings: financial institution and river bank, this rule would tell us to add two nodes to our network: a node specifying that bank1 referred to a financial institution, and a node specifying that bank1 referred to a river bank. These two nodes would have arcs pointing to the node for (word-inst bank1 bank). The conditional probability matrix for the latter would be drawn up to reflect that the two possible senses of the word are related as exclusive causes. That is to say, the probability that the word will be used given that one wishes to describe a financial institution or a river bank is some value determined by the word-sense rule; the probability that the word is used given one wishes to express both senses simultaneously is zero, and the probability that the word will be used given neither of these senses is meant is some default value.

We have similar rules for syntactic relations, reference resolution, and so on. Initially we used this kind of 'forward-chaining' for plan recognition as well. However, this led to an explosive growth in the belief networks. The reason for this is simple: consider how many possible reasons there are for 'going', for example. Worse yet, many of these plans which could explain going can themselves play roles in more complex plans. To provide more direction to the way the networks are expanded, another researcher at Brown University, Glenn Carroll, has developed a marker-passing search algorithm. For a discussion of the marker-passer, see Carroll and Charniak (1989). This marker-passer makes use of the probabilities in the network, hence the dashed line connecting these two components in Fig. 22.2. See Goldman and Charniak (1990) for more details of the network-building rules.

The posterior distribution for the belief network constitutes the interpretation of the input text. To see why this is the case, refer to the sample belief network given as Fig. 22.1. What we want to assess is the probability that Jack is going to buy liquor, given that we have been told 'Jack went to the liquor-store.' We can read this information off the network, once we have evaluated it so that it reflects the posterior distribution—the probability of each node given the evidence in the story. It should be noted that our representation does not tie us to the idea of solution by computing posterior distributions: we could compute a maximum a posteriori instantiation of the networks to get a best global interpretation. Indeed, Shimony and Charniak (1990) have done some preliminary work in this direction. We do not do this because we want intermediate results, and because these algorithms do not prövide better performance in practice.

We have been experimenting with a number of different algorithms for computing the posterior distribution of these networks. To do so, we have been using the IDEAL system (described in Srinivas and Breese, 1989), an environment in which one can define a belief net or influence diagram and apply to it a number of different evaluation algorithms. We find that a variant of the algorithm of Lauritzen and Spiegelhalter (1988), developed by Jensen (1989), gives the best results on networks like ours. Performance of this algorithm is improved by a clustering heuristic due to Kjaerulff (1990). Our cutsets grow too large for conditioning (this algorithm is given in Pearl, 1988) and, until now, we have not had a simulation-based approach that worked well on networks which contain extreme probabilities (Chin and Cooper, 1987, discuss this problem). However, there has been continuous progress in belief network evaluation algorithms, and we are still experimenting with new algorithms.

EXPERIMENTS WITH WIMP3

WIMP3 has been fully implemented. It is written in Common Lisp, and runs on Symbolics Lisp machines and Sun SPARCstations. We have developed a small test corpus to test its abilities, focusing on problems which have been difficult for earlier approaches. It can parse and 'understand' 25 stories which describe everyday occurrences. It correctly parses and classifies 18 of 25 'test set' stories synonymous to these 25 stories, and 24 of 25 after small-scale augmentation of its knowledge base and lexicon.

The test and training set story pairs were developed by the second author, partitioned randomly, and kept hidden from the first author during his development and debugging of WIMP3. The easiest story is

Bill drank a milkshake with a straw.

To understand this story it is necessary to determine from context that the word 'straw' refers to a 'drink-straw' rather than 'hay-straw', and that the straw is the instrument, rather than the co-agent of the drinking. One of the most difficult stories is

Bill took a bus to a restaurant. He drank a milkshake. He pointed a gun at the owner. He got some money from him.

which presents intertwined problems of plan recognition and pronoun reference. In particular, it is necessary to recognize Bill's plan correctly in order to realize that 'he' in the last sentence refers to Bill, and 'him' to the owner.

Appendix 22A contains the 25 training set stories, and Appendix 22B contains a sample query, used to judge WIMP3's 'comprehension' of one of these stories. Processing time for the stories in this corpus ranges from c.15 second to c.10 minutes, on a Sun SPARCstation 1 with 135M swap, 16M memory, running Sun Common Lisp, Version 4.0, code compiled but not optimized.

We developed a simple testing method to avoid some problems of wishful interpretation of our experimental results. It has been customary to demonstrate natural language understanding programs by giving a selection of texts which the program is claimed to be able to interpret. Such claims are very difficult to evaluate, and leave open the possibility that the demonstrated programs are brittle and cannot handle even slight variations from the example texts.

Our solution is a single-blind test. During the development of the WIMP3 system, the second author developed a set of 25 pairs of synonymous stories whose syntax and vocabulary were within the capabilities of WIMP3's parser and lexicon. One of each pair was randomly assigned to the test set, and the other to the training set. The first author then worked on WIMP3 until it was capable of understanding the 25 stories of the training set.

To verify that WIMP3 was able to process stories other than the ones in the training set, we developed an alignment task. The stories in the test and training sets were to be synonymous with respect to a categorizing function. This categorizing function takes the internal representation derived in the course of WIMP3's processing of the story, and yields the following information: a count of the number of top-level plans in the story and, for each top-level plan, the most specific name for that plan, the agent of that plan, and the 'generalized patient' of that plan. For example, the simple story 'Bill drank a milkshake with a straw' is categorized as: (1 (STRAW-DRINK BILL- MILK-SHAKE))—there is one top-level plan, a straw-drinking, performed by a boy named Bill, and a milkshake was the patient of this drinking. The categories of each of the training set stories are given as Appendix 22C.

As mentioned above, WIMP3 was initially able to align correctly 18 of the 25 test set stories. With some minor alterations to its knowledge base and lexicon, it was able to correctly align 24 of 25. None of the program code needed to be changed to get this improvement, rather WIMP3's database needed to be augmented so that it could handle some unexpected case relations and some additional words.

While the 'single-blind' test briefly discussed here may not seem remarkable, it is, to the best of our knowledge, without precedent in story-understanding programs. No such test appears in, for example, Norvig (1987), Cullingford (1978), Wilensky (1983), Wong (1981). For more on the difficulties of evaluating natural language systems, see Palmer and Finin (1990). The experiments are further discussed in the first author's thesis (Goldman, 1990).

SUMMARY

The contributions of this work are threefold. First of all, we have constructed a probabilistic model of story comprehension. This makes it possible to address the problem of text understanding within axiomatic probability theory. Second, we have developed a way of constructing and evaluating probabilistic models on an as-needed basis. This makes it possible to apply probability theory to problems for which a precompiled model is either not available, or impractically large. Finally, we have introduced a single-blind testing methodology for text understanding programs. While weaker than standard testing methodologies, it is suitable for this still very experimental field, and for programs developed by small research groups.

ACKNOWLEDGEMENTS

This work has been supported in part by the National Science Foundation under grants IST 8416034 and IST 8515005 and the Office of Naval Research under grant N00014-79-C-0529. The majority of the work described was done when both authors were at Brown University. The first author was supported by a summer fellowship from the Tulane University Senate Committee on Research while revising this chapter.

NOTES

1. At the same time as we developed this database architecture, Laskey and Lehner (1988) and D'Ambrosio (1988a; 1988b) reported probabilistic ATMSs that were very similar.
2. We are grateful to Jack Breese for pointing out to us the existence of this alternate sample space.

REFERENCES

Carroll, G. and Charniak, E. (1989) Finding plans with a marker-passer, in *Proceedings of the Plan-recognition Workshop*, pp. 1074–1079.
Charniak, E. (1986) A neat theory of marker passing, in *Proceedings of the Fifth National Conference on Artificial Intelligence*, pp. 584–589.
Charniak, E. and Goldman, R.P. (1988) A logic for semantic interpretation, in *Proceedings of the Annual Meeting of the Association for Computational Linguistics*, pp. 87–94.
Charniak, E. and Goldman, R.P. (1989) A semantics for probabilistic quantifier-free first-order languages, with particular application to story understanding, in *Proceedings of the 11th International Joint Conference on Artificial Intelligence*.
Charniak, E. and Goldman, R.P. (1990) Plan recognition in stories and in life, in *Uncertainty in Artificial Intelligence 5*, Henrion, M., Schachter, R.D., Kanal, L.N. and Lemmer, J.F. (eds), Amsterdam: North-Holland.
Charniak, E. and Goldman, R.P. (1991) Probabilistic abduction for plan-recognition, in *Proceedings of the 9th National Conference on Artificial Intelligence*, pp. 160–165.
Charniak, E. and McDermott, D. (1985) *Introduction to Artificial Intelligence*, Reading, MA: Addison-Wesley.
Chin, H.L. and Cooper, G.F. (1987) Stochastic simulation of Bayesian belief networks, in *Proceedings of the Workshop on Uncertainty and Probability in Artificial Intelligence*, pp. 106–113.
Cooper, G.F. (1990) The computational complexity of probabilistic inference using Bayesian belief networks. *Artificial Intelligence*, **42**, 393–405.
Cox, R.T. (1946) Probability, frequency and reasonable expectation. *American Journal of Physics*, **14**, 1–13.
Cullingford, R.E. (1978) Script application: computer understanding of newspaper stories. Technical Report, Department of Computer Science, Yale University, New Haven, CT.
D'Ambrosio, B. (1988a) A hybrid approach to reasoning under uncertainty. *International Journal of Approximate Reasoning*, **2**, 29–45.
D'Ambrosio, B. (1988b) Process, structure and modularity in reasoning with uncertainty, in *The Fourth Workshop on Uncertainty in Artificial Intelligence*, pp. 64–72.
deKleer, J. (1986) An assumption-based TMS. *Artificial Intelligence*, **28**, 127–162.
Goldman, R.P. (1990) A probabilistic approach to language understanding. Technical Report CS-90-34, Computer Science Department, Brown University.
Goldman, R.P. and Charniak, E. (1990) Dynamic construction of belief networks, in *Proceedings of the Sixth Conference on Uncertainty in Artificial Intelligence*, pp. 90–97.
Grosz, B.J. and Sidner, C. (1986) Attention, intention and the structure of discourse. *Computational Linguistics*, **12**.
Hobbs, J.R. and Martin, P. (1987) Local pragmatics, in *Proceedings of the 10th International Joint Conference on Artificial Intelligence*, pp. 520–523.
Hobbs, J.R., Stickel, M., Martin, P. and Edwards, D. (1988) Interpretation as abduction, in *Proceedings of the 26th Annual Meeting of the ACL*, pp. 95–103.
Jensen, F.V. (1989) Bayesian updating in recursive graphical models by local computations.

Technical Report R 89-15, Institute for Electronic Systems, Department of Mathematics and Computer Science, University of Aalborg, Aalborg, Denmark.

Kjaerulff, U. (1990) Triangulations of graphs—algorithms giving small total clique size. Technical Report R 90-09, Institute for Electronic Systems, Department of Mathematics and Computer Science, University of Aalborg, Aalborg, Denmark.

Laskey, K.B. and Lehner, P.E. (1988) Belief maintenance: an integrated approach to uncertainty management, in *Proceedings of the Seventh National Conference on Artificial Intelligence*, pp. 210–214.

Lauritzen, S.L. and Spiegelhalter, D.J. (1988) Local computations with probabilities on graphical structures and their application to expert systems. *Journal of the Royal Statistical Society, B,* **50**, 157–224.

Nilsson, N. (1986) Probabilistic logic. *Artificial Intelligence,* **28**, 71–88.

Norvig, P. (1987) Inference in text understanding, in *Proceedings of the Sixth National Conference on Artificial Intelligence*, pp. 561–565.

Palmer, M. and Finin, T. (1990) Workshop on the evaluation of natural language processing systems. *Computational Linguistics,* **16**, 175–181.

Pearl, J. (1988) *Probabilistic Reasoning in Intelligent Systems: Networks of Plausible Inference*, Los Altos, CA: Morgan Kaufmann.

Shimony, S. and Charniak, E. (1990) A new algorithm for finding map assignments to belief networks, in *Proceedings of the Sixth Conference on Uncertainty in Artificial Intelligence*, pp. 98–103.

Srinivas, S. and Breese, J. (1989) *IDEAL: Influence Diagram Evaluation and Analysis in Lisp*, Palo Alto, CA: Rockwell International Science Center.

Webber, B.L. (1987) The interpretation of tense in discourse, in *Proceedings of the 25th Annual Meeting of the ACL*, pp. 147–154.

Wilensky, R. (1983) *Planning and Understanding*, Reading, MA: Addison-Wesley.

Wong, D. (1981) Language comprehension in a problem solver, in *Proceedings of the 4th International Joint Conference on Artificial Intelligence*, pp. 7–12.

APPENDIX 22A. STORIES IN TRAINING SET

1. Jack went to the supermarket. He found some milk on the shelf. He paid for it.
2. Bill went to the supermarket. He paid for some milk.
3. Jack gave the busdriver a token. He got off at the supermarket.
4. Jack got off the bus at the liquor-store. He pointed a gun at the owner.
5. Jack went to the liquor-store. He found some bourbon on the shelf.
6. Bill went to the liquor-store. He pointed a gun at the owner.
7. Bill gave the busdriver a token.
8. Fred robbed the liquor-store. Fred pointed a gun at the owner.
9. Bill got a gun. He went to the supermarket.
10. Fred went to the supermarket. He pointed a gun at the owner. He packed his bag. He went to the airport.
11. Jack took the bus to the airport. He bought a ticket.
12. Bill packed a suitcase. He went to the airport.
13. Jack got on a bus. He got off at the park. Jack went to the supermarket.
14. Jack gave the busdriver a token. He got off at the park. He went to the airport. He got on a plane.

15. Fred sat down on the bus. He went to the supermarket.
16. Jack went to a restaurant. He got a milkshake.
17. Bill drank a milkshake with a straw.
18. Fred got off the bus at a restaurant. He got a milkshake.
19. Janet put a straw in a milkshake.
20. Bill got on a bus. He got off at a restaurant. He drank a milkshake with a straw.
21. Bill took a bus to a restaurant. He drank a milkshake. He pointed a gun at the owner. He got some money from him.
22. Fred gave the busdriver a token. He got off the bus at the park. He went to a restaurant. He got some money from the owner.
23. Jack took a taxi to the park.
24. Bill took a taxi.
25. Fred took a taxi to the bus-station. He got on a bus.

APPENDIX 22B. UNDERSTANDING STORIES

In order to decide when WIMP3 could be held to 'understand' the stories in the training set, we developed a set of database queries. WIMP3 was held to understand a story when it could successfully respond to the corresponding query. The query corresponding to the story

Jack went to the supermarket. He found some milk on the shelf. He paid for it.

is

```
(and (inst ?jack jack-)
     (inst ?sming superming)
     (inst ?went go-)
     (inst ?find locate-)
     (inst ?pay pay-)
     (inst ?milk milk-)
     ( = = '(agent ?sming) ?jack)
     ( = = '(go-stp ?sming) ?went)
     ( = = '(pay-stp ?sming) ?pay)
     ( = = '(bought ?sming) ?milk)
     ( = = '(loc-stp ?sming) ?find))
```

An approximate English gloss is as follows: There appear in the story a person named Jack, a supermarket-shopping plan (superming), a going event, a locating event, a paying event, and some milk. The actions mentioned in the story—the going, the locating and the paying, all play roles in the supermarket-shopping plan. Jack is the agent of the plan, and the milk is the object purchased in the plan.

Note that in order to make the last inference, WIMP3 must recognize that the pronoun 'it' refers to the milk, and not to the shelf, which is an equally good candidate at first glance. It makes this decision based on its knowledge about supermarket-shopping.

APPENDIX 22C. CATEGORIZING THE STORIES

1. (1 (SUPERMING JACK- MILK-))
2. (1 (SUPERMING BILL- MILK-))
3. (1 (SUPERMING JACK- FOOD-))
4. (1 (ROB- JACK- LIQUOR-STORE-))
5. (1 (LIQUOR-SHOP JACK- BOURBON-))
6. (1 (ROB- BILL- LIQUOR-STORE-))
7. (1 (BUS-TRIP BILL- BUS-))
8. (1 (ROB- FRED- LIQUOR-STORE-))
9. (1 (ROB- BILL- SUPER-M))
10. (1 (ROB- FRED- SUPER-M))
11. (1 (AIR-TRIP JACK- AIRPLANE-))
12. (1 (AIR-TRIP BILL- AIRPLANE-))
13. (2 (SUPERMING JACK- FOOD-) (PARK-TRIP JACK- NIL))
14. (2 (AIR-TRIP JACK- AIRPLANE-) (PARK-TRIP JACK- NIL))
15. (1 (SUPERMING FRED- FOOD-))
16. (1 (EAT-OUT JACK- MILK-SHAKE))
17. (1 (STRAW-DRINK BILL- MILK-SHAKE))
18. (1 (EAT-OUT FRED- MILK-SHAKE))
19. (1 (STRAW-DRINK JANET- MILK-SHAKE))
20. (1 (EAT-OUT BILL- MILK-SHAKE))
21. (2 (EAT-OUT BILL- MILK-SHAKE) (ROB- BILL- RESTAURANT-))
22. (2 (ROB- FRED- RESTAURANT-) (PARK-TRIP FRED- NIL))
23. (1 (PARK-TRIP JACK- NIL))
24. (1 (TAXI-TRIP BILL- TAXI-))
25. (1 (BUS-TRIP FRED- BUS-))

The application of machine learning techniques in subject classification

23

I. Kavanagh, C. Ward and J. Dunnion

INTRODUCTION

SIMPR (Structured Information Management: Processing and Retrieval) is a project in the ESPRIT II programme of the Commission of the European Community. The SIMPR system provides software support for the creation, management and querying of very large information bases on CD-ROM. The information stored will typically be technical manuals, libraries of technical reports or other **full-text** documents. A full-text document is one with no prerequisites on its content or format. Each of these documents is composed of a number of **texts**. Each text is processed in two stages. It is first indexed to extract words and phrases with a high meaning content. Then, the subject(s) of the text are identified and appropriate **classificators** are attributed to the text.

The classification work is supported by the Subject Classification Management System which creates and maintains the classification scheme. Our contribution to the project, and the work described in this chapter, is the design of an intelligent agent to help classify texts. It is expected that, as texts are processed, the SIMPR system (or more specifically, the Subject Classification Intelligent System (SCIS)) will suggest suitable classificators, and as the number of processed texts increases, the suggested classificators will become more accurate.

This chapter describes the work of University College, Dublin (UCD) on the role of learning in subject classification. The next section describes the SIMPR project in more detail. The third section introduces the topic of machine learning. The fourth presents the classification algorithm and describes the learning module which will be incorporated into this algorithm. The chapter concludes by suggesting directions in which the research might usefully develop.

THE SIMPR PROJECT

The aim of SIMPR is to develop techniques for the management of large information bases. The documents which constitute the information base are first morphologically

Artificial Intelligence Frontiers in Statistics: AI and statistics III. Edited by D.J. Hand. Published in 1993 by Chapman & Hall, London. ISBN 0 412 40710 8

and syntactically analysed. They are then indexed by the indexing module of SIMPR, which extracts **analytics** from texts. An analytic is a word or sequence of words that accurately represents the information content of a text (Smart, 1990).

The SIMPR system is intended to help in the management of very large documents, prepared by teams of authors working under the supervision of editors. It will help the editors to use, index, validate and store information so that a reader can search the resultant information base to find the answer to a query. The SIMPR information base is built by adding texts to it until all the texts have been processed. A text in this case is defined to be the information between two headings, no matter at what level the headings occur (Smart, 1989).

UCD's role in the project is to identify classificators for a text and then to try and improve on these classificators using machine learning techniques from artificial intelligence. These classificators form the basis of one of the retrieval mechanisms of the SIMPR project. Our area of interest in the project is the application of machine learning techniques to subject classification.

In SIMPR, texts are classified using a **faceted classification scheme**. A faceted classification scheme is a type of systematic classification using the analytico-synthetic principles introduced by S.R. Ranganathan in 1933 (Foskett, 1977). All knowledge is divided into classes, which are then divided into subclasses. Ranganathan imposes for most main classes the citation.order PMEST:

- Personality: key facet depending on the subject
- Matter: relates to materials
- Energy: relates to processes, activities or problems
- Space: represents geographical areas
- Time: represents time periods (Aitchison and Gilchrist, 1987).

These terms can appear several times in a subject classification if necessary, for instance, if two materials are used both may be represented.

In **subject classification** subjects are analysed into their component concepts. Compound subjects may be built up synthetically from elementary concepts. The concepts are arranged in groups called **facets**. Each concept is allocated a notational symbol derived from the facets which make up the concept. These concepts can then be combined or synthesized, using their notations, to form more complex subjects (Sharif, 1989).

In order to subject-classify a document, an indexer examines a document, selects appropriate index terms and decides how they are related (Foskett, 1977). These relationships can be represented using the notation mentioned above. For example, if the indexer selected the terms *music* and *instrument* as the index terms for a text, and music is represented by $[P_1]$ and instrument by $[M_1]$, then the text would be represented by the term $[P_1 \ M_1]$.

In SIMPR the number and type of basic facet groups will depend on the domain in question. For example, in experimentation with texts from car manuals, index terms were selected from ten facet groups including car-type, model, date, component, operation and fault. Other domains will yield different facet groups.

The next section introduces the topic of machine learning, describing and comparing different machine learning techniques.

MACHINE LEARNING

Learning is an inherent part of human intelligence. The manner in which a person acquires new knowledge, learns new skills and, in particular, his ability to improve with practice are all part of the human learning process. Therefore, the ability to learn must be part of any system that would exhibit general intelligence. Various researchers in artificial intelligence (AI) argue that one of the main roles of current research in AI is understanding the very nature of learning and implementing learning capabilities in machines (McCarthy, 1983; Schank, 1983). In recent years machine learning (ML) has developed into a separate subfield within AI.

DIFFERENT TYPES OF LEARNING

Originally, learning systems were classified depending on the kind of knowledge represented and manipulated by the system. Three distinct criteria developed: neural modelling and decision-theoretic techniques; symbolic concept acquisition (SCA); and knowledge-intensive, domain-specific learning (KDL). The different strategies differ in the amount of a priori knowledge initially built into the system, how this knowledge is represented and the way it is modified (Luger and Stubblefield, 1989). Learning in neural networks tends to be numerical in nature, in contrast with the more conceptual nature of SCA and KDL. For a more detailed discussion, see Michalski (1986).

In the early 1980s two distinct approaches to ML developed: **similarity-based learning** and **explanation-based learning**. More recently, hybrid systems have emerged which combine different aspects of each.

One of the first areas explored in ML was that of learning from examples, i.e. learning by discovery or similarity-based learning (SBL). Given an adequate description language, a training set of examples, generalization and specialization rules and other background knowledge, an SBL algorithm induces a list of concept descriptions or class membership predicates which cover all the examples in the given set of examples.

An alternative approach to SBL is explanation-based learning (EBL)—see DeJong (1983) and Ellman (1989). Unlike SBL, the system can learn new concept descriptions from a single training example. Given an adequate description language, a single positive example of a concept, generalization and specialization rules and an extensive domain theory, EBL adopts a two-step approach: an explanation is first derived for the given training example, and this 'explanation' is then generalized to obtain a more general and more useful description of the object or set of objects.

COMPARISON OF SBL AND EBL METHODOLOGIES

Each technique has inherent advantages and disadvantages over the other (Kavanagh, 1989). The major differences between the two techniques are as follows:

1. Size of the training set. Use of SBL techniques necessitates a large training set, preferably containing a number of 'near misses'. The SBL approach depends largely on similarities and differences between a large number of examples to determine a concept description. In contrast, the EBL approach requires only a single training example to discover a concept description. In addition, the SBL approach usually considers both negative and positive examples of the concept to be learned. In contrast, an EBL system usually considers a single positive example.
2. Amount of background knowledge required by each approach. SBL uses background knowledge to constrain the infinite number of possible concept descriptions which can be deduced from the initial input set. EBL techniques depend heavily on an extensive background knowledge, in particular to build the explanation to explain what is an example.
3. The main theoretical difference between the two approaches is that EBL is said to be **truth-preserving** while SBL is only **falsity-preserving**. This means applying EBL techniques ensures that if a property is true in an input example, then this property will be true in the output concept description. Applying an SBL approach to the same problem does not ensure that the property will still hold. This is entirely due to its dependence on the 'aptness' of the training set of examples (for determining feature significance). All negative assertions in the input example are guaranteed to remain so in the output example.
4. The two approaches are complementary. SBL techniques exploit the inter-example relationship, while EBL techniques exploit the intra-example relationships (Michalski, 1986).
5. In general the descriptions inferred using induction techniques are more comprehensible to the user. This is important especially in cases where human experts may be asked to validate the induced inferences.

HYBRID SYSTEMS

Many ML researchers are currently investigating methods of combining the analytical techniques of EBL with the more empirical techniques of SBL. Hybrid systems are usually categorized as one of the following:

- Systems which use explanations to process the results of empirical learning. In such systems, SBL methods are first applied to the training set to derive concept descriptions, which are then refined or discarded by the EBL process.
- Systems which use empirical methods to process the results of an explanation phase. Here the results of the EBL process are subjected to SBL to derive a general concept definition.
- Systems which integrate explanation-based methods with empirical methods. An example of this method is where SBL and domain-independent heuristics are used to extend the domain theory whenever EBL methods fail.

All learning systems must address problems of **clustering**, i.e. given a set of objects, how one identifies attributes of the objects in such a manner that the set can be divided into subgroups or clusters, such that items in the same cluster are similar to each other. Previous research has concentrated on numerical methods to establish a measure of similarity between objects and concepts. Recently this emphasis has changed, and the emphasis is now on conceptual clustering with particular importance on the development of goal-oriented classifications. The main advocates of this approach are Stepp and Michalski (1986).

Classification is an inherent part of any learning system, in that any form of learning involves the invention of a meaningful classification of given objects or events. It is clear that developments in ML will have a knock-on effect on automatic classification research, and vice versa.

THE SUBJECT CLASSIFICATION INTELLIGENT SYSTEM (SCIS)

The SIMPR Subject Classification System comprises two parts, the Subject Classification Maintenance System (SCMS) and the Subject Classification Intelligent System (SCIS). The SCMS creates and maintains the subject classification system. Among its tasks are the maintenance of the facet scheme, the **heading hierarchy** and the Facet Attribution File (FAF). A heading hierarchy is a dynamically built-up table of contents. The FAF is a file which stores the list of texts processed so far and the facets which have been attributed to each text. The SCIS consists of a classification algorithm incorporating a learning module. In this section we describe both the classification algorithm and the learning module.

To incorporate learning into the classification process, it was necessary to start with a classification algorithm. Initial work thus concentrated on the development of an adequate classification algorithm. Nine strategies were identified for inclusion in this algorithm and are detailed below. Learning strategies were then incorporated into this algorithm. These learning strategies are described later. The nine strategies of the classification algorithm are:

Strategy 1: For each text derive the Normalized Evidence Word Tokens (NEWTs) (Smart, 1990). NEWTs are the result of a simple word-count technique. To generate the NEWTs for a text, the original text is first normalized, the frequency of occurrence of the normalized words is calculated and words with a frequency value above a certain threshold value form the NEWTs. The classificators for a text are generated from the set of NEWTs for the text.

Strategy 2: Examination of the FAF entries of texts which are related in the heading hierarchy to identify discriminatory (and possibly common) facets between the related texts. In particular, facet attributions were investigated for parent and child texts and for sibling texts.

Strategy 3: Examination of words which occur in the heading of a text. From investigations with sample texts (Ward et al., 1990), it is clear that words which occur in the heading of a text and which also occur in the text itself usually provide good guidelines when determining candidate subject classificators for a text.

Strategy 4: Examination of FAF entries for referenced texts to identify if patterns of facet attribution exist. A referenced text is one which is referred to in the current text. Information about cross-references is stored in the FAF.

Strategy 5: The 'type' of a text—for example, diagnostic, explanatory, etc.—can sometimes be discerned from the structure of the sentences. If a lot of sentences in the text begin with the word 'If', then the text is probably diagnostic. If many sentences start with a verb then the text is probably instructive. The type of the text may also be discerned from the presence of key phrases such as 'This is how to...' or 'We now describe...'. This strategy attempts to use information about the type of a text to suggest classificators for the text.

Strategy 6: Examination of the occurrence of synonyms of words in the text, in particular, synonyms of words which occur in the heading of the text.

Strategy 7: Early research (Ward et al., 1990) indicated that emphasized (i.e. emboldened, underlined or italicized) text is important in deciding the subject matter of a text.

Strategy 8: Examination of the occurrence of proper nouns in a text to decide if they provide useful guidelines in determining the subject of a text.

Strategy 9: Examination of the singular occurrence of words, i.e. words which occur only once in a text, to decide if they are relevant in determining the 'aboutness' of a text.

THE LEARNING MODULE

Incorporating learning into the classification algorithm results in modifications to the background knowledge. These modifications improve future facet attribution, either by improving the efficiency of the same facet attribution or by suggesting improved alternative classificators for the same text. Alternative classificators are classificators which would not have been suggested by the system had learning not occurred. There are three ways in which learning is incorporated into the classification process. The learning module consists of these three learns.

LEARN ONE: GENERALIZATION

If a text contains a number of different entries from the same facet group, it may be useful to generalize the facet terms. For example, if a text contains five entries from the component facet group, then it may be useful to generalize the five terms. Generalization must be useful and feasible. Feasibility is determined by the relationship between the facet terms. The relationship between two terms in a facet group is determined by their relative location in the tree representing the facet group. It is feasible to generalize two facet terms in any of the following cases:

- One facet term is the parent of the other, or the two facet terms have a common parent.
- The two facet terms have a common grandparent.
- One facet term is a sibling of the parent of the other facet term.

It is feasible to generalize a set of facet terms from the same facet group, if the terms are all at the same level or are at adjacent levels in the classification scheme. Generalizing facet terms which are otherwise related will result in classificators which are too general.

Generalization of facet groups is incorporated into the classification algorithm in three places. Generalization must be appropriate to be useful. The circumstances under which generalization is appropriate are enumerated below. In each case generalization must be feasible as well as appropriate.

During NEWT generation in Step 1, if all of the individual occurrences of facet terms from a particular facet group are below the NEWT generation threshold value, and the sum of occurrences of the individual facet terms is above this threshold value, then generalization is appropriate. The facet group is then examined to see if generalization is feasible. For example, in a sample text from the car domain, if 'grind', 'rebore' and 'reface' each occur below the NEWT threshold value they are not generated as NEWTs. These three words are facet terms and occur in the same facet group for the car domain. Their common parent in the facet scheme is 'service'. Generalization results in the facet term 'service' being added to the list of NEWTs for the text. This is one way of introducing new facet terms as candidate classificators. New terms are defined as facet terms attributed by the classification process which do not occur in the original text.

Generalization is applied during Step 3 of the classification algorithm to reduce the list of possible classificators for a text. The list of facet sets generated by grouping together facet terms which co-occurred in the original text is examined to see if generalization is appropriate. Generalization is appropriate if the number of facet sets in the list which differ by a single facet term from the same facet group is above a certain threshold value (the generalization threshold value). If generalization is appropriate then the discriminatory facet terms in each of the relevant facet sets are generalized and the relevant facet sets in the list of possible classificators are replaced by the 'generalized' facet set. For example, if the list of possible classificators generated by Step 3 of the classification algorithm for a text from the car domain contained the facet sets '(ford crankcase remove)', '(ford cylinder remove)', '(ford bearing remove)' and the generalization threshold value was 2, then the three facet sets would be replaced by '(ford engine remove)'. 'Engine' is the parent of 'crankcase', 'cylinder' and 'bearing' in the facet scheme. It is not appropriate to generalize the list of facet terms before they are grouped, i.e. generalization is applied after Step 3, not after Step 2, otherwise the resultant facet sets may be too general.

Generalization is also applied during Step 4 of the classification process, when extending the list of possible classificators to include terms other than NEWT. Instead of extending the list to include all facet terms other than NEWTs which co-occurred with any of the NEWTs, the list is extended by examining the facet terms which co-occur with each NEWT and seeing if generalization is appropriate. Generalization is appropriate if the number of occurrences of terms from the same facet group is above the generalization threshold value. If the generalization threshold value is greater than or equal to the NEWT threshold value, generalization during Step 4 will not extend the list of classificators, as terms which would be added as a result of generalization during

Step 4 will have been included as a result of generalization during the generation of NEWTs in Step 1.

Assertions inferred using generalization are not necessarily valid, i.e. classificators generated using generalization may be invalid. Recall that a classificator is invalid if it violates any of the facet-set or facet-specific rules applicable to that classificator. The facet-specific and facet-set rules are applied to all classificators generated using generalization.

LEARN TWO: MATCHING

Two classificators may match exactly or inexactly. Identical classificators are said to match exactly. Classificators which are identical except for a single facet term match inexactly. For example, '(ford bearing clearance check)' matches inexactly with '(ford bearing clearance test)'.

Two cases exist when classificators match inexactly: the discriminatory term in each classificator is from the same facet group or from different facet groups. If the discriminatory terms are from different facet groups then the match returns a classificator with both facet terms as elements. For example, matching, '(ford bearing clearance instructive test)' with '(ford bearing clearance test plastigage)' gives '(ford bearing clearance instructive test plastigage)'. If the discriminatory terms are from the same facet group the matching classificators are examined to see which is the most general. A classificator is defined to be more general than another classificator, if the discriminatory term in the first classificator is more general than the discriminatory term in the second classificator. Facet terms can only be compared for generality if they belong to the same facet group. A facet term is more general than another if it is located higher in the tree representing the facet group. In the first example, 'check' and 'test' both belong to the same facet group. The second facet attribution is more general than the first because 'test' is a parent of 'check' in the tree representing the facet group.

Matching is incorporated into the classification process at two different stages:

1. As each text is classified:

    -   each classificator in the list of possible classificators generated during Step 3 of the classification algorithm is matched against the other classificators in this list;
    -   each classificator is matched against previous attributions for all other texts, related or otherwise;
    -   each classificator is matched against the classificators previously attributed to the parent text;
    -   each classificator is matched against previous attributions for referenced texts.

2. At fixed intervals the FAF is examined to determine if a pattern of facet attribution exists between related texts. This is a form of incremental learning.

Only inexact matches with the discriminatory terms from the same facet group are considered when matching classificators in the list generated during Step 3 of the classification algorithm. If the discriminatory terms are from the same facet group and generalization is appropriate, i.e. the number of inexactly matching classificators is above the generalization threshold value, then the relevant classificators will already be generalized as described in Learn One, case 2.

Only exact matches are also considered when matching classificators for the current text against all other previous attributions. If the classificator was previously attributed to more than one other text, then the other attributions for these texts are examined to see if any other common classificator(s) exist. Patterns of facet attribution are detected in this way. The fact that two classificators are always attributed together would be identified in this way. If a pattern of attribution is detected then the background knowledge is modified to account for this, i.e. the facet-specific rules associated with the terms in the relevant classificators are extended.

When matching classificators for the current text against previously attributed classificators for the parent text, inexact matches are considered. If a facet attribution for a child text is more general than a matching facet attribution for the parent text, then the more general facet attribution is removed from the list of facet attributions for the child text. That is, child facet attributions must be more specific than parent attributions. This approach minimizes the problem of recall and relevance associated with retrieval by ensuring that a parent text is always more general than a child text.

Matching the classificators attributed to the current text against previous facet attributions for related, unrelated and referenced texts is incorporated into the classification algorithm after Step 7. In addition, at fixed intervals during the classification of a large document, the facet attributions associated with related texts are examined. Any new facet-set and facet-specific rules identified in this way are generated as a result of incremental learning.

During incremental learning, classificators for sibling texts which match exactly or inexactly are considered. If a facet attribution matches a facet attribution in each sibling text, then the matching attributions are generalized and the new attribution is attributed to the parent text. The simplest case is where an attribution is common to all sibling texts. The common attribution is attributed to the parent text and removed from the list of attributions associated with each of the sibling texts. In all other cases, the discriminatory facet terms in the matching classificators are generalized in the same way as facet terms are generalized in Learn One.

The match operation is similar to the approach Kodratoff (1988) uses in structural matching. For example, the way classificators for sibling texts are matched using the match operation is analogous to the way Kodratoff structurally matches two formulae. Kodratoff's (1988) algorithm for structural matching takes two formulae that are identical except for the constants and the variables that instantiate their predicates, and uses domain-specific knowledge to transform the formulae into equivalent formulae. These equivalent formulae are generalized by retaining all the bindings common to each formula and dropping all other bindings. In the SCIS the match operation examines a set of classificators for sibling texts that are identical except for a single facet term, and transforms each classificator using domain-specific knowledge. The resultant classificators are generalized by retaining all facet terms that are common to each classificator and dropping all other terms.

LEARN THREE: FEEDBACK

The human editor, if she chooses to reject a particular classificator, is asked to 'explain' her reason(s) for doing so. This is done by presenting her with a standard set of reasons. The reasons why a human editor may reject a classificator are numerous and domain-specific. A domain expert or human editor would have to assist in enumerating the most useful reasons why a person would reject a suggested classificator for a text. The domain expert would also have to indicate the appropriate action to take. For example, if a suggested classificator is rejected because it is too general then that classificator should not be suggested for any texts lower than it in the heading hierarchy. Such information would then form part of the background knowledge for the system. Identifying a useful set of reasons for rejecting a suggested classificator is a knowledge acquisition problem and was not addressed during the implementation of the SCIS. This learn has not yet been implemented.

A sample training set, comprised of 31 texts from the car domain, was examined to determine if, given a set of texts and their associated facet attributions, it is possible to learn the associated facet-specific and facet-set rules for that domain. C4.5, an implementation of ID3, was used in this investigation. A number of facet-set rules for the car domain were identified. However, facet-set rules identified were very general. In the case of the car domain, it was easier for the domain expert to specify the facet-set rules than to characterize negative examples of classificators which violate these rules. As C4.5 requires negative examples of the concept to be learned, it was not considered suitable for identifying facet-set or facet-specific rules for the car domain. As was the case for facet-set rules, it is easier to specify the facet-specific rules for the car domain than to characterize examples which violate these rules.

THE LEARNING/CLASSIFICATION ALGORITHM

The aim of the classification algorithm is to examine a text and to suggest a list of valid classificators for that text. By definition, a valid classificator is a set of facet terms which do not violate any of the facet attribution constraints. The facet attribution constraints govern the way facet terms are combined and ordered in a given facet set. A text may have more than one classificator associated with it.

The seven steps for our classification algorithm are outlined below. These incorporate both the classification strategies and the learns which were described above. For each text the classification algorithm generates classificators for validation by the human editor as follows:

Step 1: From the frequency analysis, generate the list of NEWTs for the text. When generating NEWTs, words in the heading and proper nouns are given an extra weighting in the count. Synonyms are replaced by their preferred terms, as defined in the domain thesaurus. This step incorporates strategies 1, 3, 6 and 8.

Step 2: The list of NEWTs is reduced to include only terms which occur in the facet scheme. The non-NEWT facet terms are generated.

Step 3: Terms in this list are grouped if they co-occurred in the original text. Such information is available from the co-occurrence statistics. Each group of terms represents a possible classificator for the text.

Step 4: This list of possible classificators is extended to include facet terms other than NEWTs which co-occurred with NEWTs in the original text. Such terms were not generated as NEWTs because their frequency count in the original text was below the current NEWT threshold value.

Step 5: Terms which were grouped as analytics are also grouped in the list of possible classificators.

Step 6: The relevant facet-specific rules are applied to the facet terms in each of the possible classificators. The list of possible classificators is modified accordingly.

Step 7: The facet-set rules are applied to each of the possible classificators. The list of possible classificators is modified accordingly. This step incorporates strategy 5 as the classification scheme for a domain is designed to incorporate a facet group relating to the 'type' of a text. Facet-set rules relating to this group employ information about the 'type' of a text. The resulting list of classificators is presented to the human editor for validation.

All of the classification strategies except 2, 4 and 7 are incorporated into the algorithm. Strategy 7 is not incorporated into the algorithm because information relating to emboldening, underlining and italicizing is removed during preprocessing of the text. Strategies 2 and 4 are incorporated into the matching operation which forms part of the learning module described in the next section. Aspects of strategy 9 are incorporated into Steps 1 and 4 of the algorithm. That is, substituting synonyms with their preferred terms and including terms which occur below the NEWT threshold value, but which co-occur with NEWTs, may introduce terms which occurred only once in the original text into the list of possible classificators.

IMPROVEMENTS AND FUTURE WORK

The quality of the classificators generated by the SCIS is dependent on the quality of the classification scheme designed by the domain expert. One way in which the system could assist the domain expert in ensuring that the classification scheme is representative of the domain is by maintaining a list of all non-facet terms generated as NEWTs. The human editor may or may not select terms from this list for inclusion in the facet scheme. Ideally, as the number of texts processed by the system increases, the number of terms which need to be added to the facet scheme should reduce to zero. In other words, the classification scheme would become fully representative of the domain.

The SCIS has been tested on a set of texts from the car domain. A domain expert also classified the same set of texts so that the quality of the classificators could be determined. The quality of the set of classificators was determined by the number of candidate classificators the domain human editor added and rejected. After learning was incorporated into the algorithm the quality of the classificators improved (Kavanagh *et al.*, 1991). It is not possible to test the SCIS on another domain without designing a classification scheme, facet-specific rules, facet-set rules and a domain thesaurus for the new domain. The SCIS was designed in a modular fashion in order to minimize the effort required to move from one domain to another. If the SCIS is used to process texts from another domain, ID3 may be useful in identifying facet-specific and facet-set rules for that domain. Its usefulness will depend both on the number of such rules in the domain and whether it is easier for the domain expert to specify the rules or to characterize negative examples of the rules.

Future work could also include the development of the match operation. Relaxing the conditions for matching may improve the classificators generated by the SCIS. There are a number of ways in which the conditions for matching could be relaxed, e.g. matching classificators which differ by more than one facet term. Relaxing the conditions for matching will have implications for the generalization step in the match operation. For example, if two classificators match when they differ by two facet terms, then it is necessary to climb the facet trees in order to generalize the matching classificators.

The SCIS has not yet been evaluated in terms of retrieval. Future work in this area is to implement a retrieval module and evaluate the quality of the SCIS when used for retrieval.

CONCLUSIONS

Research in classification and machine learning indicated that the greater the domain-specificity of a classification algorithm the more useful the resultant classificators. One of the aims of the SIMPR system was to design a system which was as domain-independent as possible. With this in mind, the SCIS was designed using classification techniques applicable in any technical domain together with mechanisms for using exchangeable packages of domain-specific knowledge, thus minimizing domain-specificity.

A classification algorithm was implemented and augmented with machine learning techniques. The quality of the classificators generated by the classification algorithm for a set of sample texts improved when learning was incorporated into the classification algorithm, thus illustrating that machine learning techniques may successfully be applied to subject-classify full-text documents.

REFERENCES

Aitchison, J. and Gilchrist, A. (1987) *Thesaurus Construction*, London: ASLIB.

DeJong, G. (1983) An approach to learning from observation, in *Machine Learning*, 1, Michalski, R.S., Carbonell, J.G. and Mitchell, T.M. (eds), Palo Alto, CA: Morgan Kaufmann.

Ellman, T. (1989) Explanation-based learning: a survey of programs and perspectives. *ACM Computing Surveys*, **21**(2).

Foskett, A.C. (1977) *The Subject Approach to Information*, 3rd edn., London: Bingley.

Kavanagh, I. (1989) Machine Learning: A Survey. University College Dublin, SIMPR-UCD-1989-8.03, October.

Kavanagh, I., Ward, C. and Dunnion, J. (1991) The SCIS. University College Dublin, SIMPR-UCD-1991-16.09, July.

Kodratoff, Y. (1988) *Introduction to Machine Learning*, London: Pitman.

Laird, J.E., Rosenbloom, P.S. and Newell, A. (1986) Chunking in SOAR: the anatomy of a general learning mechanism. *Machine Learning*, **1**(1), 11–46.

Luger, G.F. and Stubblefield, W.A. (1989) *Artificial Intelligence and the Design of Expert Systems*, Menlo Park, CA: Benjamin/Cummings.

McCarthy, J. (1983) President's quarterly message: AI needs more emphasis on basic research, *AI Magazine*, **4**.

Michalski, R.S. (1983) A theory and methodology of inductive learning, in *Machine Learning*, Michalski, R.S., Carbonell, J.G. and Mitchell, T.M. (eds), Palo Alto, CA: Morgan Kaufmann, pp. 83–134.

Michalski, R.S. (1986) Understanding the nature of learning, in *Machine Learning, An Artificial Intelligence Approach*, Vol. 2, Michalski, R.S., Carbonell, J.G. and Mitchell, T.M. (eds), Palo Alto, CA: Morgan Kaufmann.

Quinlan, J.R. (1983) Learning efficient classification procedures and their application to chess end games, in *Machine Learning*, Michalski, R.S., Carbonell, J.G. and Mitchell, T.M. (eds), Palo Alto, CA: Morgan Kaufmann, pp. 463–482.

Quinlan, J.R. (1987) Simplifying decision trees. *International Journal of Man–Machine Studies*, **27**.

Rosenbloom, P.S. and Newell, A. (1986) The chunking of goal hierarchies: a generalized model of practice, in *Machine Learning: An Artificial Intelligence Approach*, Vol. 2, Michalski, R.S., Carbonell, J.G. and Mitchell, T.M. (eds), Palo Alto, CA: Morgan Kaufmann.

Schank, R.C. (1983) The current state of AI: one man's opinion, *AI Magazine*, **4**.

Sharif, C. (1989) Subject classification research, University of Strathclyde, September.

Simon, Herbert, A. (1983) Why should machines learn? in *Machine Learning*, Vol. 1, Michalski, R.S., Carbonell, J.G. and Mitchell, T.M. (eds), Palo Alto, CA: Morgan Kaufmann.

Smart, G. (1989) *SIMPR: An Introductory Description*, CRI/AS, Bregnerodvej 144, DK-3460, Birkerød, Denmark, March.

Smart, G. (1990) *Year One: The Results so far*, CRI/AS, Bregnerodvej 144, DK-3460, Birkerød, Denmark, March.

Stepp, R.E. and Michalski, R.S. (1986) Conceptual clustering: inventing goal oriented classifications of structured objects, in *Machine Learning: An Artificial Intelligence Approach*, Vol. 2, Michalski, R.S., Carbonell, J.G. and Mitchell, T.M. (eds), Palo Alto, CA: Morgan Kaufmann.

Ward, C., Dunnion, S. and Gibb, F. (1990) The heading exercise. Department of Computer Science, University College Dublin, SIMPR-UCD-1990-16.4.4, April.

PART SIX
Other areas

A statistical semantics for causation

24

J. Pearl and T.S. Verma

We propose a model-theoretic definition of causation, and show that, contrary to common folklore, genuine causal influences can be distinguished from spurious covariations following standard norms of inductive reasoning. We also establish a sound characterization of the conditions under which such a distinction is possible. Finally, we provide a proof-theoretical procedure for inductive causation and show that, for a large class of data and structures, effective algorithms exist that uncover the direction of causal influences as defined above.

THE MODEL

We view the task of causal modelling as an identification game which scientists play against nature. Nature possesses stable causal mechanisms which, on a microscopic level, are deterministic functional relationships between variables, some of which are unobservable. These mechanisms are organized in the form of an acyclic schema which the scientist attempts to identify.

Definition 24.1
A **causal model** over a set of variables U is a directed acyclic graph (DAG) D, the nodes of which denote variables, and the links denote direct binary causal influences.

The causal model serves as a blueprint for forming a 'causal theory'—a precise specification of how each variable is influenced by its parents in the DAG. Here we assume that nature is at liberty to impose arbitrary functional relationships between each effect and its causes and then to weaken these relationships by introducing arbitrary (yet mutually independent) disturbances. These disturbances reflect 'hidden' or unmeasurable conditions and exceptions which nature chooses to govern by some undisclosed probability function.

Definition 24.2
A **causal theory** is a pair $T = \langle D, \Theta_D \rangle$ containing a causal model D and a set of

Artificial Intelligence Frontiers in Statistics: AI and statistics III. Edited by D.J. Hand. Published in 1993 by Chapman & Hall, London. ISBN 0 412 40710 8

parameters Θ_D compatible with D. Θ_D assigns a function $x_i = f_i[pa(x_i), \varepsilon_i]$ and a proba-bility measure g_i, to each $x_i \in U$, where $pa(x_i)$ are the parents of x_i in D and each ε_i is a random disturbance distributed according to g_i, independently of the other εs and of $\{x_j\}_{j=1}^{i-1}$. (Variables are ordered such that all arcs point from lower to higher indices.)

The requirement of independence renders the disturbances 'local' to each family; disturbances that influence several families simultaneously will be treated explicitly as 'latent' variables (see Definition 24.3 below).

Once a causal theory T is formed, it defines a joint probability distribution $P(T)$ over the variables in the system, and this distribution reflects some features of the causal model (e.g., each variable must be independent of its grandparents, given the values of its parents). Nature then permits the scientist to inspect a select subset O of 'observed' variables, and to ask questions about $P_{[O]}$ the probability distribution over the observables, but hides the underlying causal theory as well as the structure of the causal model. We investigate the feasibility of recovering the topology of the DAG from features of the probability distribution.[1]

MODEL PREFERENCES (OCCAM'S RAZOR)

In principle, with no restriction on the type of models considered, the scientist is unable to make any meaningful assertions about the structure of the underlying model. For example, he/she can never rule out the possibility that the underlying model is a complete (acyclic) graph; a structure that, with the right choice of parameters, can *mimic* (see Definition 24.4 below) the behaviour of any other model, regardless of the variable ordering. However, following the standard method of scientific induction, it is reason-able to rule out any model for which we find a simpler, *less expressive* model, equally consistent with the data (see Definition 24.6, below). Models that survive this selection are called **minimal** models and, with this notion, we construct our definition of **inductive causation**: 'A variable X is said to have a direct causal influence on a variable Y if a unidirected edge exists in all minimal models consistent with the data'.

Definition 24.3
A **latent structure** is a pair $L = \langle D, O \rangle$ containing a causal model D over U and a set $O \subseteq U$ of observable variables.

Definition 24.4
$L = \langle D, O \rangle$ is **preferred** to $L' = \langle D', O \rangle$, written $L \preceq L'$, if and only if D' can *mimic* D over O, i.e. for every Θ_D there exists a $\Theta'_{D'}$ such that $P_{[O]}(\langle D', \Theta'_{D'} \rangle) = P_{[O]}(\langle D, \Theta_D \rangle)$. Two latent structures are **equivalent**, written $L' \equiv L$, if and only if $L \preceq L'$ and $L \succeq L'$.

Definition 24.5
A latent structure L is **minimal** with respect to a class $\mathcal{L}$ of latent structures if and only if for every $L' \in \mathcal{L}, L \equiv L'$ whenever $L' \preceq L$.

Definition 24.6
$L = \langle D, O \rangle$ is **consistent** with a sampled distribution $\hat{P}$ over O if D can accommodate some theory that generates $\hat{P}$, i.e. there exists a Θ_D such that $P_{[O]}(\langle D, \Theta_D \rangle) = \hat{P}$.

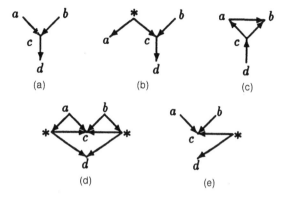

Figure 24.1 Examples of causal models.

Definition 24.7 (Induced Causation)
Given $\hat{P}$, a variable C has a **direct causal influence** on E if and only if a path from C to E exists in every minimal latent structure consistent with $\hat{P}$.

We view this definition as normative, because it is based on one of the least disputed norms of scientific investigation: Occam's razor in its semantical casting. However, as with any scientific inquiry, we make no claims that this definition is guaranteed always to identify stable physical mechanisms in nature; it identifies the only mechanisms we can plausibly induce from non-experimental data.

As an example of a causal relation that is identified by the definition above, imagine that observations taken over four variables $\{a, b, c, d\}$ reveal only two vanishing dependencies: 'a is independent of b' and 'd is independent of $\{a, b\}$ given c' (plus those that logically follow from the two). This dependence pattern would be typical, for example, of the following variables: $a = $ *having cold*, $b = $ *having hay-fever*, $c = $ *having to sneeze*, $d = $ *having to wipe one's nose*. It is not hard to show that any model which explains the dependence between c and d by an arrow from d to c, or by a hidden common cause between the two, cannot be minimal, because any such model would be able to outmimic the minimal modes shown in Figs 24.1(a) and (b). We conclude, therefore, that the observed dependencies imply a direct causal influence from c to d. Some minimal models (a and b) and non-minimal models (c and d) consistent with the observations are shown above. However, model (e) is inconsistent because it cannot account for the observed marginal dependence between b and d. (For theory and methods of reading independencies from graphs see Pearl, 1988.)

PROOF THEORY

It turns out that while the minimality principle is sufficient for forming a normative and operational theory of causation, it does not guarantee that the search through the vast space of minimal models would be computationally practical. If nature truly conspires to conceal the structure of the underlying model she could annotate that model with a distribution that matches many minimal models, having totally disparate structures. To facilitate an effective proof theory, we rule out such eventualities, and impose a restriction on the distribution called **stability**. It conveys the assumption that all vanishing dependencies are structural, not formed by incidental equalities of numerical parameters.[2]

Definition 24.8
Let $I(P)$ denote the set of all conditional independence relationships embodied in P. A probability distribution $\hat{P}$ is stable if there exists a dag D such that $\hat{P}$ precisely embodies the independencies dictated by D, i.e. there exists a set of parameters Θ_D such that for any other set Θ'_D we have:

$$I(P(\langle D, \Theta_D \rangle)) \subseteq I(\hat{P}) \subseteq I(P(\langle D, \Theta'_D \rangle))$$

With the added assumption of stability, every distribution has a unique causal model (up to equivalence), as long as there are no hidden variables (Verma and Pearl, 1990). The search for the minimal model then boils down to recovering the structure of the underlying DAG from probabilistic dependencies that perfectly reflect this structure (see Pearl and Verma, 1987; and Pearl, 1988) for a characterization of these dependencies). This search is exponential in general, but simplifies significantly when the underlying structure is sparse (for such algorithms, see Verma and Pearl, 1990; Spirtes and Glymour, 1991).

RECOVERING LATENT STRUCTURES

When nature decides to 'hide' some variables, the observed distribution $\hat{P}$ need no longer be stable relative to the observable set O, i.e. $\hat{P}$ may result from many equivalent minimal latent structures, each containing any number of hidden variables. Fortunately, rather than having to search through this unbounded space of latent structures, it turns out that for every latent structure L, there is an equivalent latent structure called the *projection* of L on O in which every unobserved node is a root node with exactly two observed children.

Definition 24.9
A latent structure $L_{[O]} = \langle D_{[O]}, O \rangle$ is a **projection** of another latent structure L if and only if

(1) every unobservable variable of $D_{[O]}$ is a parentless common cause of exactly two non-adjacent observable variables;
(2) for every stable distribution P generated by L, there exists a stable distribution P' generated by $L_{[O]}$ such that $I(P_{[O]}) = I(P'_{[O]})$.

Theorem 24.1
Any latent structure has at least one projection (identifiable in linear time).

(Proofs can be found in Verma, 1992).

It is convenient to represent projections by bidirectional graphs with only the observed variables as vertices (i.e. leaving the hidden variables implicit). Each bidirected link in such a graph represents a common hidden cause of the variables corresponding to the link's end-points.

Theorem 24.1 renders our definition of induced causation (Definition 24.7) operational; we will show (Theorem 24.2 below) that if a certain link exists in a distinguished projection of any minimal model of $\hat{P}$, it must indicate the existence of a causal path in every minimal model of $\hat{P}$. Thus the search reduces to finding a projection of any minimal model of $\hat{P}$ and identifying the appropriate links. Remarkably, these links can be identified by a simple procedure, the IC algorithm, which is not more complex than that which recovers the unique minimal model in the case of fully observable structures.

IC Algorithm (Inductive Causation)

Input: $\hat{P}$, a sampled distribution.
Output: $core(\hat{P})$, a marked hybrid acyclic graph.

(1) For each pair of variables a and b, search for a set S_{ab} such that (a, S_{ab}, b) is in $I(\hat{P})$, namely a and b are independent in $\hat{P}$, conditioned on S_{ab}. If there is no such S_{ab}, place an undirected link between the variables.
(2) For each pair of non-adjacent variables a and b with a common neighbour c, check if $c \in S_{ab}$.
    If it is, then continue.
    If it is not, then add arrowheads pointing at c (i.e. $a \to c \leftarrow b$).
(3) Form $core(\hat{P})$ by recursively adding arrowheads according to the following two rules:
    If $\overline{ab}$ and there is a strictly directed path from a to b then add an arrowhead at b.
    If a and b are not adjacent but $\overrightarrow{ac}$ and c—b, then direct the link $c \to b$.[3]
(4) If $\overline{ab}$ then mark every unidirected link $b \to c$ in which c is not adjacent to a.

The result of the IC algorithm is a substructure called $core(\hat{P})$ in which every marked unidirected arrow $X \to Y$ stands for the statement: 'X is a direct cause of Y (in all minimal latent structures consistent with the data)'. We call these relationships **genuine** causes (e.g., $c \to d$ in Fig. 24.1(a)).

Theorem 24.2
If every link of a directed path from C to E is marked in $core(\hat{P})$ then C has a causal influence on E according to $\hat{P}$.

Theorem 24.3
If $core(\hat{P})$ contains a bidirectional link $E_1 \leftrightarrow E_2$, then there is a common cause X influencing both E_1 and E_2, and no direct causal influence between the two, in every minimal latent structure consistent with $\hat{P}$.

SUMMARY AND INTUITION

For the sake of completeness we now present explicit definitions of potential and genuine causation, as they emerge from Theorem 24.2 and the IC algorithm. Additional conditions, sufficient for the determination of spurious and genuine causes, with and without temporal information, can be found in Pearl and Verma (1991).

Definition 24.10 (Potential Cause)
A variable X has a **potential causal influence** on another variable Y (**inferable** from $\hat{P}$), if

(1) X and Y are dependent in every context.
(2) There exists a variable Z and a context S such that
    (i) X and Z are independent given S
    (ii) Z and Y are dependent given S

Definition 24.11 (Genuine Cause)

A variable X has a **genuine causal influence** on another variable Y if there exists a variable Z such that either:

(1) X is a potential cause of Y and there exists a context S satisfying:
    (i) Z is a potential cause of X.
    (ii) Z and Y are dependent given S.
    (iii) Z and Y are independent given $S \cup X$.
(2) X and Y are in the transitive closure of the relation defined by Part 1, that is, there exists $k \geqslant 1$ variables, $W_1, \ldots, W_k$ such that X has a genuine causal influence on W_1 and W_i has a genuine causal influence on W_{i+1} for all $k > i \geqslant 1$ and W_k has a genuine causal influence on Y, all defined by Part 1.

Definition 24.10 was formulated in Pearl (1990) as a relation between events (rather than variables) with the added condition $P(Y|X) > P(Y)$ in the spirit of Suppes (1970). Condition (1.i) in Definition 24.11 may be established either by statistical methods (per Definition 24.10) or by other sources of information, e.g., experimental studies or temporal succession (i.e. that Z precedes X in time). When temporal information is available, as it is assumed in the formulations of Suppes (1970), Granger (1988) and Spohn (1983), then every link constructed in step 1 of the IC algorithm corresponds to a potential cause (genuine or *spurious cause* in Suppes terminology). In such cases, Definition 24.11 can be used to distinguish genuine from spurious causes without the usual requirement that all causally relevant background factors be measurable.

The intuition behind our definitions (and the IC recovery procedure) is rooted in Reichenbach's (1956) 'common cause' principle stating that if two events are correlated, but one does not cause the other, then there must be causal explanation to both of them, an explanation that renders them conditionally independent. As it turns out, the pattern that provides us with information about causal directionality is not the 'common cause' but rather the 'common effect'. The argument goes as follows: If we create conditions (fixing S_{ab}) where two variables, a and b, are each correlated with a third variable c but are independent of each other, then the third variable cannot act as a cause of a or b; it must be either their common effect, $a \rightarrow c \leftarrow b$, or be associated with a and b via common causes, forming a pattern such as $a \leftrightarrow c \leftrightarrow b$. This is indeed the eventuality that permits our algorithm to begin orienting edges in the graph (step 2), and assign arrowheads pointing at c. Another explanation of this principle appeals to the perception of 'voluntary control' (Pearl, 1988, p. 396). The reason why people insist that the rain causes the grass to become wet, and not the other way around, is that they can find other means of getting the grass wet, totally independent of the rain. Transferred to our chain $a-c-b$, we can preclude c from being a cause of a if we find another means of potentially controlling c without affecting a, namely b.

The notion of genuine causation also rests on the 'common effect' principle: Two causal events do not become dependent simply by virtue of predicting a common effect. Thus, a series of spurious associations, each resulting from a separate common cause, is not transitive; it predicts independence between the first and last variables in the chain. For example, if I hear my sprinklers turn on, it suggests that my grass is wet, but not that the parking lot at the local supermarket is wet even though the latter two events are highly correlated by virtue of a common cause in the form of rain.[4] Therefore, if correlation is measured between my sprinkler and the wetness of the parking lot then there ought to be a non-spurious causal connection between the wetness of my grass and that of the parking lot (such as the water saturating my lawn, running off into the gutter and into the parking lot).

CONCLUSIONS

The results presented in this chapter dispel the claim that statistical analysis can never distinguish genuine causation from spurious covariation (Otte, 1981; Cliff, 1983; Holland, 1986; Gardenfors, 1988; Cartwright, 1989). We show that certain patterns of dependencies dictate a causal relationship between variables, one that cannot be attributed to hidden causes lest we violate one of the basic maxims of scientific methodology: the semantical version of Occam's razor.

On the practical side, we have shown that the assumptions of model minimality and 'stability' (no accidental independencies) lead to an effective algorithm for recovering causal structures, transparent as well as latent. Simulation studies conducted at our laboratory show that networks containing 20 variables require less than 5000 samples to have their structure recovered by the algorithm. Another result of practical importance is the following: Given a proposed causal theory of some phenomenon, our algorithm can identify those causal relationships (or the lack thereof) that could potentially be substantiated by observational studies, and those whose directionality might require determination by controlled, manipulative experiments.

From a methodological viewpoint, our results should settle some of the ongoing disputes between the descriptive and structural approaches to theory formation (Freedman, 1987). It shows that the methodology governing path-analytic techniques is legitimate, faithfully adhering to the traditional norms of scientific investigation. At the same time, our results also explicate the assumptions upon which these techniques are based, and the conditions that must be fulfilled before claims made by these techniques can be accepted.

ACKNOWLEDGEMENTS

This work was supported, in part, by NSF grant IRI-88-2144, and NRL grant N000-89-J-2007. T.S. Verma was supported by an IBM graduate fellowship.

NOTES

1. This formulation employs several idealizations of the actual task of scientific discovery. It assumes, for example, that the scientist obtains the distribution directly, rather than events sampled from the distribution. This assumption is justified when a large sample is available, sufficient to reveal all the dependencies embedded in the distribution. Additionally, we assume that the observed variables actually appear in the original causal theory and are

not some aggregate thereof. Aggregation might result in feedback loops, which we do not discuss in this chapter. Our theory also takes variables as the primitive entities in the language, not events which permits us to include 'enabling' and 'preventing' relationships as part of the mechanism.
2. It is possible to show that, if the parameters are chosen at random from any reasonable distribution, then any unstable distribution has measure zero (Spirtes *et al.*, 1989). Stability precludes deterministic constraints, as well as aggregated variables.
3. $\overline{ab}$ denotes adjacency, $\overrightarrow{ab}$ denotes either $a \rightarrow b$ or $a \leftrightarrow b$.
4. Apparently this lack of transitivity has not been utilized by path analysts.

REFERENCES

Cartwright, N. (1989) *Nature Capacities and their Measurements*. Clarendon Press: Oxford.
Cliff, N. (1983) Some cautions concerning the application of causal modeling methods. *Multivariate Behavioral Research*, **18**, 115–126.
Freedman (1987) As others see us: a case study in path analysis (with discussion). *Journal of Educational Statistics*, **12**, 101–223.
Gardenfors, P. (1988) Causation and the dynamics of belief, in *Causation in Decision, Belief Change and Statistics II*, Harper, W.L. and Skyrms, B. (eds), Dordrecht: Kluwer Academic Publishers, pp. 85–104.
Glymour, C., Scheines, R., Spirtes, P. and Kelly, K. (1987) *Discovering Causal Structure*, New York: Academic Press.
Granger, C.W.J. (1988) Causality testing in a decision science, in *Causation in Decision, Belief Change and Statistics I*, Harper, W.L. and Skyrms B. (eds), Dordrecht: Kluwer Academic Publishers, pp. 1–20.
Holland, P. (1986) Statistics and causal inference. *Journal of the American Statistical Association*, **81**, 945–960.
Otte, R. (1981) A critique of Suppes' theory of probabilistic causality. *Synthese*, **48**, 167–189.
Pearl, J. (1988) *Probabilistic Reasoning in Intelligent Systems*, San Mateo, CA: Morgan-Kaufmann.
Pearl, J. (1990) Probabilistic and qualitative abduction, in *Proceedings of AAAI Spring Symposium on Abduction*, Stanford, 27–29 March, 155–158.
Pearl, J. and Verma, T.S. (1987) The logic of representing dependencies by directed acyclic graphs. *Proceedings of AAAI-87*, Seattle, Washington, pp. 347–379.
Pearl, J. and Verma, T.S. (1991) A theory of inferred causation, in *Principles of Knowledge Representation and Reasoning: Proceedings of the Second International Conference*, Allen, J.A., Fikes, R. and Sandwall, E. (eds), San Mateo, CA: Morgan Kaufmann, pp. 441–452.
Reichenbach, H. (1956) *The Direction of Time*, Berkeley: University of California Press.
Simon, H. (1954) Spurious correlations: a causal interpretation. *Journal American Statistical Association*, **49**, 469–492.
Spirtes, P. and Glymour, C. (1991) An algorithm for fast recovery of sparse causal graphs. *Social Science Computer Review*, **9**(1), 62–72.
Spirtes, P., Glymour, C. and Scheines, R. (1989) Causality from probability. Technical Report CMU-LCL-89-4, Department of Philosophy, Carnegie-Mellon University.
Spohn, W. (1983) Deterministic and probabilistic reasons and causes. *Erkenntnis*, **19**, 371–396.
Suppes, P. (1970) *A Probabilistic Theory of Causation*, Amsterdam: North-Holland.
Verma, T.S. (1992) Causal modeling: A graph-theoretic approach. PhD dissertation, UCLA Computer Science Department, Los Angeles, CA (in preparation).
Verma, T.S. and Pearl, J. (1990) Equivalence and synthesis of causal models, in *Proceedings of the 6th Conference on Uncertainty in Artificial Intelligence*, Cambridge, MA, pp. 220–227; Amsterdam: North-Holland, pp. 255–268.

Admissible stochastic complexity models for classification problems

25

P. Smyth

INTRODUCTION AND MOTIVATION

Consider in a very general sense the problem of building a classifier given a set of training data. Given no prior information about the underlying distributions describing the domain features and the class variable, it is increasingly common to design the classifier using a so-called *non-parametric* model, such as a decision tree or a neural network. For each type of model a variety of well-understood algorithms exist to select the best model and its parameters, given the training data. Traditionally these algorithms rely on a goodness-of-fit measure as evaluated on the training data to select the best model—for example, the classification error rate or the mean-squared error. However, empirical experience has universally shown that relying on goodness of fit alone to select a model can be misleading—one can argue from a variety of viewpoints that a quantitative version of Occam's razor is required to select a model which appropriately trades off goodness of fit with model complexity. Simpler models stand a better chance of generalizing well to new data than more complicated models do. Hence, for example, in decision tree design, 'pruning' techniques are widely employed to prune back the irrelevant branches of an initially large tree.

A very general technique, independent of any particular class of models, by which to control the trade-off of goodness of fit and complexity is that of minimum description length (MDL). In this chapter we examine in a general sense the application of MDL techniques to the problem of selecting a good classifier from a large set of candidate models or hypotheses. Pattern recognition algorithms differ from more conventional statistical modelling techniques in the sense that they typically choose from a very large number of candidate models to describe the available data. The problem of searching through this set of candidate models is frequently a formidable one, often approached in practice by the use of greedy algorithms. In this context, techniques which allow us to eliminate portions of the hypothesis space are of considerable interest. We will show in this chapter that it is possible to use the intrinsic

Artificial Intelligence Frontiers in Statistics: AI and statistics III. Edited by D.J. Hand. Published in 1993 by Chapman & Hall, London. ISBN 0 412 40710 8

structure of the MDL formalism to eliminate large numbers of candidate models given only minimal information about the data. Our results depend on the very simple notion that models which are obviously too complex for the problem (e.g., models whose complexity exceeds that of the data itself) can be discarded from further consideration in the search for the most parsimonious model.

BACKGROUND ON STOCHASTIC COMPLEXITY THEORY

GENERAL PRINCIPLES

Stochastic complexity (Rissanen, 1984; Wallace and Freeman, 1987) prescribes a general theory of inductive inference from data, which, unlike more traditional inference techniques, takes into account the *complexity* of the proposed model in addition to the standard goodness of fit of the model to the data. In this manner estimation and classi-fication algorithms can consider families of models, rather than just optimizing the parameters of a single model. For a detailed rationale the reader is referred to the work of Rissanen (1984) or Wallace and Freeman (1987) and the references therein. Note that the minimum description length (MDL) technique (as Rissanen's approach has become known) is closely related to maximum a posteriori (MAP) Bayesian estimation techniques if cast in the appropriate framework. In particular, as pointed out by Buntine (Chapter 15, this volume), the MDL approach can be viewed as an approximation to the optimal Bayes solution—we will not dwell on this point in this chapter, except to point out that the reader should be aware of this implicit relation. We also make the point that the Bayesian/MDL approach is a *theory-based* model-selection approach which seeks to make the most efficient use of the available data by use of prior knowledge. In contrast, *data-dependent* model-selection approaches such as cross-validation require no prior knowledge but do not necessarily make the most efficient use of the available data. Each of these approaches has its role and uses in the modeller's 'toolbox' of inductive techniques. In this chapter we will focus only on the MDL approach, a theory-based method.

MINIMUM DESCRIPTION LENGTH AND STOCHASTIC COMPLEXITY

Following the notation of Barron and Cover (1991), we have N data points, described as a sequence of tuples of observations $\{x_i^1, \ldots, x_i^K, y_i\}$, $1 \leqslant i \leqslant N$. For brevity we will refer to the pair $\{\mathbf{x}_i, y_i\}$, meaning the entire collection of $\mathbf{x}_i$ data points plus the class data y_i. The $\mathbf{x}_i$ correspond to values taken on by the K random variables X^k (which may be continuous or discrete (finite alphabet)). For the purposes of this chapter, the y_i are elements of the finite alphabet of the discrete m-ary class variable Y. Let $\Gamma_N = \{M_1, \ldots, M_{|\Gamma_N|}\}$ be the family of candidate models under consideration. Note that by defining Γ_N as a function of N, the number of data points, we allow the possibility of considering more complicated models as more data arrives. For each $M_j \in \Gamma_N$ let $C(M_j)$ be non-negative numbers such that

$$\sum_j 2^{-C(M_j)} \leqslant 1 \tag{25.1}$$

The $C(M_j)$ can be interpreted as the cost in bits of specifying model M_j—in turn, we have that

$$p(M_j) = \frac{2^{-C(M_j)}}{\sum\limits_{i=1}^{|\Gamma_N|} 2^{-C(M_i)}} \tag{25.2}$$

where $p(M_j)$ is the prior probability assigned to model M_j (suitably normalized).

For example, let Γ_N be the class of all binary decision trees which can be generated using K binary attributes. A particular balanced tree of depth d could be coded using

$$C(M_d) = \sum_{i=0}^{d} 2^i \log (K - i)$$

bits, where $\log (K - i)$ bits are required to specify the node attribute at each of the 2^i nodes at depth i. We will assume $\log_2$, or units of bits, throughout.

Let us use $\mathscr{C} = \{C(M_1), \dots, C(M_{|\Gamma_N|})\}$ to refer to a particular coding scheme for Γ_N. Hence the total **description length** of the data plus a model M_j is defined as

$$L(M_j, \{x_i, y_i\}) = C(M_j) + C(\{y_i\} | M_j(\{x_i\}))$$

$$= C(M_j) + \log \left(\frac{1}{p(\{y_i\} | M_j(\{\mathbf{x}_i\}))} \right) \tag{25.3}$$

i.e. we first describe the model and then the class data relative to the given model (as a function of $\{\mathbf{x}_i\}$). Typically for classification problems we assume that $p(\{y_i\} | M_j(\{\mathbf{x}_i\})) = \prod_{i=1}^{N} p(y_i | \mathbf{x}_i)$, i.e. that the data represents independent samples from the environment. This assumption is not strictly necessary but makes the notation considerably easier to deal with.

The **stochastic complexity** of the data $\{\mathbf{x}_i, y_i\}$ relative to $\mathscr{C}$ and Γ_N is the minimum description length

$$I(\{\mathbf{x}_i, y_i\}) = \min_{M_j \in \Gamma_N} \{L(M_j, \{\mathbf{x}_i, y_i\})\} \tag{25.4}$$

Hence, we define the **minimum complexity** model M^* of the data relative to $\mathscr{C}$ and Γ_N as

$$M^* = \arg \{I(\{\mathbf{x}_i, y_i\})\} \tag{25.5}$$

The problem of finding M^* is intractable in the general case. However, since any inductive procedure of merit which considers multiple models suffers the same limitation, this does not limit the usefulness of the approach. In particular, it is usually sufficient to find a good local minimum (a relatively simple model which fits the data well) for most practical applications.

ADMISSIBLE STOCHASTIC COMPLEXITY MODELS

DEFINITION OF ADMISSIBILITY

We will find it useful to define the notion of an *admissible* model for the classification problem. Note that the notion of admissible *models*, as we will describe it, is quite distinct from the well-known notion in estimation of admissible *estimators*.

Definition 25.1
The set of admissible models Ω_N is defined as all models whose complexity is such that there exists no other model whose description length is known to be smaller.

In other words, we are saying that inadmissible models are those which have complexity in bits greater than any *known* description length—clearly they cannot be better than the best known model in terms of description length and can be eliminated from consideration. Hence, Ω_N is defined dynamically and is a function of how many description lengths we have already calculated in our search. Typically Γ_N may be predefined, such as the class of all decision trees of maximum depth k, or all 3-layer feedforward neural networks with particular node activation functions. We would like to restrict our search for a good model to the set $\Omega_N \subseteq \Gamma_N$ as far as possible (since non-admissible models are of no practical use). In practice, it may be difficult to determine the boundaries of Ω_N, particularly when $|\Gamma_N|$ is large (with decision trees or neural networks, for example). One could view the notion of admissibility as a generalization to the stochastic case of the deterministic techniques prevalent in artificial intelligence approaches to learning algorithms. Typically, these algorithms assume (unrealistically) that perfect goodness of fit is achievable and eliminate from consideration any model which cannot perfectly account for the data.

RESULTS FOR ADMISSIBLE MODELS

Simple techniques for eliminating obvious non-admissible models are of interest:

Theorem 25.1
For the classification problem a necessary condition that a model M_j be admissible is that

$$C(M_j) \leqslant NH(Y) \leqslant N \log(m) \tag{25.6}$$

where $H(Y)$ is the entropy of the m-ary class variable Y.

The proof of this result follows directly from the definition of description length and the fact that we can always imagine a coding scheme $\mathscr{C}$ where we assign very low complexity to the 'null model', where we leave the data in its original state and use no model. Hence, the obvious interpretation in words is that any admissible model must have complexity less than that of the data itself. It is easy to show, in addition, that the complexity of any admissible model is upper-bounded by the parameters of the classification problem: given $p(M_j)$, and without knowing anything about the data except the number of samples and (possibly) the entropy of the class variable, we have

Corollary 25.1

If $p(M_j) < 2^{-NH(Y)} < 2^{-N\log(m)}$ then $M_j \notin \Omega_N$. The proof of this corollary is a direct consequence of Equations 25.2 and 25.6. Hence, the size of the space of admissible models can also be bounded:

Corollary 25.2

$$|\Omega_N| \leqslant 2^{NH(Y)} \leqslant 2^{N\log(m)} \qquad (25.7)$$

The proof is straightforward. By Theorem 25.1 we have that

$$\sum_{i=1}^{|\Omega_N|} 2^{-C(M_i)} \geqslant 2^{-NH(Y)}|\Omega_N| \qquad (25.8)$$

and since by Equation 25.2 we have that the left-hand side is less than 1, we get the desired result.

Note that although the bound on $|\Omega_N|$ is exponential in $NH(Y)$, one must remember that the size of many model spaces can grow *super*exponentially as a function of their parameters due to basic combinatorics. Consider again the case of complete binary decision trees of depth d—there are of the order of 2^{2^d} such trees. In terms of Corollary 25.2, this in turn implies that for the class of admissible trees, d can only grow as a function of $\log(NH(X))$.

These results are interesting in the sense that they run counter to statements in the MDL and Bayesian literature that countably infinite sets of models may be considered for a particular inference problem. Our approach suggests that, for classification at least, once we know N and the number of classes m, there are strict limitations on how many *admissible* models we can consider. In practice, of course, one may not know these quantities exactly in advance, yet it seems reasonable that one can at least provide upper bounds at the time one is constructing the set of candidate hypotheses or models.

In decision tree design algorithms, for example, there is often either an implicit set of candidate models (trees) in use, or else a prespecified hypothesis space (tree descriptions) with codelengths. In practice, then, for many problems, the original set of models Γ may be very much larger than Ω_N. Of course, the theory does not state that considering a larger subset will necessarily result in a less optimal model being found. However, it is difficult to argue the case for including large numbers of models which are clearly too complex for the problem. At best, such an approach will lead to an inefficient search, whereas at worst a bad model will be chosen either as a result of a poor coding scheme for the unnecessarily large hypothesis space or by a chance fit of model to noise in the data.

DYNAMIC BOUNDS ON ADMISSIBLE MODELS

We can also calculate bounds on the extent of Ω_N dynamically during our model search. Imagine that we index the models by j such that we search the models in the order $M_1, \ldots, M_j, M_{j+1}, \ldots$. Let L_j (min) be the model of least description length which we have found so far, i.e. among models 1 through j. We can then define $\Omega_{j,N}$ as the admissible set of models at this point of the search. Naturally there is an ordering

$$\Omega_{j,N} \subseteq \Omega_{j-1,N} \subseteq \cdots \subseteq \Omega_{1,N} \Omega_N$$

Hence, the size of the admissible model space can shrink as we start to evaluate the description lengths of particular models. The equivalent statement to Theorem 25.1 is that for all models M_k, $k > j$,

$$\text{if } C(M_k) \geq L_j(min) \quad \text{then} \quad M_k \notin \Omega_{j,N}$$

For example, consider a coding scheme such that the model complexity term is $\frac{k}{2} \log N$—this is a commonly-used term in MDL applications, where k is the number of parameters in the model. In this case we need only ever consider models such that

$$k \leq \frac{2L_j(min)}{\log N}$$

In the next section we will see how this approach can be applied to particular applications.

ADMISSIBLE MODELS AND BAYES RISK

The Bayes risk for a given classification problem is the minimum attainable mean classification accuracy for the problem. For discrete-valued feature variables we can relate description lengths and P-yes risk for particular problems. The following definition is useful:

Definition 25.2
Let M_B be any model (not necessarily unique) which achieves the optimal Bayes risk (i.e. minimizes the classifier error) for the classification problem.

In particular, $C(\{y_i\}|M_B(\{x_i\}))$, the goodness-of-fit term, is not necessarily zero—indeed, in most practical problems of interest it is non-zero, due to the ambiguity in the mapping from the feature space to the class variable. In addition, M_B may not be defined in the set Γ_N, and hence, M_B need not even be admissible. If, in the limit as $N \to \infty$, $M_B \notin \Gamma_\infty$ then there is a fundamental *approximation* error in the representation being used, i.e. the family of models under consideration is not flexible enough to represent optimally the mapping from X^k to Y.

Let p_B be the classification error rate of the optimal Bayes model M_B. In principle this is only achievable with an infinite amount of data while in practice we have a finite amount of training data. Hence, we define $\hat{p}_B(N)$ as the minimum attainable error achievable on the training data, the finite training data analogue of the Bayes error rate. If N is sufficiently large relative to the dimensionality of the feature space then $\hat{p}_B(N) \simeq p_B$, while if N is too small then $\hat{p}_B(N) = 0$ (it is possible to 'memorize' the training data perfectly). Using this notion of $\hat{p}_B(N)$, we have a lower bound on attainable accuracy:

Theorem 25.2

$$C(\{y_i\}|M(\{\mathbf{x}_i\})) = \log\left(\frac{1}{p(\{y_i\}|M(\{\mathbf{x}_i\}))}\right) \geqslant 2N\hat{p}_B(N) \qquad \forall M \qquad (25.9)$$

Proof
It has been shown by Kovalevsky (1980) and others that

$$H(Y|X) \geqslant 2p_e \qquad (25.10)$$

where $H(Y|X)$ is the average conditional entropy of the discrete random variable Y given knowledge of the discrete random variable X, and p_e is the mean classification accuracy obtainable if we were optimally to use X to guess Y (i.e. knowing both the value of X and the conditional probabilities $p(y|x)$). Given any particular training data set, and a particular model M, imagine estimating the conditional probabilities $p(Y|M(X))$. Then one can calculate both $\hat{H}(Y|M(X))$, the empirical conditional entropy, and $\hat{p}_e(M(X))$, the minimum classification accuracy attainable using the estimated probabilities to predict Y. Since Kovalevsky's result is true in general for arbitrary channels, then the bound must hold for the channel defined by the empirically estimated probabilities. Hence we get that

$$\hat{H}(Y|M(X)) = \log\left(\frac{1}{p(\{y_i\}|M(\{\mathbf{x}_i\}))}\right) \geqslant 2N\hat{p}_e(M(X)) \qquad (25.11)$$

Since, by definition, $\hat{p}_e(M(X)) \geqslant \hat{p}_B(N)$, the proof is complete. ∎

Theorem 25.2 expresses the fact that there is a lower limit to the **compressibility** of the class data in terms of being described by the feature data. In practice, of course, it is difficult to estimate $\hat{p}_B(N)$ for a given problem. However, if we have a lower bound on $\hat{p}_B(N)$, namely $l(\hat{p}_B(N))$, then we can tighten the admissibility criterion stated in Theorem 25.1:

Corollary 25.3
Any admissible model $M_j \in \Omega_N$ must satisfy

$$C(M_j) \leqslant N(H(Y) - 2l(\hat{p}_B(N))) \leqslant N(\log(m) - 2l(\hat{p}_B(N))) \qquad (25.12)$$

APPLICATIONS OF ADMISSIBLE MODELS

We now consider the use of admissible models for specific applications. In general the applications can be divided into two general types. First, there is the use of

Theorem 25.1 to predefine an admissible model set Ω when initially choosing a set of candidate models. Second, there is the use of the results stated in the previous section to shrink dynamically the boundaries of Ω_N and hence to reduce the search space during the model selection phase. Naturally, the latter could be embedded within a general 'branch-and-bound' type of search algorithm—in this chapter we will not explore in any detail the connection between the particular approach presented here and generic branch-and-bound techniques.

MARKOV RANDOM FIELD MODELS

A recent application of using MDL to discover the appropriate order for Markov random field (MRF) texture models was reported by Smith and Miller (1990), with applications to the problem of segmenting electron-microscope autoradiography (EMA) images. Under their scheme, the description length per pixel for model M_j is calculated as

$$L(M_j, \{y_i\}) = -\frac{1}{N} \sum_{i=1}^{N} \log \hat{p}(y_i | S_i(j)) + \frac{k_j}{2N} \log N \qquad (25.13)$$

where N is the number of training pixels, y_i is the ith pixel, $S_i(j)$ is the Markov neighbourhood of y_i defined under model j, $\hat{p}(y_i | S_i(j))$ is the conditional probability estimate from the training data, and k_j is the number of parameters in the model (in this case k_j is the number of conditional probabilities which must be defined). Smith and Miller essentially considered three particular models, corresponding to the cases of no model, a first-, and a second-order model. The authors found that the best model (that which minimized $L(M_j, \{y_i\})$) for both the mitochondria and background textures, was the first-order model. This model had 1024 conditional probabilities. Given that the EMA images used had four grey levels, it is easy to show (from Theorem 25.1) that a necessary condition that a model M_j be admissible (i.e. $M_j \in \Omega_N$) can be written as

$$k \leqslant \frac{4N}{\log N} \qquad (25.14)$$

With the four 128×128 training images used in the experiment, this translates into a requirement that $k < 23\,637$. Since the second-order model requires over ten times this number of parameters, while the first-order model only needs 1024, we can conclude that the first-order model and the 'null' model are the *only* admissible models. Hence, the experimental results are entirely to be expected given the model of order 0 was unlikely to be competitive. The natural conclusion would be that, within the class of admissible models, other intermediate Markov models are worth considering in order to take advantage of the MDL approach, for example, models which do not require the full specification of *all* conditional probabilities. Goodman *et al.* (1992) describe the use of such 'sparse' Markov models as an alternative to standard fully-specified *n*-gram language models.

DECISION TREE CLASSIFIERS

Wallace and Patrick (1990) have recently proposed an extension to Quinlan and Rivest's (1989) original paper on using MDL to find good decision tree classifiers. The MDL approach seems well suited to the decision tree problem, which has typically been plagued by the fact that no well-defined criterion for controlling the size of the tree seems to exist. If we run through the analysis for Wallace and Patrick's scheme we find, for example, that the class of admissible trees must have no more leaves l than

$$l \leqslant \frac{2 + N(H(Y))}{1 + \log K} \leqslant \frac{2 + N \log m}{1 + \log K} \tag{25.15}$$

where $H(Y)$ is the entropy of the class variable and K is the number of attributes available. Similarly, with Rissanen's (1989, p. 165) proposed coding scheme for decision tree classifiers, we get that for all admissible models the number of leaves l must obey the following inequality:

$$\log\left(\frac{2l-1}{l}\right) + L^*(l) + \frac{l}{2}\log K \leqslant 1 + NH(Y) \tag{25.16}$$

where $L^*(\cdot)$ is Rissanen's recursive log function. For typical values of N, K, and $H(Y)$ these bounds are quite loose and, hence, of limited practical utility. However, the appearance of $H(Y)$ is a common thread across admissible classification models, that is, low-entropy problems permit fewer models to be considered than high-entropy problems, for fixed N. The notion of admissible models could play a role in decision tree design by providing dynamic bounds (which shrink as we continue to grow the tree) on the allowable complexity of candidate models, thus eliminating unnecessary node expansion.

MULTILAYER FEEDFORWARD NEURAL NETWORKS

There has been considerable recent interest in the application of multilayer neural networks for non-parametric estimation and classification. Barron and Barron (1988) provide a useful overview of the field in a statistical context. A significant problem with the current generation of network training algorithms (such as the back-propagation algorithm) is that they require that a network architecture is specified a priori, independently of the data. The MDL approach clearly shows that the architecture of the network needs to be constrained by factors such as the entropy of the class variable $H(Y)$ and the size of the data set N. For low-entropy problems, we only need simple architectures, or equivalently, for a fixed simple architecture, we need little data to support it. High-entropy problems may require more complex architectures, which in general require more data to support them. Hence, it is easy to see how the practice of fixing the network architecture in advance may cause problems if network complexity is not appropriately matched to the complexity of the problem.

Work on developing an MDL or Bayesian approach to neural network architecture selection is already underway (Mackay, 1992; Buntine and Weigend, 1991). As pointed out by Cybenko (1990) and Smyth (1991), there is a subtle problem in estimating the description length (complexity) of neural models, namely, that it is difficult to estimate the true dimensionality of the parameter space. The **precision** of the connection strengths appears to play a key role in network complexity estimation. As pointed out by Gish (Chapter 18, this volume), it is possible to formulate the network design problem within a likelihood framework—hence, in principle at least, optimal weight precision (and, as such, codelengths for the weights) can be calculated in a *data-dependent* manner such as that described in Wallace and Freeman (1987). However, it remains to be seen whether the computational complexity of such an exercise is feasible for practical network training algorithms. From the point of view of admissible models, the fact that the resulting 'codelengths' are completely data-dependent makes it difficult to apply any general bounds.

HYBRID RULE-BASED NETWORKS

Goodman *et al.* (1992) have described a hybrid rule-based neural network architecture for classification problems. In this model the hidden units correspond to boolean combinations of discrete input variables. The link weights from hidden to output (class) nodes are proportional to log-conditional probabilities of the class given the activation of a hidden node. The output nodes form estimates of the posterior class probabilities by a simple summation followed by a normalization. The implicit assumption of conditional independence is ameliorated in practice by the fact that the hidden units are chosen in a manner to ensure that the assumption is violated as little as possible.

The complexity penalty for the network is calculated as being $\frac{1}{2}\log N$ per link from the hidden to output layers, plus an appropriate coding term for the specification of the hidden units. Hence, the description length of a network with k hidden units would be

$$L_k = -\sum_{i=1}^{N} \log\left(\hat{p}(y_i|\mathbf{x}_i)\right) + \frac{k}{2}\log N - \sum_{i=1}^{k} \log \pi(o_i) \qquad (25.17)$$

where o_i is the order of the ith hidden node and $\pi(o_i)$ is a prior probability on the orders, and $\hat{p}(y_i|\mathbf{x}_i)$ is the network's estimate of the probability of y_i given $\mathbf{x}_i$. Using this definition of description length we get that the number of hidden units in the architecture will be bounded above by

$$k \leqslant \frac{L(min)}{0.5\log N + \log K + 1} \leqslant \frac{NH(Y)}{0.5\log N + \log K + 1} \qquad (25.18)$$

where K is the number of binary input attributes and $L(min)$ is the smallest description length we have found in the search so far.

Let us examine the calculation of the admissible models criteria for this particular set of rule-based networks when applied to a real-world medical diagnosis problem (also treated in Goodman *et al.*, 1992). We will not describe the details of the problem here—the interested reader is referred to Wolberg and Mangasarian (1990). There are

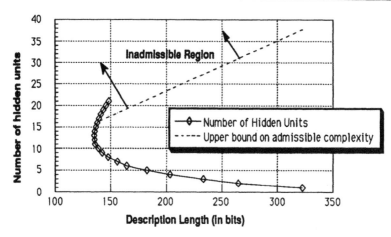

Figure 25.1 Inadmissible region as a function of description length.

two classes (*benign* and *malignant*), and nine attributes, each of which can take on any of ten discrete values. The training data consists of 439 labelled feature vectors. Given that the prior class entropy is almost 1 bit, one can immediately state from Equation (25.18) that networks with more than 51 hidden units are inadmissible. Furthermore, as we evaluate different models we can narrow the region of admissibility using the results stated earlier. Figure 25.1 gives a graphical interpretation of this procedure showing how the algorithm's search procedure causes a shrinking of the admissible model set, running on the medical diagnosis database.

The algorithm in effect moves up the left-hand axis, adding hidden units in a greedy manner. Initially the description length (the lower curve) decreases rapidly as we capture the gross structure in the data. For each model for which we calculate a description length, we can in turn calculate an upper bound on admissibility (the upper curve)—this bound is linear in terms of description length. Hence, for example, by the time we have 5 hidden units we know that any models with more than 21 hidden units are inadmissible. Finally, a local minimum of the description length function is reached at 12 units, at which point we know that the optimal solution can have at most 16 hidden units. Knowing that more complicated models are inadmissible gives us a certain degree of confidence that the model discovered by the algorithm (in the form of a set of probabilistic rules) is at least of the appropriate order of complexity given the data.

RELATED TOPICS AND OPEN RESEARCH ISSUES

A number of interesting related issues arise from this general framework. Let us define $H(\Omega_N) = -\sum_j p(M_j) \log (p(M_j))$ as the **prior** model entropy—it describes our initial uncertainty in terms of which models we think are likely to describe the data. How does this prior model uncertainty affect the complexity of the search problem? One expects that if the prior entropy is large then the search should be more difficult since all models are almost equally likely. Conversely, for low-entropy problems, the search should be easy if our prior intuition is correct—naturally, though, if we have placed a large amount of prior probability mass on models which are unsupported by the data, it may take a considerable amount of searching through the numerous unlikely (from a prior standpoint) models to find a good model.

The problem of **incremental** learning also brings up some interesting points. Consider an agent which receives data over time, in batches of size k, where k can be as small as 1. Clearly, if the agent wishes to adhere to the principles of admissible models, he must allow the size of his hypothesis space to grow at a reasonable rate as a function of the amount of data he has seen so far. Asymptotic results for this type of problem are available in the work of Barron and Cover (1991) (for the MDL framework) and White (1989) (for the method of sieves). However, there is little work in the literature (as far as the author is aware) devoted to more practical, non-asymptotic results for this class of problems. A naive conjecture might be that the agent should only consider very simple memory-based models initially (such as nearest-neighbour classifiers) and gradually extend his model space to more sophisticated representations as justified by the data (an obvious analogy exists with animal learning behaviour and the onset of generalization phenomena in memory-based mechanisms). By embedding the MDL model selection technique within a general automated **hypothesis generation** framework, the agent should be able to display more interesting and (ultimately) more powerful behaviour. How the automated hypothesis generation mechanism might work is another issue—one can certainly imagine heuristic approaches tailored for specific domains as have been explored in a simple manner by AI researchers in automated discovery (Shrager and Langley, 1990) and machine learning (Mehra *et al.*, 1989). Of course, this 'gradually expanding hypothesis space' approach produces even more open questions from a statistical viewpoint. Within such a general framework, issues such as admissible models become a very important consideration.

CONCLUSION

It should not be construed from this chapter that consideration of admissible models is the major factor in inductive inference—certainly the choice of description lengths for the various models and the use of efficient optimization techniques for seeking the parameters of each model remain the cornerstones of success in finding good models from data. Our analysis of various classes of models (Markov models, trees, neural networks) and specific coding schemes revealed that admissible models are more useful for certain problems than others. None the less, our results suggest that the structure of the MDL formulation may be taken advantage of for search purposes and are practical to the extent that they provide a 'sanity check' for model selection in MDL.

ACKNOWLEDGEMENTS

The author would like to thank the anonymous referees for their constructive comments on an original version of this chapter. The research described in this chapter was performed at the Jet Propulsion Laboratory, California Institute of Technology, under a contract with the National Aeronautics and Space Administration, and was supported in part by DARPA under grant number AFOSR—90—0199.

REFERENCES

Barron, A. and Barron, R. (1988) Statistical learning networks: a unifying view. Paper presented at the 1988 Symposium on the Interface: Statistics and Computing Science.

Barron, A.R. and Cover, T.M. (1991) Minimum complexity density estimation. *IEEE Transactions on Information Theory*, **37**, 1034–1054.

Buntine, W.L. and Weigend, A.S. (1991) Bayesian back-propagation. Technical Report FIA-91-22, RIACS and NASA Ames Research Center, Moffett Field, CA.

Cybenko, G. (1990) Complexity theory of neural networks and classification problems, in *Neural Networks*, Almeida, L. (ed.), Springer Lecture Notes in Computer Science, pp. 26–45.

Goodman, R.M., Smyth, P., Higgins, C. and Miller, J.W. (1991) Rule-based networks for classification and probability estimation. To appear in *Neural Computation*.

Kovalevsky, V.A. (1980) *Image Pattern Recognition*, translated by A. Brown, New York: Springer-Verlag, p. 79.

Mackay, D. (1992) A practical Bayesian framework for backprop networks. To appear in *Neural Computation*.

Mehra, P., Rendell, L.A., Wah, B.W. (1989) Principled constructive induction, in *Proceedings of IJCAI 1989*, San Mateo, CA: Morgan Kaufmann, pp. 651–656.

Quinlan, J.R. and Rivest, R. (1989) Inferring decision trees using the minimum description length principle. *Information and Computation*, **80**, 227–248.

Rissanen, J. (1984) Universal coding, information, prediction, and estimation. *IEEE Transactions on Information Theory*, **30**, 629–636.

Rissanen, J. (1989) *Stochastic Complexity in Statistical Inquiry*, Teaneck, NJ: World Scientific.

Shrager, J. and Langley, P. (eds) (1990) *Computational Models of Scientific Discovery and Theory Formation*, San Mateo, CA: Morgan Kaufmann.

Smith, K.R. and Miller, M.I. (1990) A Bayesian approach incorporating Rissanen complexity for learning Markov random field texture models, in *Proceedings of the IEEE International Conference on Acoustics, Speech, and Signal Processing*, New York: IEEE Press.

Smyth, P. (1991) On stochastic complexity and admissible models for neural network classifiers, in *Advances in Neural Information Processing 3*, Touretzky, D., Lippman, R. and Moody, J. (eds), San Mateo, CA: Morgan Kaufmann, pp. 818–824.

Wallace, C.S. and Freeman, P.R. (1987) Estimation and inference by compact coding, *Journal of the Royal Statistical Society*, B, **49**, 240–251.

Wallace, C.S. and Patrick, J.D. (1990) Coding decision trees. To appear in *Machine Learning*.

White, H. (1989) Learning in artificial neural networks: a statistical perspective. *Neural Computation*, **1**, 425–464.

Wolberg, W.H. and Mangasarian, O.L. (1990) Multisurface method of pattern separation for medical diagnosis applied to breast cytology. *Proceedings of the National Academy of Sciences, U.S.A.*, **87**, 9193–9196.

Combining the probability judgements of experts: statistical and artificial intelligence approaches

26

L.A. Cox, Jr.

INTRODUCTION

Suppose that N agents must jointly select one from a finite set A of possible acts. Agents can be human decision-makers, knowledge sources sharing inferences and expertise via a blackboard, or expert systems distributed throughout a complex system and exchanging messages via an internal communication network. Unless all N of them agree on a choice from A within a certain amount of time, a 'disagreement outcome' occurs by default. After an act has been chosen by consensus, 'nature' chooses a state from a set S of possible states. A consequence from a set C of possible consequences then results. In the simplest causal model, the consequence corresponding to a particular act–state pair is uniquely determined and the causal correspondence between choices and consequences can be represented by a consequence function $c: A \times S \Rightarrow C$ giving the consequences for each agent of each act–state pair (a, s). Elements of C are typically represented as N-ary 'payoff vectors' for the agents (Rosenschein and Breese, 1989). Let $c(a, s)$ denote the consequence obtained when the group chooses act a in A and 'nature' chooses act s in S. Many apparently more complicated decision problems can be reduced to this 'normal form' by suitable definition of A and S; see, for example, Luce and Raiffa (1957, especially Chapter 3) for a classic treatment of this framework for decision analysis.

If agents initially disagree about the probable consequences of acts in A, how can they resolve their differences and reach a mutually agreed decision? That is the subject of this chapter. Agents may initially disagree on which element of A to choose because of differences in their preferences for outcomes in C, differences in their attitudes toward risk, or different understandings of what the acts in A entail (Bacharach, 1975; Bonduelle, 1987). This chapter concentrates on disagreements arising from differences in beliefs about the probable consequences of alternative acts in A. Thus, agents are assumed to constitute a **team** in that they have the same preferences for final consequences in C and for lotteries over C (and thus for the acts in A when a probability

Artificial Intelligence Frontiers in Statistics: AI and statistics III. Edited by D.J. Hand. Published in 1993 by Chapman & Hall, London. ISBN 0 412 40710 8

measure $p(s)$ over states s in S is common knowledge and each act can be defined with enough precision to be viewed as a deterministic mapping from $A \times S$ to C). Without the team assumption, game-theoretic issues from preference revelation, mechanism design, and bargaining theory arise that are beyond the scope of this chapter. The papers in Roth (1985) and in Gasser and Huhns (1989) give excellent discussions of these issues for human and machine agents, respectively, especially for the case where A involves allocations of a limited resource among group members.

The problem of resolving conflicting beliefs about causal mechanisms—here represented by disagreement over the probable consequences caused by different choices of acts—is a fundamental one for applied artificial intelligence. For example, some expert systems use an architecture in which multiple expert 'knowledge sources' or 'critics' (typically implemented as alternative rule bases) are brought to bear on the same user-supplied problem or query. When they give conflicting answers or recommendations, the problem of reconciling the results arises. Similarly, in plausible reasoning systems, different lines of plausible argument or plausible inference may support conflicting conclusions. The problem of combining evidence across multiple sources can also sometimes be usefully viewed from the standpoint of reconciling disagreements in the opinions of different agents. If an agent acts purely as a source of beliefs (with no other preferences or strategies to confound its behaviour) then interpreting knowledge sources, plausible inference rules and arguments, or pieces of evidence as the 'agents' whose beliefs are to be reconciled casts each of these well-known problems in applied AI as a problem in resolving the initially conflicting beliefs of different agents in a team.

STATISTICAL APPROACHES TO COMBINING PROBABILITY JUDGEMENTS

The problem of belief resolution for multiple knowledge sources with beliefs represented by probability measures has been extensively studied in the statistics and management science literatures, often in the context of a single decision-maker being advised by experts who disagree. This section briefly surveys statistical approaches to formulating and solving the resolution problem. This provides the background for an examination of alternative 'knowledge-based' approaches.

RESOLUTION BY AGGREGATION: AN UNSUCCESSFUL STRATEGY

A fully adequate theory of belief resolution must answer three basic questions. First, how do agents *form* their beliefs based on observations? Second, how are beliefs *represented* for purposes of analysis? Third, how are beliefs *revised* during the resolution process? These questions may be answered differently in different domains (for instance, in contexts where other intelligent agents are involved as against contexts where only natural causal laws operate). The remainder of this chapter focuses on beliefs about the relation between acts in A and consequences in C. Beliefs about intents, plans, goals, and likely behaviours of other agents are not treated except inasmuch as they fall within this framework.

In Bayesian statistics, beliefs are represented by probability measures over subsets of S and are revised by conditioning. (How they are formed prior to conditioning on observations is left open, despite a variety of maximum-entropy schemes and other proposals. However, they can in principle be measured, even if they cannot be derived, by reference to subjective betting odds.) A classical approach to belief resolution is **axiomatic aggregation of probabilities** (Bacharach, 1975; Bordley, 1982; Morris, 1986). Each agent's beliefs are represented by a subjective probability measure over S. Let $p(j)$ denote the probability measure representing the beliefs of agent j and let $p(s; j)$ denote the probability density assigned to state s by this measure. An aggregation function is a function that maps tuples of probability measures onto probability measures. If f is such a function and $[p(1), \ldots, p(N)]$ is an N-tuple of individual probability measures, then the 'aggregate' probability measure associated with this tuple by f is denoted by $f[p(1), \ldots, p(N)]$. For example, a simple aggregation function is $f[p(1), \ldots, p(N)] = w(1)p(1) + \cdots + p(N)w(N)$, where the weights $w(j)$ satisfy $w(1) + \cdots + w(N) = 1$. This weighted average aggregation procedure assigns probability density $p(s; f) = w(1)p(s; 1) + \cdots + w(N)p(s; N)$ to state s when the agents $j = 1, \ldots, N$ assign probability densities $p(s; 1), \ldots, p(s; N)$ to it. $p(s; f)$ denotes the probability density assigned to state s by the aggregated probability measure.

Rather than using arbitrary aggregation functions, it is natural to seek sets of normative axioms for probability aggregation that will single out a unique aggregation function. Despite its intuitive appeal and some interesting mathematical results, this line of research does not produce a useful theory of belief resolution. Consider the types of axiom that might seem reasonable to impose on aggregation functions. Candidates include the following:

Axiom 1 (Unanimity)
If $p(1) = p(2) = \cdots = p(N) = p$, so that all agents have identical beliefs, then it should be the case that $f[p(1), \ldots, p(N)] = p$.

Axiom 2 (Continuity)
The aggregation function is continuous (i.e. $p(s; f)$ varies continuously with respect to $p(s'; j)$ for any two states s and s' and any agent j).

Axiom 3 (Monotonicity)
If all agents agree that state s is at least as likely as state s', then the aggregate probability measure should preserve this relative ordering, i.e. $p(s; j) \geqslant p(s'; j)$, for $j = 1, \ldots, N$, implies $p(s; f) \geqslant p(s'; f)$.

The problems with using such axioms to obtain an aggregation function as an approach to belief resolution are as follows. First, the axioms are not consistent with each other in general (unless there are only two possible states, or one agent's beliefs are selected ignoring the rest, or other similarly extreme restrictions are imposed). That is, no general probability aggregation function exists that simultaneously satisfies all the apparently reasonable axioms (Barrett and Pattanaik, 1987). Unique aggregation formulae are produced by some subsets of axioms (see, for example, Barrett and Pattanaik, 1987, for axiomatic derivations of additive and other aggregation formulae and the papers by Winkler, Lindley, Schervish, Clemen, French, and Morris for an excellent introduction to axiomatic and Bayesian expert aggregation). However, all such formulae violate one or more normative axioms. Barrett and Pattanaik (1987) present details of several 'impossibility theorems' identifying subsets of normative axioms that are mutually inconsistent.

Second, aggregation does not solve the resolution problem, although it may produce an agreement on an act if the agents have committed themselves to abide by the results of the aggregation process (and if the consequent incentives to misrepresent their true beliefs can be ignored). By 'resolution' of differing beliefs, we mean a process by which the agents come to agree on a set of shared beliefs. An aggregation function identifies a single probability measure, but fails to provide an adjustment process by which agents will (or rationally should) come to share it.

Third, the axioms are not necessarily reasonable if agents have dependent information—i.e. if the empirical bases for their beliefs overlap or interact—(Bordley, 1982; Winkler, 1981). The following example illustrates this last point.

Example 26.1: Unanimity is not always desirable for aggregation
Consider an axiom stating that if all agents assign the same probability to a state, then the aggregate probability measure should also do so. Now suppose that three experts are asked about the probability that a certain multi-component reliability system will function. Assume that the structure of the system is common knowledge and is as follows: first, the system functions if any of its three components functions; and second, the prior probability that each component will function is 0.5, independent of the states of the other components. Finally, suppose that an expert can tell whether a component will function by inspecting it. If expert 1 inspects component A, expert 2 inspects component B, and expert 3 inspects component C, and if each expert then reports that, in her opinion, the system has a 75% chance of functioning, then it might seem desirable to require that the probability assigned to the state 'system will function' by any aggregation of these three opinions should also be 75%. However, a slightly more knowledge-based approach reveals the defect in such reasoning. An expert who examines exactly one component but not the other two concludes that whole system has a 75% chance of working only if the component she has observed is non-functioning. (Otherwise, she would conclude that the system has a 100% chance of functioning.) Therefore, given the reports of the three experts, it is clear that the system has a 0% chance of working, rather than a 75% chance. Similarly, if only two experts had inspected components and if each had independently reported a 75% chance that the system will function, it would follow that the true probability was 50%. Thus, any naive aggregation formula (e.g., weighted averaging) that preserves unanimously believed probabilities would fail in this example.

RESOLUTION BY ITERATIVE COMMUNICATION OF PROBABILITIES

In Example 26.1, the discrepancy between the unanimous expert opinions about the probability of an event (namely, that the system is functional) and the probability for this event that can be inferred from these opinions would disappear if the experts shared their findings with each other before presenting their final opinions. This suggests the following very general resolution procedure for beliefs represented by probability measures:

The common-knowledge procedure

Let each agent in turn announce his posterior probability measure conditioned on all announcements made so far (where all conditioning is accomplished using Bayes's rule). Repeat this cycle until all agents have the same posterior probability measure and until this posterior probability measure is common knowledge.

Remarkably, under conditions discussed below, this simple procedure is guaranteed to terminate in a finite number of iterations, leaving all N agents with a shared posterior probability measure that is common knowledge (McKelvey and Page, 1986; 1990; Bacharach, 1985). By 'common knowledge' is meant not just that all agents know the final probability measure, but that all agents know that all agents know it, know that all agents know that all agents know it, etc. (Notice that a fact becomes common knowledge if it is announced publicly in a context where everyone knows that everyone is listening.)

To sharpen the statement of this result, we adopt the following probabilistic model for the beliefs of agents. Each agent has a (possibly noisy) **channel** through which it observes (i.e. receives private information about) the state of the world. A channel is a probabilistic mapping from states in S onto a set of potentially observable signals, Y; i.e. it is a set of conditional probability density functions $p(y|s)$ over signals in Y given states in S. Assume that the channels through which agents observe the state are common knowledge, that a prior probability measure over states in S is also common knowledge, and that S is finite. However, each agent's observations, and therefore his posterior beliefs about the state given these observations, are initially his own prior knowledge. Since different agents have different information (perhaps reflecting different channel characteristics as well as different observed signals) they generally disagree in their opinions about the probabilities of states in S—and hence about the probable consequences $c(a, s)$ of acts a in A. In this model, beliefs are *formed* by updating the common knowledge prior probability measure with private information. Beliefs are *represented* as probability measures. And beliefs are *resolved* by the common knowledge procedure. The fundamental theorem of the economics literature on common knowledge is that in this model of agent beliefs, successive announcements of their posterior probabilities by the agents will lead them back to a common-knowledge posterior on which they all agree—a true resolution of their initially conflicting beliefs. What is surprising is that the agents need not directly reveal to each other what they saw (their private information): exchanges of probabilities suffice.

Example 26.2: An example of the common-knowledge procedure
Consider again the reliability system from Example 26.1, and suppose now that it works if and only if at least two out of its three components work. Suppose also that this time there are only two experts: expert 1, who inspects A and B, and expert 2 who inspects B and C. Each expert has a deterministic channel mapping the eight possible states of the system (three components, each with two possible states) onto one of four possible signals (telling which of the inspected components works). The prior probability measure assigns probability $\frac{1}{8}$ to each of the system states and is common knowledge. Suppose that expert 1 announces a posterior probability that the system works (after observing his private information, i.e. the states of components A and B) of 0.5. Expert 2 will realize that expert 1 must have observed exactly one defective component—either A or B. Moreover, expert 2 will know whether B is defective from his own direct observation. Therefore, given expert 1's announced posterior probability, he can deduce what expert 1 actually saw, i.e. which of the two components A and B is defective. Combining this knowledge with his own observations, he will be able to deduce whether the system has probability 0 probability 1 of functioning. When he announces the result, expert 1 will be able to deduce what expert 2 has seen, and a common-knowledge posterior will have been reached.

In this example, convergence of beliefs to a common-knowledge posterior occurs in a single round of probability exchanges. More generally, especially with noisy channels, there may need to be many rounds of updating before common knowledge is achieved. In fact, examples can be constructed for any positive integer, k, such that for k rounds the agents exchange probabilities that remain constant, and then in round $k + 1$ both probabilities adjust to a new final value as common knowledge is finally achieved (Geanakoplos and Polemarchakis, 1982). The common-knowledge procedure depends in general on having the agents reason quite subtly about the possible private information that other agents can have that is consistent with their announcements and with Bayesian updating. It is one of the earliest and most fruitful results of the 'reasoning about knowledge' line of research initially explored by mathematical statisticians and economists and now attracting great interest in the AI community (Rosenschein, 1985).

BAYESIAN COMBINATION OF EXPERT PROBABILITIES

By allowing direct communication among the experts, the common-knowledge procedure generalizes and in some important ways improves upon a class of **Bayesian aggregation procedures** that have been studied in the statistics and management science literatures as an alternative to axiomatic aggregation of probabilities (Morris, 1977; Winkler, 1986; Bonduelle, 1987). In these procedures, experts are viewed as noisy measurement devices or channels. A single decision-maker advised by the experts receives their announced probabilities for an event as inputs and conditions his own beliefs on these announcements and on his beliefs about the experts themselves (their reliabilities, calibrations, etc.). To this framework, the common-knowledge paradigm contributes the important ideas of treating the decision-maker as just another expert or knowledge source for the purposes of belief resolution (since all experts will come to agree after their private information has been combined via successive rounds of Bayesian conditioning to give a common-knowledge posterior); and of allowing the Bayesian updating of beliefs based on the announcements of others to proceed for more than one round, so that a common-knowledge equilibrium can be reached.

The strengths and limitations of the Bayesian approach are illustrated by the following result. Suppose that each expert is asked to assess the probability of an event E and let p_i be the probability assigned to E by expert i. For algebraic convenience, let $L_i = p_i/(1 - p_i)$ denote the corresponding odds in favour of E, as assessed by i. Assume that the source of i's expertise is observation of a binary signal whose value (0 or 1) is correlated with the occurrence of E; specifically, if E occurs, then the probability that i sees a '1' is f_{i1}, while if E does not occur, then the probability that i sees a '1' is f_{i2}. Provided that f_{i1} and f_{i2} are different, i's observation contains some information about the occurrence of E. Finally, denote the prior odds favouring E by L_{prior}, which is assumed to be common knowledge. Then it follows from Bayes's rule that the decision-maker's posterior odds for E, after receiving the experts' posterior probabilities $\{p_i\}$ (or, equivalently, their individual posterior odds $\{L_i\}$) should be given by the simple multiplicative formula $L_{post} = L/[L + (L_{prior})^{1-N}]$, where $L = L_1 L_2 \cdots L_N$ is the product of the posterior odds for the N experts. This combination rule is independent of the detailed information (determined by f_{i1} and f_{i2}) for the experts. However, no similar formula can be derived if there are three or more states (rather than just 'E' and 'not-E') whose probabilities are to be assessed; in general, Bayesian combination of individual posterior probabilities requires knowledge of the individual channel characteristics (the source of each expert's 'expertise' in this framework). Without such knowledge, the decision-maker must make judgements about how well calibrated and reliable each expert is and about correlations among different experts. Bordley (1982) discusses these types of judgement for a more general multiplicative combination formula obtained by axiomatic methods.

KNOWLEDGE-BASED APPROACHES TO BELIEF RESOLUTION

The model of agent beliefs used in the above theory is limited in several ways—for instance, by its assumptions that agent beliefs are represented by probability measures; that agents share common knowledge of the prior probability measure and of each others' channels; and that all reasoning is accomplished by Bayes's rule. The remainder of this chapter examines some alternatives suggested by current paradigms in artificial intelligence.

DEDUCTIVE MODELS OF BELIEF FORMATION AND REVISION

In the economic model, the source of an agent's opinions and expertise is private information that he obtains via observations. A more general model of agent beliefs stresses the fact that agents may know not only different empirical information, but also different rules for reasoning from known facts and probabilistic information to deduced conclusions. More specifically, suppose that each agent's knowledge can be decomposed into three parts: a **rule base** K consisting of causal rules of the form

(act a causes consequence c) if (assertions $a_1, a_2, \ldots, a_k$ are all true)

and logical rules or logical constraints such as

(assertion a_j is true) if (assertions $a_{j1}, \ldots, a_{jn}$ are all true);

a **fact base** or **empirical information set**, I, consisting of concrete facts of the form (assertion a_j is true) or (assertion a_j is false); and a **statistical information set**, Z, consisting of statements of the form (assertion a_j has prior probability p_j). The logical language used to express causal relations, logical constraints and concrete facts may be as simple as propositional calculus restricted to Horn clauses or as complex as full predicate calculus augmented with modal operators and other expressive devices. The choice of language depends on the domain being discussed and on the desired trade-offs between expressive power and representational adequacy versus computational tractability and decidability (Brachman and Levesque, 1985). Without formally specifying the particular languages to be used to represent an agent's knowledge, we shall assume that the partitioning of agent j's knowledge into sets $K(j)$, $I(j)$, and $Z(j)$ for rules, concrete facts, and prior statistical information, respectively, makes sense. $K(j)$ may be interpreted as agent j's **causal model** of how the world works. $I(j)$ may be interpreted as j's **empirical information**, and $Z(j)$ as his **statistical information** about the state of the world. Facts in $I(j)$ might typically be derived from direct observations, while statements in $Z(j)$ represent prior statistical information and evidence ('beliefs') about the world from unspecified sources.

For simplicity, the prior probabilities in $Z(j)$ are assumed to be common knowledge, and hence are identical across agents, i.e. $Z(j) = Z^*$ for all j, where Z^* is the common-knowledge prior statistical information. Of course, updating these shared prior statistical expectations with the private information in their individual information sets $I(j)$ in general gives the agents different beliefs about the truth probabilities of the assertions $\{a_j\}$ in their knowledge bases. The assumption of a common-knowledge prior is therefore innocuous and not to be taken literally, since all agents may have different empirical information sets at the time they are first brought together to make a decision. However, it allows a theoretically clean explanation of differences in agent beliefs in terms of differences in information.

The intuitive idea behind our knowledge-based approach to belief resolution is that agents may disagree in their opinions about the probable consequences of acts in A not only because of differences in their empirical information—the sole source of disagreement in the common-knowledge model of agent beliefs—but also because they have different rules or causal models relating their empirical information to implied probabilistic relations between acts in A and consequences in C. To resolve this 'knowledge-based' (as opposed to 'information-based') source of disagreement, we first assume that each agent's causal model is *sound*, meaning that the rules in it are correct (i.e. they correctly describe the ways in which the world works and constraints on possibilities). However, they may be *incomplete*, meaning that an agent may not know (have in his causal model) all the relevant rules. Making these rough concepts precise requires a more formal treatment of the languages in which rules are expressed. Here, we concentrate on intuitive ideas. The chief idea is that to resolve conflicting opinions about probabilities, it is sufficient for the agents to build a **shared causal model** and then to combine their empirical information. Then all N agents will calculate the same posterior probabilities with respect to the merged rule set, say K^*, and information set, say I^*. This resolution procedure involves direct merging of the statements representing agents' knowledge and beliefs. Provided that the single-agent problems involved in inferring the beliefs generated by such a merged set of statements can be solved, and that the communication capabilities required to accomplish the merge can be provided, knowledge-based resolution provides a quite general approach to resolving conflicting opinions.

Example 26.3: Knowledge-based belief resolution
Suppose that two experts have the following knowledge about a two-component reliability system with components A and B:

Expert 1's knowledge:
K_{11}:   (system is functional) if (A is functional and B is functional)
K_{12}:   (system is not functional) if (A is defective and B is defective)
I_{11}:   (A is functional)
Z_1^*:   (P(A functional) $= P$(B functional) $= 0.5$)
Z_2^*:   Failures of components occur statistically independently of each other
K_1^*:   (if system functions when a set T of its components functions, then system functions when any superset of T functions)
K_2^*:   (if system fails when a set F of its components fails, then system fails when any superset of F fails)
K_3^*:   (system functions for some state of its components)
K_4^*:   (every component is essential, i.e. it makes the difference between whether or not the system functions for at least one set of component states)

Expert 2's knowledge:
I_{21}: (B is not functional)
$Z_1^*, Z_2^*, K_1^*-K_4^*$.

Here, K_{ij} denotes the jth causal rule for agent i; I_{ij} denotes the jth fact for agent i; and starred statements are assumed to be common knowledge. Rules $K_1^*-K_4^*$ express the meta-knowledge that the system is 'coherent', in the terminology of systems reliability theory (Barlow and Proschan, 1975). They (and also Z_2^*) are obviously drawn from a much more powerful language than the other causal rules, which can be adequately expressed in propositional logic. They reflect a meta-assumption that the system has been designed by an intelligent designer to be coherent. The statistical knowledge in Z_1^* refers only to **prior** information; it must be modified by experts 1 and 2 to account for the specific facts I_{11} and I_{21}, respectively.

What is the probability that the system is functional? As judged by expert 1 in isolation, given only the knowledge $(K(1), I(1), Z^*)$, the two constraints K_{11} and K_{12} do not determine a unique probability. In contrast to the management science and economic models previously described, agents' beliefs are represented not by complete probability measures over the four states in S but by explicit statements. Thus, agents are allowed to have **incomplete probability models** (Cox, 1990). In particular, expert 1 knows that there are exactly two possible completions of his causal model that are logically consistent with what he already knows: the system functions if and only if both A and B function; and the system functions if and only if at least one of A or B functions. (Possibilities such as that the system functions if and only if A functions are eliminated by the coherence requirements.) The probabilities that the system will function under each of these two alternative model completions are 0.5 and 1, respectively. Thus, for expert 1, the possible probabilities are $\{0.5, 1\}$. Which one is correct depends on which set of logically consistent assumptions is correct about the two missing causal rules in $K(1)$—i.e. the rules determining the state of the system for (A defective, B functional) and for (A functional, B defective), respectively. Nothing in expert 1's knowledge allows a particular assumption set to be singled out, so a unique probability for the system functioning is logically underdetermined.

Note that one subjectivist school of modern decision analysis would argue that the two completions of the model should be treated as if they were equally likely, for instance, because this is a 'maximum-entropy' or 'non-informative' prior distribution representing ignorance of the true model. By this reasoning, expert 1 would have a probability of 0.75 for the event 'system functions'. Our view is that a distribution does not represent ignorance, but rather a default assumption similar to those made in some non-monotonic logics (Etherington, 1988). However, unlike default logic, this initial default assumption cannot be gracefully retracted when new information is acquired. Bayesian conditioning on subsequent information can never lead to a conclusion that the prior distribution was wrong, given what was known. In the Bayesian paradigm, current beliefs must always be an update of initial beliefs conditioned on subsequent observations. Thus, an initial default assumption is not innocuous: in a sense, it can never be retracted. For this reason, we use a set of *possible probabilities* rather than a single probability number based on a default prior to account for ignorance due to incompleteness of the causal model. In summary, we resist combining 'epistemic probabilities' expressing speculations about model rules with 'evidential probabilities' expressing beliefs about facts.

For expert 2, the two possible completions of his empty causal model that satisfy the coherence requirements are also (system works if and only if both A and B work) and (system works if and only if at least one of A or B works). (Notice that coherence alone implies both of the causal rules in expert 1's knowledge base.) The probabilities that the system will work under these two model completions, given that B does not work, are 0 and 0.5, respectively.

Knowledge-based resolution of the probability opinions 0.75 from expert 1 (if he uses a uniform prior over model completions, or $\{0.5, 1\}$ if he reports possible probabilities) and 0.25 (or $\{0, 0.5\}$) from expert 2 proceeds as follows. First, the statements in the two knowledge bases are merged, giving the common knowledge base

1. (system is functional) if (A is functional and B is functional)
2. (system is not functional) if (A is defective and B is defective)
3. (A is functional)
4. (B is not functional).

$K_1^*-K_4^*$

The statistical information in Z_1^* and Z_2^* is superseded by empirical information, and is no longer needed. The possible probabilities are now 0 and 1, or 0.5 if a uniform prior distribution over models is used. The opinions of both experts have been fully resolved. They will agree that $\{0, 1\}$ is the possible probability set; moreover, they will agree on the probability that the system will function based on any assumed complete causal model. Finally, given *any* additional fact or causal rule, they will be able to determine whether the system will function. Although the conclusion in this simple example may seem somewhat disappointing—either the system works or it does not and nothing more can logically be concluded without further information—the resolution procedure produces less trivial sets of possible probabilities in more complex cases. The main point is that resolution has been achieved and the most specific conclusion warranted by the combined knowledge of the two experts has been drawn. If a decision-maker insists that he or she needs a single probability number to make a decision, the experts can truthfully report that there is insufficient objective information and knowledge available to them to justify such a single probability. Any further reduction of the set of possible probabilities for the convenience of the decision-maker (absent further knowledge or information) would not improve the objective basis for decision-making, but only conceal its weakness. In short, realistic decision-making must often confront the fact of inadequate information. Using incomplete probability models is one approach to force this confrontation.

KNOWLEDGE-BASED BELIEF RESOLUTION IN INFLUENCE DIAGRAMS

The strategy we have advocated of merging agents' knowledge and then recalculating probabilities with respect to the merged knowledge base has been extensively developed and applied to the special case of influence diagrams by Bonduelle (1987). Influence diagrams (Smith, 1989) are generalizations of the belief networks currently being studied in artificial intelligence and expert systems technology (Pearl, 1988). The main differences between the present framework and Bonduelle's influence diagram approach are that we do not require agents to have complete causal or probabilistic models; and that we have emphasized deterministic rather than stochastic causal rules. All rules in an influence diagram can be represented by conditional probability statements of the form 'The probability that variable j has value x when the values of variables $k, l, \ldots, m$ are $w, y, \ldots, z$ is $f(x|w, y, \ldots, z)'$. Other types of knowledge requiring different representation languages (e.g., 'the system being modelled is coherent') cannot be represented explicitly within the influence diagram framework. On the other hand, the influence diagram formalism provides a variety of effective computational techniques for representing probabilistic knowledge, propagating new information through probabilistic relations, converting disagreements about models into disagreements about probabilities, and so forth (Bonduelle, 1987). Ongoing research at US WEST Advanced Technologies in Boulder, Colorado, is exploring extensions of these ideas to contexts in which statements from more expressive knowledge-representation languages are needed to represent agents' beliefs.

CONCLUSIONS AND OPEN QUESTIONS

The idea of resolving conflicting expert opinions by first combining the bases for their opinions has often been considered but dismissed as being impractical for many applications (Barrett and Pattanaik, 1987). However, there are many multi-agent decision and control applications in which the time and vocabulary to build shared causal knowledge and information bases are available. In such cases, resolution of expert knowledge can give different (and demonstrably better-informed) results than axiomatic or Bayesian aggregation of expert probabilities.

The theory of common knowledge provides an alternative to direct sharing of facts and rules. However, it requires complete beliefs, represented by probability measures, and restrictive assumptions about what is initially common knowledge (e.g., the channels by which different players observe the state of the world). Moreover, the common-knowledge posterior probabilities obtained by successive exchanges of opinions may in principle differ from the ones that would result from direct sharing of observations, although such an outcome is unlikely (Geanakoplos and Polemarchakis, 1982). As in the case of influence diagrams, the types of knowledge that can be represented in the common-knowledge model are very limited. Meta-knowledge about causal processes cannot be explicitly represented.

Resolution via merging of expert knowledge is conceptually appealing but raises many questions, especially about computational tractability of inferences, that have not been addressed here. Moreover, major questions about the explicit dynamics of model formation, application and revision remain to be answered, as is also the case for single decision-making agents. A final example will illustrate this type of problem. Suppose that a third expert were added to the setting in Example 26.3. Expert 3 has only one item of knowledge, namely, 'system functions if A functions and B is defective'. As long as the experts do not share their knowledge, this creates no difficulties. But if this rule is pooled with the rules in the causal models held by experts 1 and 2, it becomes clear that the models are mutually inconsistent (coherence is violated: either B is an entirely superfluous component or fixing a defective component can cause the system to fail). In this case, what should the agents conclude? It is clear that some of their 'knowledge' must be wrong, but it is not clear what alternative models might be appropriate. Belief formation, revision and resolution with potentially incorrect or inconsistent models must be better studied for single agents before a useful approach in this area for multiple agents can be created.

In summary, knowledge-based approaches to expert resolution appear to offer some promising enhancements to the purely probabilistic approaches to evidence combination and expert resolution. Communication that allows agents to exchange **explanations** for their opinions, rather than only numerical **strengths** of opinions (or of supports for beliefs, etc.) can improve both the efficiency of the consensus-building process and the quality of the common knowledge base finally achieved.

ACKNOWLEDGEMENTS

This chapter is an updated version of Cox (1991), with new material added on the Bayesian approach. I thank Patrick Beatty, Peter Craig and James Wilson for renewing my interest in applications of Bayesian methods for dealing with uncertainties in scientific forums where currently only probability numbers are being exchanged. I also thank an anonymous reviewer whose comments helped to clarify the exposition and make more explicit the point of Example 26.3. Finally, I appreciate several stimulating conversations with Al Drake of MIT on common knowledge and the epistemic interpretation of probability measures.

REFERENCES

Bacharach, M. (1975) Group decisions in the face of differences of opinion. *Management Science*, **22**, 182–191.

Bacharach, M. (1985) Some extensions of a claim of Aumann in an axiomatic model of knowledge. *Journal of Economic Theory*, **37**, 167–190.

Barlow, R.E. and Proschan, F. (1975) *Statistical Theory of Reliability and Life Testing*, New York: Holt, Rinehart and Winston.

Barrett, C.R. and Pattanaik, P.K. (1987) Aggregation of probability judgements. *Econometrica*, **55**, 1237–1241.

Bonduelle, Y. (1987) Aggregating expert opinions by resolving sources of disagreement. Ph.D. thesis, Department of Engineering-Economic Systems, Stanford University.

Bordley, R.F. (1982) A multiplicative formula for aggregating probability assessments. *Management Science*, **28**, 1137–1148.

Brachman, R. and Levesque, H. (1985) *Reading in Knowledge Representation*, Los Altos, CA: Morgan Kaufmann.

Clemen, R.T. (1986) Calibration and the aggregation of probabilities, *Management Science*, **32**, 312–314.

Cox, L.A., Jr (1990) Incorporating statistical information into expert classification systems to reduce classification costs. *Annals of Mathematics and Artificial Intelligence*, **2**, 93–108.

Cox, L.A., Jr (1991) Knowledge-based resolution of conflicting expert opinions. *Journal of Applied Statistics*, **18**, 23–34.

Etherington, D. (1988) *Reasoning with Incomplete Information*, Los Altos, CA: Morgan Kaufmann.

French, S. (1986) Calibration and the expert problem, *Management Science*, **32**, 315–320.

Gasser, L. and Huhns, M.N. (1989) *Distributed Artificial Intelligence Volume II*, San Mateo, CA: Morgan Kaufmann.

Geanakoplos, J.D. and Polemarchakis, H. (1982) We can't disagree forever. *Journal of Economic Theory*, **28**, 192–200.

Lindley, R.V. (1986) Another look at an axiomatic approach to expert resolution, *Management Science*, **32**, 303–305.

Luce, R.D. and Raiffa, H. (1957) *Games and Decisions*, New York: Wiley.

McKelvey, R.D. and Page, T. (1986) Common Knowledge, consensus, and aggregate information. *Econometrica*, **54**, 109–127.

McKelvey, R.D. and Page, T. (1990) Public and private information: an experimental study of information pooling. *Econometrica*, **58**, 1321–1340.

Morris, P.A. (1977) Combining expert judgments: a Bayesian approach. *Management Science*, **23**.

Morris, P.A. (1986) Observations on expert aggregation. *Management Science*, **32**, 321–328. ·

Pearl, J. (1988) *Probabilistic Reasoning in Intelligent Systems: Networks of Plausible Inference*, San Mateo, CA: Morgan Kaufmann.

Rosenschein, S. (1985) Formal theories of knowledge in AI and robotics. *New Generation Computing*, **3**, 345–357.

Rosenschein, J.S. and Breese, J.S. (1989) Communication-free interactions among rational agents: a probabilistic approach, *Distributed Artificial Intelligence Volume II*, Gasser, L. and Huhns, M.N. (eds), San Mateo, CA: Morgan Kaufmann, Chapter 5.

Roth, A.E., ed. (1985) *Game-Theoretic Models of Bargaining*, New York: Cambridge University Press.

Schervish, M.J. (1986) Comments on some axioms for combining expert judgements, *Management Science*, **32**, 306–311.

Smith, J.Q. (1989) Influence diagrams for statistical modeling. *Annals of Statistics*, **17**, 654–672.

Winkler, R.L. (1981) Combining probability distributions from dependent information sources. *Management Science*, **27**, 479–488.

Winkler, R.L. (1986) Expert resolution. *Management Science*, **32**, 298–302.

Randomness and independence in non-monotonic reasoning

27

E. Neufeld

INTRODUCTION

If **randomness** is at the heart of Kyburg's (1971) theory of epistemological probability, **independence** (especially conditional independence) is at the heart of the work of Pearl (1988), Lauritzen and Speigelhalter (1988) and Shachter (1986) on causal probabilistic nets. Both ideas extend the range of inferences that can be made from a set of probabilities (or statistical statements), but in very different ways.

Kyburg's randomness is a formal inductive tool for choosing among conflicting statistics when making inferences about an individual. If we only know that 80% of birds fly, randomness lets us ignore the fact that *this* bird happens to be a sparrow for which class we have no statistics. If this bird is an emu, randomness chooses the narrowest class. On the other hand, Pearl's conditional independence is a purely deductive tool that lets us derive new relative frequencies (or measures of belief) from old ones given certain knowledge about conditional independence. Given point values for all conditional probabilities represented directly on a Bayes net, one can compute a point value for any other relevant conditional probability.

Related ideas appear widely in the non-monotonic reasoning literature. Figure 27.1 depicts a simple well-known example. A link $a \to b$ means 'as are typically bs', $a \Rightarrow b$ means 'all as are bs' and arrows with crosses indicate negation of the heads. We refer to such statements as **defaults** and **facts**, respectively.

Non-monotonic reasoning might be viewed as the problem of discovering domain-independent justifications for combining statements of typicality as well as the problem of attaching properties to individuals, although it is not usually framed in these terms.

A typical combination problem is sound transitive inference. It seems reasonable to chain from *emu* to *bird* to *has-wings*, but the chain from *emu* to *bird* to *fly* seems completely unreasonable. The transitive inference from *bird* to *fly* to *airborne* involving two non-monotonic links seems to fall in the middle—it is not as strong as *emus have wings* yet not as absurd as *emus fly*. Many (Loui, 1987) view the task of non-monotonic inference as arbitrating among such chains.

Artificial Intelligence Frontiers in Statistics: AI and statistics III. Edited by D.J. Hand. Published in 1993 by Chapman & Hall, London. ISBN 0 412 40710 8

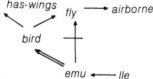

Figure 27.1 The bird world.

We can see the non-monotonic inference problem in another light if we try to apply a generic statement to an individual. Suppose we redundantly know Tweety is a *bird* and an *emu*. We might use the *birds fly* link to conclude she flies or the *emus don't fly* link to conclude she doesn't. How do we choose the right statement to apply? Note that this remains a problem even if we forbid chaining.

Thus it is easy to confuse the problem of combining generic statements with the problem of attaching the appropriate generic statement to an individual. For example, if Tweety is an *emu*, do we conclude she has wings by chaining from *emu* to *bird* to *has-wings* or by deducing that she is also a bird and applying the *bird → has-wings*?

If both approaches always gave the same answer, this would not be a problem. However, in more complex domains, this confusion produces peculiar behaviour.

Here we wish to distinguish carefully the two kinds of inference by comparing them to their probabilistic counterparts. In particular, combining generic statements is closely related to the inferences permitted by knowledge about conditional independence and the problem of attaching generic statements to individuals is closely related to statistical inference.

We will do this partly by simulating certain non-monotonic inferences within a probabilistic framework by perturbing the definitions of randomness and independence. The observation of interest is that although these perturbations justify conclusions that seem natural in the non-monotonic logic setting, the assumptions they entail (about sampling procedures) are hard to swallow from the probabilistic perspective. To highlight the issues, we confine ourselves to queries about simple individuals based on generalized knowledge encoded as an inheritance hierarchy with exceptions (Horty *et al.*, 1987) such as Figure 27.1. We conclude by considering the implications for non-monotonic reasoning formalisms.

NON-MONOTONIC INFERENCE

We will avoid the details of a formal logic and just sketch a probabilistic formalism to use as a benchmark. Similar to previous work (Neufeld, 1990), let the link $a \rightarrow b$ mean $p(b|a) - p(b|\neg a) > 1 - \varepsilon$ where a and b are sets and ε is a domain constant that may be large in unusual cases. Thus a link indicates both high probability and a relation of effect.[1] Strict links $a \Rightarrow b$ mean $p(a|b) = 1 > p(a|\neg b)$. The problem is, given evidence b (some of which may be irrelevant) about an individual i, is i typically a?

We first need to know or need to make an assumption about how i was chosen from the general population. The notion is randomness, and is defined for an individual i, evidence b, a property a and a body of statistical knowledge.

Definition 27.1

(Kyburg). We assume an individual i is chosen *randomly* from, or is a *random member* of a set b with respect to a property a, if $i \in b$ and we know $p(a|b)$ and we do not know that $i \in b^* \subset b$ where the probability $p(a|b^*)$ is known and differs from $p(a|b)$.

This formalizes the statistician's saw that we make an inference by matching an individual with the most representative sample.

A separate issue is that of deducing new statistics from old ones, and a useful tool is independence.

Definition 27.2

(Pearl). Corresponding to every proposition on such a graph (say, *grey*), there is a variable *Grey* with outcomes *grey*, $\neg grey$, and we distinguish the variable from its corresponding proposition by capitalizing the name of the variable. If A, B, C, are variables, then we say A is *conditionally independent* of B given C if $p(A|BC) = p(A|B)$ (Pearl, 1988), for all outcomes of the variables when the mentioned conditional probabilities are defined.

Here we assume that a node is independent of all predecessors given its immediate predecessors. This entails other independency relationships (see Pearl, 1988, for details).

Now we provisionally define 'typically' over sets and over individuals.

Definition 27.3

as are typically bs if $p(a|b) - p(a|\neg b) = 1 - O(\varepsilon)$ where $O(\varepsilon)$ tends to zero as ε. An individual i is typically an a if i is a random member of b, and bs are typically as.

Two examples illustrate some difficulties.

Example 27.1

Consider the simple 'chain' of approximate inference where i is an *emu* and we want to decide whether i has wings (see Fig. 27.1). Does i typically have wings because i is a random bird and birds typically have wings and we have no specific information about emus, or does i typically have wings because i is a random emu and we *know Has-Wings* is conditionally independent of *Emu* given *Bird* and that $p(h|e) - p(h|\neg e) = 1 - O(\varepsilon)$? (This is an easy proof.)

Now consider j, a *bird*. We want to wonder whether j is typically *airborne*. We could use independence knowledge if applicable, but randomness cannot be applied without generalizing the definition to include 'partial' subsets.

Thus randomness licenses $p(h|b, e) = p(h|b)$, whereas independence also entails $p(h|b) = p(h|b, \neg e)$, $p(h|\neg b) = p(h|\neg b, \neg e) = p(h|\neg b, \neg e)$. The ideas seem closely related.

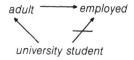

Figure 27.2 Adult students graph.

Example 27.2

Suppose we redundantly known Tweety is both a *bird* and an *emu*. Do we conclude typically $\neg fly$ from the definition of randomness, or from an 'evidence combination rule' such as

$$p(f \mid be) = \frac{p(f \mid e) - p(f \mid \neg be)p(\neg b \mid e)}{p(b \mid e)} \tag{27.1}$$

If we use randomness, then we are guided inductively to our conclusion by randomness and the logical equivalence of **emu** and **bird ∧ emu** in the antecedents of the target conditional probabilities. If we think in terms of the preceding 'evidence combination rule', we are guided by a strictly deductive principle.

This has subtle implications when the link from *emu* to *bird* is not strict. This appears frequently in the literature as the 'unemployed student' example (Fig. 27.2) given the query 'are adult students unemployed?'.

The usual answer is 'no', but why? In the non-monotonic reasoning literature, this example is usually lumped with the emu-bird example and justified by a graph-theoretic preference for an otherwise undefined inductive notion of 'specificity'. In Pearl's ε-calculus, which interprets defaults as statements of high probability, this arises deductively as a probabilistic constraint, an interesting generalization of 'Jeffrey's rule'. However, if we interpret defaults as statistical majorities, or as shifts in belief, the deductive inference is not possible.

REITER'S DEFAULT INFERENCES

Normally our confidence in probabilities diverges when we combine them. However, as in the case of chaining, we can exploit conditional independence to combine probabilities in a very precise way.

Non-monotonic logicians use **logical** independence in a similar manner to combine defaults. (Propositions are logically independent if they do not entail an inconsistency.) For example, consider Reiter's (1980) implementation of the *bird* → *fly* default as 'if ⊢ *bird(i)*, and it is consistent to assume *fly(i)*, then ⊢ *fly(i)*'.

Indeed, for the important special cases in Reiter (1980), including the instance of the 'birds fly' problem where no emus fly, logical independence is enough to solve neatly all our problems. It permits interesting transitive inferences: if birds fly and flying things are airborne, we infer that birds are airborne. It permits us to ignore irrelevant knowledge: if *i* is a *blue bird*, we can still infer *i* flies. No emus fly. From these examples, it is not surprising that this simple idea seemed like a promising solution to the 'probabilistic bottleneck'. So we might be tempted to say we might assume i is a **Reiter-random** member of **any** set appearing as an antecedent to a default, whenever this assumption does not produce an inconsistency.

But if some emus fly, and i is an emu, we can consistently either assume i is a random emu that typically does not fly or that i is a random bird that typically flies. This is neither probabilistically nor logically satisfying, but it is not completely absurd if we imagine a sampling procedure where emus might be evenly distributed among the birds. Then it might be possible that a particular individual i was randomly chosen both from the set of birds and from the set of emus.

Then a solution must be found by restricting the kinds of assumption we make about the method of selection. Etherington (1987) offers the solution of **seminormal defaults**. In the case of Fig. 27.1, we implement 'birds fly' as, 'if i is a bird and it is consistent to assume $\neg\, emu \wedge fly$, then conclude fly'. This is different from knowing $\neg\, emu(i)$, and it is easy enough to see that this is equivalent to 'hand-coding' Kyburgian randomness for this set–superset hierarchy. We note Etherington uses this method on the adult student problem, thus generalizing randomness to 'partial' subsets.

THEORIST'S DEFAULT AXIOMS

The Theorist reasoner (Poole *et al.*, 1987) implements 'birds fly' as a default axiom $bird(X) \rightarrow fly(X)$ with the single inference rule that a default conclusion can be explained by applying any number of logically independent default axioms to an individual.

Since the defaults $bird \rightarrow fly$ and $\neg fly \rightarrow \neg bird$ are equivalent, we simulate this kind of default in a probabilistic framework by changing the definition of typicality to

Redefinition 27.1
If as are typically bs then $\neg\, b$s are typically $\neg\, a$s.

To incorporate Theorist's inference patterns fully into our probabilistic framework, we must also change the definition of randomness:

Redefinition 27.2
For any set a, an individual i is a random member of a or a random member of $\neg\, a$.

This is a key observation! Theorist, unlike Reiter's defaults, permits proof by cases. To emulate this behaviour, assumptions about the method of selection must mimic all the rules of logic. In particular, random set membership, like simple set membership, must obey the law of the excluded middle!

Then it is no wonder that the qualitative lottery paradox is so devastating in tightly connected databases[2] since it seems straightforward to construct examples that permit us to assume i is a representative member of any set.

Poole (1988) shows how to achieve the equivalence of seminormal defaults by the method of 'constraints'. These more freewheeling defaults require a more complex machinery to restrain them, but the underlying idea is closely related to seminormal defaults and randomness. Perhaps a more general solution is the idea of **dialectics**

(Poole, 1990), where we accept an argument just when there are no counterarguments against it, where a counterargument is essentially one that rests on more specific knowledge.

ε-SEMANTICS

There is an interesting duality between probabilistic and consistency-based approaches to non-monotonic inference. Where the latter get into trouble because of 'too much inferencing', the former illustrate the 'probabilistic bottleneck'. Since we cannot generally combine probabilities, probabilistic non-monotonic inference seems restricted to a table look-up. For example, the probability that birds fly tells us nothing about the probability of *blue* birds flying. Secondly, the probability that those who are both parachutists and firewomen are brave is completely unconstrained by the probabilities that firewomen are brave and parachutists are brave. (These are usually called the problems of **irrelevant knowledge** and **multiple inheritance**, respectively.)

Pearl's (1988) ε-semantics interprets defaults as constraints on probability distributions. Thus *bird→fly* represents the constraint that

$$p(fly|bird) = 1 - \varepsilon$$

and any other inference $A \to B$ is supported just when

$$p(B|A) = 1 - O(\varepsilon)$$

and $O(\varepsilon)$ tends to 0 as ε.

Substituting into Equation (27.1), Pearl shows adult students are typically employed, seeming to generalize randomness to partial ('almost all') subsets. But he observes we still cannot conclude emus have wings since $p(wings|emu)$ is completely unconstrained by our other knowledge. The same applies for irrelevant information and multiple inheritance.

Thus an inductive assumption is still needed, and Pearl (1989) considers several 'adventurous' extensions to the conservative core of ε-semantics including maximum entropy and a graph-theoretic notion of **irrelevance** similar in syntactic notation to independence but similar in semantic spirit to randomness.

Note that this problem also arises in our work on inference graphs (Neufeld, 1990b). These graphs interpret links as above, including the interpretation of the topology of the graph as encoding all the independence knowledge of a Bayes network. In that paper, we made inferences about individuals using the heuristic of **conditioning on all observations**. This permits many interesting inferences, but an inductive assumption is needed to factor out irrelevant observations and permit 'multiple inheritance'.

SUMMARY AND CONCLUSIONS

Problems arise in the non-monotonic reasoning formalisms partly because of a lack of clear distinction between the inductive and deductive aspects of non-monotonic inference. Here we have argued that combination of generic statements (statements of typicality) is a deductive problem; that is, domain-independent inferences about typical properties of classes should be objective. However, an individual often is a member of many classes, and the problem of matching an individual with an appropriate generic statement an inductive one since we may later discover that the individual is a member of a subclass that is exceptional in some way.

A counterargument might be that combination of generic statements might also be viewed as inductive assumption-making. In the probabilistic framework, we might view the conditional independence relationships defined by a Bayes network as objective facts or inductive assumptions. However, the problem of attaching probabilities to individuals remains even after we verify all conditional independence assumptions.

It has been sufficient for the purpose of illustration to restrict attention to simple systems and examples. However, related problems arise in more complex systems and examples. For example, refer to Fig. 27.1 again, and interpret *lle* as 'looks like an emu'. Then, in the system we have described, we conclude that if *i* looks like an emu, then *i* typically does not fly, and *i* typically is not airborne (see Neufeld, 1991, for a proof). Now if we learn of a strict subclass of *emu* called *flemu* that typically flies and add the appropriate links, we can still infer that *lle*s typically do not fly. Oddly, had we learned instead that flemus typically do not fly and added a redundant link, we could conclude nothing about *lle*s. This is because the additional links introduce new probabilistic dependencies between *lle* and *fly*. The links on inference graphs are perhaps 'overloaded' in that we cannot add subset links without also introducing dependencies that may or may not have side-effects. Something similar happens with Geffner's (1988) logic. If *i* looks like an emu and is not airborne, we cannot deduce that *i* cannot fly. Here the problem is not a lack of an inductive assumption like randomness, but one that differs in the very precise sense of defining a method of selection. (See Poole, 1990: 74, for details on this example.)

Many other interesting comparisons are possible. For example, Neufeld (1991) shows that within the hypothetical reasoning framework, randomness becomes a kind of 'universal conjecture'. It can also be shown that randomness is equivalent to interesting special cases of **circumscription of abnormality** (McCarthy, 1986).

Discussing the relationship between non-monotonic inference and qualitative probabilities, Wellman (1990) states a belief that the solution is not purely probabilistic, and we agree. However, once we have settled on the issues of what sets to store information about and on the idea that defaults have *some* probabilistic interpretation, it seems that the problem of sound non-monotonic inference is closely related to statistical inference. We hope that the distinctions described here offer a useful perspective from which to view non-monotonic inference as statistical inference without numerical probabilities.

ACKNOWLEDGEMENTS

This research has been supported by Natural Science and Engineering Research Council grant OGP0099045. The author is also a member of the Institute for Robotics and Intelligence Systems and wishes to acknowledge the support of the Networks of Centres of Excellence Program of the Government of Canada, NSERC and the participation of PRECARN Associates Inc. Thanks to all readers for comments on earlier drafts of this chapter.

NOTES

1. Let $p(a|b) = v$ denote the conditional probability that the relative frequency of as among bs is $0 < v \leqslant 1$. To stress our main point, the links of the graph should be considered as observed statistics.
2. The paradox arises as follows. Imagine the set *bird* consists of n mutually exclusive and exhaustive subclasses (s_j) such that for every relevant property p_j, $bird \rightarrow p_j$ and $s_j \rightarrow \neg p_j$. By contraposing, we have for all j that a bird is typically not an s_j. But then for any k, we can conclude a bird is typically $\bigwedge_{j \neq k} \neg s_j$ and by cases conclude the bird must typically be s_k and hence has the properties typical of that subclass.

REFERENCES

Etherington, D. (1987) Formalizing nonmonotonic reasoning systems. *Artificial Intelligence*, **31**, 41–85.

Geffner, H. (1988) A logic for defaults, in *Proceedings of the Seventh National Conference on Artificial Intelligence*, 449–454.

Horty, J., Touretzky, D.S. and Thomason, R.H. (1987) A clash of intuitions: the current state of nonmonotonic multiple inheritance systems, in *Proceedings of the Tenth International Joint Conference on Artificial Intelligence*, 476–482.

Kyburg, H.E., Jr (1971) *Logical Foundations of Statistical Inference*, Dordrecht: Kluwer Academic.

Lauritzen, S. and Speigelhalter, D. (1988) Local computations with probabilities on graphical structures and their application to expert systems. *Journal of the Royal Statistical Society, B*, 157–224.

Loui, R.P. (1987) Defeat among arguments: a system of defeasible inheritance. *Computational Intelligence*, **3**, 100–106.

McCarthy, J. (1986) Applications of circumscription to formalizing commonsense knowledge. *Artificial Intelligence*, **28**, 89–118.

Neufeld, E. (1990) A probabilistic commonsense reasoner. *International Journal of Intelligent Systems*, **5**, 565–594.

Neufeld, E. (1991) Notes on 'A clash of intuitions'. *Artificial Intelligence*, **48**, 225–240.

Neufeld, E. (1991) The abnormality predicate. *21st IEEE Conference on Multiple Valued Logic*, pp. 218–224.

Pearl, J. (1988) *Probabilistic Reasoning in Intelligent Systems*. San Mateo, CA: Morgan Kaufmann.

Pearl, J. (1989) Probabilistic semantics for nonmonotonic reasoning: a survey, in *Proceedings of the First International Conference on Principles of Knowledge Representation and Reasoning*, 505–516.

Poole, D.L. (1988) A logical framework for default reasoning. *Artificial Intelligence*, **36**, 27–48.

Poole, D.L. (1990) Dialectics and specificity—preliminary report, in *Proceedings of the Eighth Biennial Conference of the CSCSI*, 69–78.

Poole, D.L., Goebel, R. and Aleliunas, R. (1987) Theorist: a logical reasoning system for defaults and diagnosis, in *The Knowledge Frontier: Essays in the Representation of Knowledge*, N. Cercone and G. McCalla (eds), New York: Springer-Verlag.

Reiter, R. (1980) A logic for default reasoning. *Artificial Intelligence*, **13**, 81–132.

Shachter, R.D. (1986) Evaluating influence diagrams. *Operations Research*, **34**, 871–882.

Wellman, M.P. (1990) Fundamental concepts of qualitative probabilistic networks. *Artificial Intelligence*, **44**, 257–303.

Consistent regions in probabilistic logic when using different norms

28

D. Bouchaffra

INTRODUCTION

Ordinary logic is very often used when dealing with certainty, but in many fields when we have to treat uncertainty or 'fuzziness' boolean logic appears a very poor concept. Several artificial intelligence (AI) or linguistic applications require the ability to reason with uncertain information. The semantic generalization of logic in which the truth values of sentences are probability values between 0 and 1 seems to be necessary, assignment of probability values could be consistent or inconsistent.

A lot of mathematical work on probabilistic inference already exists (Suppes, 1966; Zadeh, 1975). For instance, the MYCIN expert system handled uncertain knowledge, and the PROSPECTOR system used a method based on Bayes's rule (Duda *et al.*, 1976). In this chapter, our interest is in providing a new way of extending Nilsson's (1986) probabilistic logic by exploring different metrics. We adopt a geometric approach and discuss the relative notion of an extreme vector.

The second section defines the notion of consistency and the probability concept. The third section presents the algebraic and geometric interpretation. The next four sections present several ways of obtaining the consistent region according to metrics. We explore the infinite norm, the L_1 norm, the Euclidean norm, and also their corresponding isometries. The interpretation of Bayes's theorem when using the L_1 norm is discussed in the eighth section.

CONSISTENCY AND PROBABILITY

We define these concepts by an example. Let us consider the set of sentences $\mathscr{S} = \{P, \neg P \vee Q, Q\}$, so that we expect to obtain among all possible worlds (2^3) just 2^2 consistent possible worlds; other possible worlds are inconsistent.

Many methods exist for determining the sets of consistent truth values, given a set of sentences. A basic method is that called the **binary semantic tree** (Kleene, 1987).

Artificial Intelligence Frontiers in Statistics: AI and statistics III. Edited by D.J. Hand. Published in 1993 by Chapman & Hall, London. ISBN 0 412 40710 8

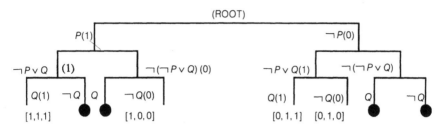

Figure 28.1 The binary tree.

We create nodes by branching left or right, depending on whether or not we assign one sentence a value of TRUE or FALSE respectively.

At each node, we branch left or right, depending on whether or not we assign one of the sentences in $\mathcal{S}$ a value of TRUE or FALSE, respectively. Just below the root, we branch the truth value of the next sentences of our set $\mathcal{S}$, and the procedure continues until the last sentence of $\mathcal{S}$. We obtain at the end a set of paths corresponding to unique assignments and we close off those paths corresponding to inconsistent valuations, as illustrated in Fig. 28.1. The black circles correspond to the closed paths (inconsistent evaluation), the numbers '1' and '0' at the right of each sentence correspond, respectively, to the evaluation TRUE and FALSE of the sentence. Thus we can write the matrix corresponding to the diagram, which consistent matrix we will denote C:

$$C = \begin{bmatrix} 1 & 1 & 0 & 0 \\ 1 & 0 & 1 & 1 \\ 1 & 0 & 1 & 0 \end{bmatrix}$$

We can choose the probability of any sentence S as $P(S) = \sum_i P(\mathbf{V}_i)^* \psi_{V_i}(S)$ where $P(\mathbf{V}_i)$ is the probability that the actual world $\mathbf{V}_a$ is equal to $\mathbf{V}_i$, the characteristic function ψ is defined as:

$$\psi_{V_i}(S) = \begin{cases} 1 & \text{if } S \text{ is true in } \mathbf{V}_i \\ 0 & \text{if } S \text{ is false in } \mathbf{V}_i \end{cases}$$

$\mathbf{V}_i$ are the column vectors of the matrix C, and i belongs to the set $\{1, 2, 3, 4\}$. This definition of the probability associated with a sentence gives a mathematical and geometric interpretation of the consistent area in the space of the three sentences.

THE ALGEBRAIC AND GEOMETRIC INTERPRETATION

Let us consider the following mapping between the space of possible worlds and the one of sentences. This mapping is the homomorphism associated with the consistent matrix C called Ψ. Let X and Y denote, respectively, the space of possible words and that of sentences. We can plot the following diagram corresponding to our example:

$$X \xrightarrow{\hspace{5.5cm}} \Psi \xrightarrow{\hspace{5.5cm}} Y$$

$$\mathbf{V} = \begin{bmatrix} P(\mathbf{V}_1) \\ P(\mathbf{V}_2) \\ P(\mathbf{V}_3) \\ P(\mathbf{V}_4) \end{bmatrix} \quad \text{------------------------------} \quad \rightarrow \Psi(\mathbf{V}) = \boldsymbol{\Pi} = \begin{bmatrix} P(P) \\ P(\neg P \vee Q) \\ P(Q) \end{bmatrix}$$

The system $\Psi(\mathbf{V}) = \boldsymbol{\Pi}$ is equivalent to:

$$P(\mathbf{V}_1) + P(\mathbf{V}_2) = P(P)$$
$$P(\mathbf{V}_1) + P(\mathbf{V}_3) + P(\mathbf{V}_4) = P(\neg P \vee Q)$$
$$P(\mathbf{V}_1) + P(\mathbf{V}_3) = P(Q)$$

Such that $\sum_{i=4}^{4} P(\mathbf{V}_i) = 1$. By doing some algebraic transformations, one obtains:

$$P(\neg P \vee Q) + P(P) - 1 - P(Q) = P(\mathbf{V}_1) + P(\mathbf{V}_3) + P(\mathbf{V}_4)$$
$$+ P(\mathbf{V}_1) + P(\mathbf{V}_2) - 1 - P(\mathbf{V}_1) - P(\mathbf{V}_3)$$
$$= - P(\mathbf{V}_3) \leqslant 0$$

Then, one can write:

$$P(\neg P \vee Q) + P(P) - 1 \leqslant P(Q)$$

One also notices that:

$$P(Q) - P(\neg P \vee Q) = P(\mathbf{V}_1) + P(\mathbf{V}_3) - P(\mathbf{V}_1) - P(\mathbf{V}_3) - P(\mathbf{V}_4) = - P(\mathbf{V}_4) \leqslant 0$$

In conclusion, we obtain:

$$P(\neg P \vee Q) + P(P) - 1 \leqslant P(Q) \leqslant P(\neg P \vee Q).$$

Remark 28.1
The bounds relative to $P(Q)$ are not found by a geometric approach, but by an algebraic approach.

The geometric interpretation is based on the following theorem.

Theorem 28.1
Let X and Y be two normed spaces and Ψ a linear mapping from X onto Y. If $\{\mathbf{V}_i\}$ is a set of extreme vectors according to the norm defined in X and Y, then the set $\{\Psi(\mathbf{V}_i)\}$ is also a set of extreme vectors according to the same norm.

Proof
Ψ is a linear operator. ∎

THE USE OF THE INFINITE NORM: $L_\infty(x) = \max |x_i|$

EXTREME VECTORS

We express in the space X (normed space of possible worlds) the extreme vector concept by the fact that the norm of any vector $\mathbf{V}$ is equal to 1, which can be written as $\|\mathbf{V}\|_\infty = 1$. Thus, we can write $\max |V_i| = 1$ which is equivalent to $\max |P(V_i)| = 1$, $\forall i \in \{1, 2, 3, 4\}$. However, since $\sum P(V_i) = 1$, we know that if one component of the vector $\mathbf{V}$ is equal to 1 then the others must be 0. Finally, we have obtained four extreme vectors (according to $\|\cdot\|_\infty$) corresponding to the standard basis for $\mathbb{R}^4$ (the classical Euclidean space):

$$
\mathbf{V}_1 = \begin{bmatrix} 1 \\ 0 \\ 0 \\ 0 \end{bmatrix} \quad
\mathbf{V}_2 = \begin{bmatrix} 0 \\ 1 \\ 0 \\ 0 \end{bmatrix} \quad
\mathbf{V}_3 = \begin{bmatrix} 0 \\ 0 \\ 1 \\ 0 \end{bmatrix} \quad
\mathbf{V}_4 = \begin{bmatrix} 0 \\ 0 \\ 0 \\ 1 \end{bmatrix}
$$

By using the linear relation due to the product of the consistent matrix C by the column vector of possible worlds probabilities $\mathbf{V}$, we can write $\Psi(\mathbf{V}) = \Pi$, which is equivalent to $C\mathbf{V} = \Pi$. Using Theorem 28.1, $\Psi(\mathbf{V})$ is also extreme in the space of sentences according to the same norm.

This algebraic property enables us to plot the consistent region according to the infinite norm. As noted above, the extreme vectors in the space of possible worlds correspond to the standard basis for $\mathbb{R}^4$, so the extreme vectors in the space of sentences are the column vectors of the matrix C (consistent):

$$
\Pi_1 = \begin{bmatrix} 0 \\ 1 \\ 0 \end{bmatrix} \quad
\Pi_2 = \begin{bmatrix} 0 \\ 1 \\ 1 \end{bmatrix} \quad
\Pi_3 = \begin{bmatrix} 1 \\ 0 \\ 0 \end{bmatrix} \quad
\Pi_4 = \begin{bmatrix} 1 \\ 1 \\ 1 \end{bmatrix}
$$

Geometrically, the consistent region is the inside of the convex hull of the extreme vectors $\psi(\mathbf{V})$. Since these extreme vectors are the columns of the matrix C, we can plot the consistent region using $\|\cdot\|_\infty$ as in Fig. 28.2.

The points outside this volume are inconsistent. When using the mixed product in order to determine the plane equations, the constraints corresponding to the consistent

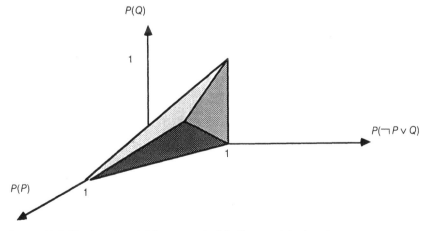

Figure 28.2 The interior of this convex hull is the consistent region.

region are easily found. Thus, we obtain the inequalities:

$$P(Q) \leqslant P(\neg P \vee Q)$$
$$P(\neg P \vee Q) + P(P) - 1 \leqslant P(Q)$$

The length of the interval corresponding to $P(Q)$ is equal to: $I_\infty = 1 - P(P)$. It depends only on the value of $P(P)$.

Remark 28.2
The bounds relative to $P(Q)$ are now found geometrically. The geometric region has been searched in order to match with the region found algebraically. The geometrical construction according to the infinite norm is a particular semantic.

THE CONSISTENT REGION DEDUCED FROM ISOMETRIES

It is of interest to see what is the consistent region when dealing with special auto-morphisms, called *isometries*.

Definition 28.1
Let $f: X \to Y$ a non-singular mapping and X and Y be two normed spaces. f is called an isometry if $\|x - y\| = \|f(x) - f(y)\| \; \forall (x, y) \in X^* Y$.

Proposition 28.1
If $C \in L(\mathbb{R}^n)$, the following assertions are equivalent:

(a) C is an isometry
(b) $\|C^* V\|_\infty = \|V\|_\infty \; \forall V \in \mathbb{R}^n$

Proof
We apply Definition 28.1 with $V = (x - y)$ and f associated with C. ∎

Remark 28.3
Any extreme vector $\Pi = C^* V$ defined in the space of sentences is an image of an extreme vector V defined in the space of possible worlds. However, if C is an isometry, the second equation of Proposition 28.1 can be written as: $\|\Pi\|_\infty = 1$, where Π is an extreme vector. Thus, the method used in order to find $P(Q)$, when $P(P)$ and $P(\neg P \vee Q)$ are known, is less complicated. This is due to the fact that any solution for V has to check equation (b) of Proposition 28.1.

THE USE OF THE L_1 NORM: $L_1(x) = \sum |x_i|$

Using the property of a linear mapping and the continuity concept, we define an overvalue relative to the norm of any vector in the space of possible worlds. The consistent region is contained inside the maximal area assigned to this overvalue.

THE MAXIMAL AREA CONTAINING THE CONSISTENT REGION

In the case of the L_1 norm, an extreme vector in the space of possible worlds is defined by $\|V\|_1 = 1$. Therefore, by this definition, all the vectors defined in the space of possible worlds become extreme. This remark is inherent to the fact that a probability must satisfy the concept of exclusivity and exhaustivity.

Additional information is required in order to obtain a square matrix C. Therefore, the problem is approached via Proposition 28.2.

Proposition 28.2
The image of a compact set by a continuous mapping is also a compact set in the space of sentences.

Proof
See Choquet (1964). ∎

As $\mathbb{R}^4$ has finite dimension, the compact image is in particular bounded, we can express the continuity of the matrix C by:

$$\|CV\|_1 \leqslant \|C\|_1 \|V\|_1 \, \forall V \in \mathbb{R}^4$$

The norm of a linear mapping in a normed space is defined by the following relation:

$$\|C\|_1 = \max \|CV\| 1/\|V\|_1$$

and the extreme condition is expressed by $\|V\|_1 = 1$. Thus, we obtain:

$$\|C\|_1 = \max \|P(P); P(\neg P \vee Q); P(Q); P(\tau)\|_1 = 4$$

In conclusion, the final result is:

$$\|CV\|_1 \leqslant \|C\|_1 = 4$$

which is equivalent to:

$$P(P) + P(\neg P \vee Q) + P(Q) + P(\tau) \leqslant 4$$

(4 is the number of sentences). The maximal area is then:

$$\mathscr{D} = \{[P(P), P(\neg P \vee Q), P(Q), P(\tau)]^T \text{ such that } P(P) + P(\neg P \vee Q) + P(Q) + P(\tau) \leqslant 4\}$$

where T indicates a vector transpose. The consistent region is strictly contained inside the region $\mathscr{D}$.

THE CONSISTENT REGION

Proposition 28.3
Any vector defined in the space of possible worlds is extreme according to this norm.

Proof
This is due to the fact that any probability vector must satisfy $\sum P(V_i) = 1$. ∎

Before giving the method and a solution, we present some topological properties.

Let $\mathcal{P} = \{\mathbf{V} = [P(\mathbf{V}_1), P(\mathbf{V}_2), \ldots, P(\mathbf{V}_n)]^T / \sum P(V_i) = 1 \text{ and } \forall i, \ P(\mathbf{V}_i) = P_i \geqslant 0\}$

Lemma 28.1

$\mathcal{P}$ is the space of possible worlds and it is a convex and compact set in $\mathbb{R}^n$.

Proof

$\mathcal{P}$ is convex and closed because it is the intersection of closed convex sets; it is also bounded because it is contained inside the unit sphere $\|\mathbf{V}\|_1 = 1$. ∎

Theorem 28.2

Let U be a linear operator in $\mathbb{R}^n$, $\mathbf{C} = (C_{ij})$ its associated matrix relative to the standard basis for $\mathbb{R}^n$. If $\sum C_{ij} = n \, \forall j$ ($n-1$ is the number of sentences in the set of beliefs), $C_{ij} \geqslant 0$, then we can write the following assertions:

(a) $U(\mathcal{P}) \subset n\mathcal{P}$
(b) $\|\boldsymbol{\varPi}\|_1 = n\|\mathbf{V}\|_1$.

Proof

By using the hypothesis of this theorem, one can write:

$$\|\mathbf{CV}\|_1 = \sum_{i=1}^{n} \sum_{j=1}^{n} C_{ij}\mathbf{V}_j = \sum_{j=1}^{n} \sum_{i=1}^{n} C_{ij}\mathbf{V}_j = \sum_{j=1}^{n} \mathbf{V}_j \sum_{i=1}^{n} C_{ij} = n \sum_{j=1}^{n} \mathbf{V}_j$$

and both of the relations (a) and (b) are proved. ∎

Proposition 28.4

The sets $\mathcal{P}, U(\mathcal{P})/n, U^2(\mathcal{P})/n^2, \ldots, U^k(\mathcal{P})/n^k, \ldots$, form a decreasing sequence with an intersection D different from $\varnothing$.

Proof

$U^k(\mathcal{P})/n^k$ is a decreasing sequence when applying Theorem 28.2, so any finite intersection of the squence is different from $\varnothing$. Now, for all k, U^k is continuous because it is a linear operator in a finite dimension space, thus, $U^k(\mathcal{P})/n^k$ is compact, and $\cap U^k(\mathcal{P})/n^k = D \neq \varnothing$. ∎

Proposition 28.5

$U(D) = D$ and for each probability vector $\mathbf{V}$, $U(\mathbf{V}) \in D$.

Proof

If $\mathbf{V} \in D$, for all $k \geqslant 0$, $\mathbf{V} \in U^k(\mathcal{P})/n^k$, therefore $U(\mathbf{V}) \in U^{k+1}(\mathcal{P})/n^k$, e.g. $U(\mathbf{V}) \in D$. If $\mathbf{W} \in D$, by the definition of D, for all $k \geqslant 0$, $\exists \mathbf{V}_k \in U^k(\mathcal{P})/n^k$ such that $\mathbf{W} = U(\mathbf{V}_k)$. As $\mathcal{P}$ is compact, the sequence $\mathbf{V}_k$ has at least one adherence value $\mathbf{V}$, then $\mathbf{V} \in D$, since $\mathbf{V}_k \in \cap U^i(\mathcal{P})/n^i$ with $i \leqslant k$ and the $U^i(\mathcal{P})/n^i$ are closed and decreasing. In conclusion, as $U(D) \subset D$ and $D \subset U(D)$ then $U(D) = D$. ∎

Theorem 28.3

If $C(k) = (C_{ij}(k))$ is the matrix associated with U^k/n^k relative to the canonical basis is for $\mathbb{R}^n$, then a strictly increasing sequence, $k_1, k_2, k_3, \ldots$ of natural numbers exists so that, for all i and j, the sequence $(C_{ij}(k_p))$ tends to the limit matrix $\mathbf{D} = (D_{ij})$ as p tends to infinity and $L(\mathcal{P}) = \mathbf{D}$, where L is a linear operator associated to the matrix $\mathbf{D} = (D_{ij})$.

Proof

For all i, j, k, we can write: $0 \leqslant C_{ij}^k \leqslant 1/n^k$; so, for all i, j, the sequence (C_{ij}^k) possesses an adherence value (Bolzano–Wiertrass). So by setting in order the finite set of indices i and J, we can, by n^2 extractions of the sequence, find an increasing sequence of natural numbers (k_p) so that, for all i and j, the sequence $(C_{ij}(K_p))$ tends to $\mathbf{D} = (D_{ij})$ (adherence value) as p tends to infinity. For all $p \geqslant 1$, we can write, using the definition of L: $L(\mathcal{P}) \subset U^{k_p}(\mathcal{P})/n^{k_p}$, and, therefore, $L(\mathcal{P}) \subset \mathbf{D}$ because $U^i(\mathcal{P})/n^i$ is a decreasing sequence. Let $\mathbf{V} \in \mathbf{D}$; for all natural numbers p, we can find $\mathbf{V}_{k_p} \in \mathcal{P}$ such that $\mathbf{V} = U^{k_p}(\mathbf{V}_{k_p})/n^{k_p}$. If $\mathbf{W}$ is an adherence value associated to the sequence $\mathbf{V}_{kp}$, we obtain the result by using, for example, the following inequality:

$$\| U^{k_p}(\mathbf{V}_{k_p})/n^{k_p} - L(\mathbf{W}) \| \leqslant \| U^{k_p}(\mathbf{V}_{k_p})/n^{k_p} - L(\mathbf{V}_{k_p}) \| + \| L(\mathbf{V}_{k_p}) - L(\mathbf{W}) \| \qquad \blacksquare$$

Proposition 28.6

$\mathcal{P}$ is the convex hull relative to $\{\mathbf{e}_i\}$, D is the convex hull relative to $\{L(\mathbf{e}_i)\}$, where $\{\mathbf{e}_i\}$ is the canonical basis associated with $\mathbb{R}^n$.

Proof

This is due to the fact that L is a linear operator and $L(\mathcal{P}) = \mathbf{D}$. $\qquad \blacksquare$

APPLICATION OF THE PREVIOUS TOPOLOGICAL RESULTS

The original problem was $\mathbf{CV} = \mathbf{\Pi}$, where $\mathbf{C}$ is the consistent matrix, $\mathbf{V}$ is the probability vector in the space of possible worlds, and $\mathbf{\Pi}$ is the vector probability associated with the sentences. Methods using the entropy of the distribution $\mathbf{V}$ have been used, but they remain approximate and subjective. Our consistent matrix is:

$$C = \begin{bmatrix} 1 & 1 & 0 & 0 \\ 1 & 0 & 1 & 1 \\ 1 & 0 & 1 & 0 \end{bmatrix}$$

We complete the matrix $\mathbf{C}$ by adding another row so that the sum of each column will be equal to the number of sentences $+ 1 = 4 = n$, in this example. Our new matrix, $\mathbf{C}'$, is square and all its elements values are positive. It corresponds to our linear operator U used in Theorem 28.3, and is as follows:

$$C' = \begin{bmatrix} 1 & 1 & 0 & 0 \\ 1 & 0 & 1 & 1 \\ 1 & 0 & 1 & 0 \end{bmatrix}$$
$$\begin{bmatrix} 1 & 3 & 2 & 3 \end{bmatrix}$$

The problem becomes:

$$C'^{*}V = \boldsymbol{\Pi}' = [P(P), P(\neg P \vee Q), P(Q), \mathscr{A}]^{\mathrm{T}}$$

where $\mathscr{A}$ is the component deduced from the added row of the matrix C'. In order to find D we have to multiply the square new matrix $C'^{*}\, 1/n$ by itself p times; the number p of iterations used is the **precision** of the method.

We obtain the following diagram:

$$C \rightarrow C'^{*}\, 1/n \rightarrow C'^{2*}1/n^{2} \rightarrow C'^{3*}1/n^{3} \rightarrow \cdots \rightarrow C'^{p*}1/n^{p}$$

$C'^{p*}1/n^{p}$ corresponds to the limit matrix $D = (D_{ij})$ of the topological results; this matrix is associated with the linear operator L. As the approximated limit matrix $C'^{p*}1/n^{p}$ is found, then D is obtained by the convex hull $[C'^{p*}1/n^{p}](e_{i})$, where $\{e_{i}\}$ is the canonical base of our vector space. As $[C'^{*}1/n](D) = D$, inside D, which is completely defined, any point considered is consistent. However, to find $P(Q)$, we have to project to its corresponding axis. The programming of this method is easy to define, as any iterative process.

THE CONSISTENT REGION DEDUCED FROM ISOMETRIES

As each vector defined in the space of possible worlds is extreme according to this norm, when dealing with isometries, we obtain the following proposition.

Proposition 28.7
If $C \in L(\mathbb{R}^{n})$, the following assertions are equivalent:

(a) $\| C^{*}V \|_{1} = \| \boldsymbol{\Pi} \|_{1} = \| V \|_{1} = 1$
(b) $\sum C_{ij} = 1 \; \forall j \in \{1, \ldots, n\}$ and $C_{ij} \geqslant 0$.

Proof
This is Theorem 28.2 with $n = 1$. ∎

Remark 28.4
This is a special case ($n = 1$) of the work done previously. However, any vector $\boldsymbol{\Pi} = (\boldsymbol{\Pi}_{i})$ satisfies $\sum \boldsymbol{\Pi}_{i} = 1$. This case is the one where the space of sentences is a probability space. The solution for $P(Q)$ is unique and is equal to $1 - P(\neg P \vee Q) - P(P) - \mathscr{A}$.

Remark 28.5
The approximate methods using the entropy concept (Nilsson, 1986) in order to find a solution for $P(Q)$ can now be checked. We can apply them to isometries and the result for $P(Q)$ must be $1 - P(\neg P \vee Q) - P(P) - \mathscr{A}$. However, some work needs to be done in order to define all 'logical isometries'; one certainly has to change the binary representation of the truth values assigned to predicates.

THE USE OF THE L_2 NORM: $L_2(x) = (\sum x_i^2)^{1/2}$

In this section, we use the Euclidean norm to find the minimal sphere containing the 'consistent' region. We maximize a function which is defined as the image of an extreme vector in the space of possible worlds. We show that the problem is equivalent to finding the eigenvalues associated with a quadratic form.

THE MINIMAL SPHERE CONTAINING THE CONSISTENT REGION

This approach is based on weights W_i (Zadeh, 1975; 1983) given to the sentences instead of probabilities and the concept of continuity in the space of finite dimensions. Each weight W_i is assigned to a possible world. One condition is that each weight belongs to the interval [0, 1] and their sum can be greater than 1. By using this norm, we are approaching a fuzzy logic.

To make calculations easier, in order to apply some methods using spectral theory, one can add additional information τ (tautology) to fulfil the necessary hypothesis. As $\|V\|_2^2$ cannot be greater than 1, we impose the condition that the extreme vector is defined according to this norm as:

$$\|V\|_2^2 = 1, \text{ which is equivalent to } \sum W_i^2 = 1 \qquad (28.1)$$

We are interested in determining the minimal sphere containing the consistent region. The constraint corresponding to this aim is: $\exists k_0 \in \mathbb{R}^+$ such that:

$$\|\Pi\|_2^2 = K_0 \qquad (28.2)$$

The continuity of the homomorphism corresponding to the consistent matrix yields the following assertion:

$$\exists K > 0; \ \|C^*V\|_2 \leqslant K\|V\|_2 \Leftrightarrow \|\Pi\|_2 \leqslant K \quad (\text{as } \|V\|_2 = 1) \qquad (28.3)$$

Equation (28.2) is equivalent to the sphere whose equation in $\mathbb{R}^4$ is:

$$W^2(P) + W^2(\neg P \vee Q) + W^2(Q) + W^2(\tau) = k_0$$

We choose k_0 equal to inf $\{K_i\}$ where K_i is the set of positive numbers satisfying the relation (28.3), and τ is a tautology added as the last row vector in the square consistent matrix. Finally, as $C^*V = \Pi$, our optimization approach is to solve according to $W(V_i) = w_i$ the following problem:

$$\max f(w_1, w_2, w_3, w_4) = \max \left\{ (w_3 + w_4)^2 + (w_1 + w_2 + w_4)^2 + (w_2 + w_4)^2 \right.$$
$$\left. + (w_1 + w_2 + w_3 + w_4)^2 \right\}$$

subject to:

$$Q(w_1, w_2, w_3, w_4) = \sum w_i^2 = 1 \qquad i \in \{1, 2, 3, 4\}$$

The mapping Q is quadratic and positive definite, so solutions can be found by using spectral theory properties. Before giving a solution using spectral properties, one can transform our optimized problem into an equivalent one. In order to obtain a well-known function to optimize, we shall change the variables in the following manner:

$$(w_3 + w_4) = x_1; (w_1 + w_2 + w_4) = x_2; (w_2 + w_4) = x_3; (w_1 + w_2 + w_3 + w_4) = x_4$$

The determinant of the associated matrix $\mathcal{M}$ of this linear system is:

$$|\mathcal{M}| = \begin{bmatrix} 1 & 1 & 1 & 1 \\ 1 & 1 & 0 & 1 \\ 0 & 1 & 0 & 1 \\ 0 & 0 & 1 & 1 \end{bmatrix}$$

This determinant is non-zero (1 in this example), thus the solution for w_i is unique. This transformation is quite independent of the number of sentences; it can be used each time the determinant of the matrix is different from 0 (the matrix is invertible).

The problem is transformed into the following one:

$$\max f(x_1, x_2, x_3, x_4) = \max \sum x_i^2$$

subject to:

$$Q(x_1, x_2, x_3, x_4) = (x_2 - x_3)^2 + (x_4 + x_3 - x_1 - x_2)^2 + (x_4 - x_2)^2 + (x_1 + x_2 - x_4)^2 = 1$$

Let us consider $\mathbb{R}^4$ with its Euclidean structure. We associate with the quadratic form Q, for example, the symmetric automorphism U defined by: $Q(x) = (U(x), x) = (x, U(x))$, the comma denoting the scalar product. This automorphism may make the problem less complicated to solve. Therefore, the automorphism which can be associated with the quadratic form Q is not unique.

The differential of

$$g: \mathbb{R}^4 \rightarrow \mathbb{R}$$
$$x \rightarrow Q(x) - 1$$

is the following mapping:

$$h \rightarrow Q'(x) \cdot h = 2(U(x), h)$$

The differential of

$$f: \mathbb{R}^4 \rightarrow \mathbb{R}$$
$$x \rightarrow (x, x)$$

is the following mapping:

$$h \rightarrow f'(x) \cdot h = 2(x, h)$$

Theorem 28.4
Let $f: D \rightarrow \mathbb{R}$, where D is an open set of E (vector space). If f possesses a local extremum for $x_0 \in D$, then $f'(x_0) = 0$.

Conversely, if $f \in C^2$ in D, and if $x_0 \in D$ exists where $f'(x_0) = 0$ and $f''(x_0)$ is a positive definite (negative definite) quadratic form, then f possesses a local minimum (maximum) for x_0.

Proof
See Choquet (1964) and Dieudonné (1960). ∎

Proposition 28.8
Let $f, g : D \to \mathbb{R}$ differentiable in the open D of E (vector space), and let $V = \{x \in D \,|\, g(x) = 0\}$ if $f \,|\, V$ possesses an extremum for x_0. Then the forms $f'(x_0)$ and $g'(x_0)$ are proportional.

Proof
See Dieudonné (1960). ∎

Proposition 28.9
If $E = \mathbb{R}^P$, this proportionality is expressed by the proportionality of partial derivatives.

Proof
Since $f'(x_0)$ and $g'(x_0)$ are proportional, then it is easy to obtain the proportionality of the partial derivatives. ∎

Proposition 28.8 proves that there exists $\lambda \in \mathbb{R}$ such that $(U(x), h) = \lambda(x, h)$ for all h. Thus, by Proposition 28.9, we deduce:

$$\exists \lambda \in \mathbb{R} \text{ with: } U(x) = \lambda x$$

The eigenvalues are all positive (Q is positive definite), the extreme vectors are obtained in all points x which are eigenvectors of the automorphism U with $Q(x) = 1$. In those solution points for example x_0, one can write:

$$f(x_0 + h) = (x_0, x_0) + 2(x_0, h) + (h, h)$$
$$Q(x_0 + h) = Q(x_0) + 2(U(x_0), h) + Q(h)$$

with $Q(x_0) = Q(x_0 + h) = 1$. Thus, we obtain the following relations:

$$(x_0, h) = (U(x_0), h)^* 1/\lambda = - Q(h)/2\lambda$$

and thus we can write:

$$f(x_0 + h) = f(x_0) + (h, h) - Q(h)/\lambda$$

Since λ_m and λ_M are respectively the minimal and maximal eigenvalue, we have, for all h,

$$\lambda_m(h, h) \leqslant Q(h) \leqslant \lambda_M(h, h)$$

for $\lambda = \lambda_m$ we obtain, for all h,

$$f(x_0 + h) \leqslant f(x_0)$$

and thus the function f possesses a maximum, and this minimal eigenvalue is the solution to our problem. So $\lambda = \lambda_m = k_0 = \max f(x_1, x_2, x_3, x_4)$, and for $\lambda = \lambda_M$ we obtain a minimum. At any point obtained $(x_0, x_0) = (U(x_0)/\lambda, x_0) = (1/\lambda)Q(x_0) = 1/\lambda$.

We should, however, select among all possible solutions those which give a weight value in the interval [0, 1]; so we must have the following inequalities:

$$\begin{cases} 0 \leqslant x_2 - x_3 \leqslant 1 \\ 0 \leqslant x_4 + x_3 - x_1 - x_2 \leqslant 1 \\ 0 \leqslant x_4 - x_2 \leqslant 1 \\ 0 \leqslant x_1 + x_2 - x_4 \leqslant 1 \\ 0 \leqslant x_i \leqslant 1 \end{cases}$$

If the eigenvector corresponding to the minimal eigenvalue does not agree with the previous inequalities, we shall have to test the eigenvector associated to the next minimal eigenvalue. The solution of our problem is transformed to the search of the eigenvalues of the following symmetric automorphism U deduced from the previous expression of Q (Gaussian method):

$$U = \begin{bmatrix} 2 & 2 & -1 & -2 \\ 2 & 4 & -2 & -3 \\ -1 & -2 & 2 & 1 \\ -2 & -3 & 1 & 3 \end{bmatrix}$$

The characteristic polynomial of this automorphism is:

$$P(\lambda) = \lambda^4 - 11\lambda^3 + 21\lambda^2 - 11\lambda + 1$$

Many numerical algorithms exist (Ciarlet, 1983) for determining the minimal eigenvalue. As has been proved, this minimum is equal to $k_0 \leqslant 1$. The consistent region is limited by the positive part of the sphere in which the radius is equal to $(\inf \lambda_i)^{1/2}$. It can be represented in $\mathbb{R}^3$ by Fig. 28.3.

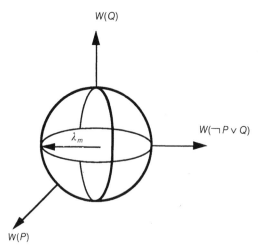

Figure 28.3 The interior of the positive part of this sphere is the minimal area containing the consistent region.

THE CONSISTENT REGION DEDUCED FROM ISOMETRIES

The idea has to do with the following question: in what conditions could the consistent region be contained inside the unit ball, which is the unit sphere in three dimensions? Mathematically this is expressed by the following equality:

$$\| C^*V \|_2^2 = \| V \|_2^2 = 1 \Leftrightarrow \| \Pi \|_2^2 = 1$$

Proposition 28.10
If $C \in L(\mathbb{R}^n)$ the following assertions are equivalent:

(a) C is an orthogonal matrix.
(b) $\| CV \|_2 = \| V \|_2$ (C is called an isometry).

Proof
$\| CV \|_2^2 = (CV, CV) = (V, C^*CV)$ as $C^*C = I$ (C is an orthonormal matrix) thus:
$\| CV \|_2^2 = (V, V) = \| V \|^2$ and $\| CV \|_2 = \| V \|_2$. ∎

Remark 28.6
Statement (a) of Proposition 28.10 is the characterization of the isometries. We know that in the case of $\mathbb{R}^2$, the rotations and the orthogonal symmetries are those matrices. In the case of $\mathbb{R}^4$, which is our example, not only are the family rotations obtained but other families as well.

Proposition 28.11
If $C \in L(\mathbb{R}^n)$ is an isometry then C^T remains an isometry. The set of all isometries is a group for the product operation.

Proof
$(C^*C)^T = I = C^T(C^*)^T = C^T(C^T)^* = I$; this proves that C^T is an orthogonal matrix which is equivalent to an isometry. In the general case, the following kind of matrices are isometries:

$$
\begin{bmatrix}
1 & 0 & 0 & 0 & \cdots & & 0...0.....0 \\
0 & 1 & 0 & 0 & \cdots & & 0 \ldots 0 ....0 \\
& & & \cdots & & & \\
0.........0 & & \cos\theta....\sin\theta.....0.....0 \\
& & \cdots & & & \\
0.....0......... & -\sin\theta....\cos\theta.........0 \\
& & \cdots & & & \\
0.....0.....0......................................1
\end{bmatrix}
\begin{matrix} p \quad q \end{matrix}
$$

where p and q are subject to the condition:

$$1 \leqslant p < q \leqslant n$$

p and q are not only indications of columns but also indications of lines, n is the

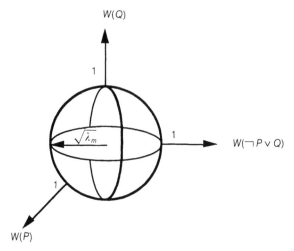

Figure 28.4 The interior of the positive part of this sphere is the minimal area containing the consistent region.

dimension of the matrix, θ is the angle of the vector of indices p and q of the rotation. The minimal area containing the 'consistent region' can be represented in $\mathbb{R}^3$ by Fig. 28.4.

A GEOMETRICAL INTERPRETATION OF BAYES'S THEOREM

Bayes inference (Duda *et al.*, 1976; Glymour, 1985) allows of easy geometrical interpretation when using the norm L_1. We show that the conditional probabilities $P(S_i/D)$ are given by the boundary of the unit ball associated to the L_1 norm.

Theorem 28.5
If the set $S_1, S_2, S_3, \ldots, S_n$ realizes a partition of the fundamental space Ω, e.g. for all $(i, j) S_i \cap S_j = \emptyset$ if $i \neq j$ and $\cup S_i = \Omega$, and $D \subset \mathcal{P}(\Omega)$ is an event, then we can write, for all i, the following formula:

$$P(S_i/D) = \frac{P(D/S_i)P(S_i)}{\sum\limits_{i=1}^{i=n} P(D/S_i)P(S_i)}$$

and the equality is true for all i in $[1, n]$.
We can express this in vector notation as:

$$
\begin{bmatrix} P(S_1/D) \\ P(S_2/D) \\ P(S_3/D) \\ \vdots \\ P(S_n/D) \end{bmatrix}
= \frac{1}{\sum\limits_{i=1}^{i=n} P(D/S_i)P(S_i)}
\begin{bmatrix} P(D/S_1)P(S_1) \\ P(D/S_2)P(S_2) \\ P(D/S_3)P(S_3) \\ \vdots \\ P(D/S_n)P(S_n) \end{bmatrix}
$$

One notices that if we call $Y = (Y)_i = P(S_i/D)$, the conditional vector associated with D, and $X = (X)_i = P(D/S_i)P(S_i)$, this equation can be written as:

$$Y = \frac{X}{\|X\|_1}$$

and thus $\|Y\|_1 = 1$. The vector Y belongs to the boundary of the unit ball associated to the L_1 norm. Other results could be deduced by this way of interpretation, the norm L_1 seems to be much more adequate to inference concept using probabilities.

CONCLUSION

In this chapter we have presented different ways of obtaining the consistent region. We pointed out that the notion of consistency is inherent in the choice of a topology. One notices that the notion of inclusion between the set of possible worlds and the set of sentences is hidden when using the infinite norm. When dealing with the 'logical isometries' (logical consistent isometries), the solution for $P(Q)$ is easy to find. The use of the L_1 norm is much more advantageous in the geometrical aspect than that of the infinite norm, because many relations exist between the linear mappings and the L_1 norm. The consistent region when using this norm can be reduced, and this is not a mathematical advantage but an experimental advantage, since we must distinguish between the possible realizations of the proposition $P(Q)$. Experimentally, the ideal case is that corresponding to a unique solution for $P(Q)$. The condition of isometry when using the L_1 norm is equivalent to the exclusivity and exhaustivity of the sentences: it is the case where the space of sentences is a probability space. The determination of the probability associated with a sentence S given the other sentences is then unique. The case of the L_2 norm requires a definition of weights in the space of possible worlds. Our approach is based on the minimal area containing the consistent region corresponding to this norm. The results obtained are interesting since we reach the spectral theory which is still a very rich area of mathematics. Some isometries have been cited; their minimal area containing the consistent region is the unit sphere. The geometric interpretation of Bayes's theorem is easily given by the boundary of the unit ball when using the L_1 norm.

REFERENCES

Choquet, G. (1964) *Cours d'analyse*, Vol. 2, Paris: Masson.
Ciarlet, P.G. (1983) *Introduction à l'analyse numérique matricielle et à l'optimisation*, Paris: Masson.
Dieudonné, J. (1960) *Foundations of Modern Analysis*, New York: Academic Press.
Duda, R.O., Hart, P. and Nilsson, N.J. (1976) Subjective Bayesian methods for rule-based inference systems, in *Proceedings of the 1976 National Computer Conference, AFIPS*, **45**, 1075–1082. Reprinted in B.W. Webber and N.J. Nilsson (eds) (1981) *Readings in Artificial Intelligence*, Palo Alto, CA: Tioga.
Glymour, C. (1985) Independence assumptions and Bayesian updating. *Artificial Intelligence*, **25**, 95–99.
Grosof, B.N. (1985) An inequality paradigm for probabilistic knowledge. In *Proceedings*

AAAI/IEEE Workshop on Uncertainty and Probability in Artificial Intelligence, Los Angeles.

Kleene, S.C. (1987) *Logique mathématique*, Paris: Jacques Gabay.

Nilsson, N.J. (1986) Probabilistic logic. *Artificial Intelligence*, **28**, 71–87.

Suppes, P. (1966) Probabilistic inference and the concept of two evidences, in J. Hintikka and P. Suppes (eds), *Aspects of Inductive Logic*, Amsterdam: North-Holland, pp. 49–65.

Zadeh, L.A. (1975) Fuzzy logic and approximate reasoning. *Synthese*, **30**, 407–428.

Zadeh, L.A. (1983) The role of fuzzy logic in the management of uncertainty in expert systems. *Fuzzy Sets and Systems*, **11**, 199–227.

A decision-theoretic approach to controlling the cost of planning 29

L. Hartman

INTRODUCTION

It has long been believed that one of the cornerstones of intelligence is the ability to plan, or simulate the future. With this ability a planner can test the usefulness of certain action sequences without exposing itself to the hazards of actually executing those sequences in a real-world environment. Planning here includes such problem-solving activities as that carried out, for example, by Strips systems (Fikes and Nilsson, 1971) or theorem provers (Green, 1969). Such systems take formal descriptions of an external environment and of a problem statement and attempt to compute **plans**, or sequences of actions, which when executed yield a state of the agent and the environment satisfying the problem statement. The computational costs are the time and space consumed by the planning process. Evidence that planning is expensive is the fact that even though in principle it is possible to solve problems, a guarantee provided by many **complete** search procedures, there are few practical applications of what has been presented in the planning literature. There are exceptions, such as reported in Vere (1983), Georgeff and Lansky (1987) and Stefik (1981a, 1981b). However, these cases seem unrepresentative of the broad coverage that the completeness property suggests is possible.

It is well known that general forms of the planning problem have high computational complexity (Plaisted, 1984; Fischer and Rabin, 1974; Canny, 1988; Hopcroft and Wilfong, 1986; Chapman, 1987; Korf, 1985). For the most part these results are obtained by encoding a problem of high complexity as a version of the planning problem. In addition to other important problems, such as finding adequate formal encodings of domain knowledge (Hayes, 1979; 1985), involved in the practical application of planning, the computational expense of general planning algorithms certainly stands as a great hindrance to the application of planning techniques to real-world problems.

An interpretation of the dissatisfaction with current planner performance (see, for example, Chapman, 1990; Brooks, 1991) is that people are unwilling to expend such

Artificial Intelligence Frontiers in Statistics: AI and statistics III. Edited by D.J. Hand. Published in 1993 by Chapman & Hall, London. ISBN 0 412 40710 8

a large amount of their computational resources to acquire solutions to these kinds of problems. That is, the solutions to such problems are not as valuable as the computational resources required to generate them. Automatic planners are presently unable to expend their resources in a way that is commensurate with the value of the problems being solved. People are able to solve the same problems in ways that are less expensive than using an automatic planner or they have the option of solving other problems.

The approach presented here uses techniques of decision theory to measure and control the computational effort involved in planning. The application of these techniques yield planning or problem-solving strategies that measure their use of computational resources and make decisions about how to use these resources appropriately to solve each problem instance. The standard decision-theoretic technique for generating an optimal action sequence to solve a problem involves an algorithm requiring time exponential in the length of the solution. The EU-strategies presented here instead efficiently compute an estimate of the global expected utility of each execution sequence that the strategy exhibits and selects among them on that basis. For these EU-strategies the computation associated with making each of these decisions requires only constant time and space. For each problem instance posed to such a strategy, the strategy is able to take into account in its decision-making essentially all computation expended to solve the instance.

Our approach assumes that agents are autonomous in that their activity is unsupervised for extended periods of time during which they are exposed to many problem instances. Such agents are able to make a greater investment of effort in acquiring and using information about how they solve problems. Information, such as statistical properties of the environment, gained in solving past problems can be used in attempting to solve future problems more quickly.

The computational sequences that a strategy is able to carry out partition the space of problem instances into equivalence classes. Two problem instances are in the same equivalence class if the strategy, when presented with each instance, carries out the same computational sequence. Members of these equivalence classes of problems (or, simply, **problem classes**) occur according to some distribution function determined by the agent's environment. That is, for the sequence of problem instances to which an autonomous agent is exposed, members of a class appear in the sequence with some probability. Whether a solution is found (the strategy's **success**) and what effort is involved (the strategy's **cost**) determine the performance of the strategy. The expected utility of a strategy is also a function of the distribution of problem classes since, in general, a strategy has a distinct success rate and distribution of cost for the members of each problem class. Because an agent employing a strategy may not be aware of how problem classes are distributed and, thus, may not be aware of the best substrategy to employ in attempting to solve a particular problem instance, there is uncertainty associated with this relationship between the strategy and the distribution problem classes. The strategies of interest here address this uncertainty by measuring statistical properties of the problem domain. Specifically, these strategies gather statistics over the agent's problem-solving history in order to approximate the expected utility of executing a substrategy and to aid in making decisions about resources.

This work extends that of Feldman and Sproull (1974) by incorporating the cost of planning into the evaluation of plans. These strategies use techniques of standard decision theory (see, for example, Raiffa, 1968) to evaluate alternative ways to compute plans. For any plan with a finite description—for instance, the plans with conditional branches as in Feldman and Sproull (1974)—there are only a finite number of outcomes so the expectation of such a plan can be computed in an efficient way, i.e. in time linear in the size of the plan. Similarly (given certain formal constraints) a strategy with a finite encoding has a well-defined expected utility. An approximation to this expectation can be obtained from statistics about the strategy's past performance. Given two strategies, it is easy to construct another strategy that gathers performance statistics for the given strategies, computes approximations to their expected utility and executes the one with the largest value. Such a decision accounts for all the computational resources that the larger strategy expends except for the constant time computation involved in making the decision itself and updating the performance statistics for the executed strategy. Since the expectations include the computational expenditure, the choice of substrategies is one that, at least to within the errors of the approximation of expected utility, involves the best investment of computational effort.

EU-STRATEGIES

We now present a more precise notion of strategy. In decision theory a strategy is basically an imperative specification of what the agent does under any circumstances (Raiffa, 1968). An **expected utility strategy**, or EU-strategy, is a particular kind of decision-theoretic strategy. For the current purposes an agent is identified by an EU-strategy and behaves only by virtue of the activity that results from the execution of its EU-strategy, such as the invocation of a motor system primitive. An agent attempts to solve a problem by executing its EU-strategy. Execution begins after a description of a problem instance, the **current** problem instance, is deposited or written into a part of the agent's memory. The point of describing an agent in limited terms such as an EU-strategy is to make the agent's use of computational resources explicit and to make it possible to attribute utility in a well-defined way to all of the computation that an agent does.

EU-strategies are composed ultimately of primitive actions that are either external motor actions or internal computational actions. The external actions, M, consist of all those primitives, both sensory and motor actions, that involve interaction with the agent's external environment. The internal actions, C, are computational actions that do not involve direct interaction with the external environment and include simple actions such as assigning a value to a cell in memory and complex actions such as executing a procedure that carries out an A^* search. The planning an agent carries out is to be encoded as computations composed of elements of C.

An EU-strategy has one of the following forms:

1. a for any single action $a \in M \cup C$.
2. $[s_1; s_2]$ for EU-strategies s_1 and s_2.
   $[s_1; s_2]$ is a sequence and is executed by executing first s_1 and then s_2.

3. $[c \rightarrow s_1, s_2]$ for EU-strategies c, s_1 and s_2.

$[c \rightarrow s_1, s_2]$ is a conditional action in which the state of the agent resulting from the execution of c determines which of s_1 and s_2 is executed. If the execution of c leaves a non-zero value in special status register then s_1 is executed. Otherwise s_2 is executed. The proposition $True(c)$ holds when the execution of c yields this non-zero value.

As part of a larger EU-strategy c, s_1 and s_2 are sub-EU-strategies. They are referred to below simply as **substrategies** since in the context of this chapter no confusion should arise.

With a function U that the agent is able to compute, it is possible to obtain approximations to $E[U, S_i]$, the expected utility of executing S_i, such that the error becomes arbitrarily small as the number of times S_i is executed increases. Let $eu(T)$ be the sample mean of U for all executions of substrategy T in the agent's problem-solving history. We can define the EU-strategy

$$[c \rightarrow S_1, S_2] \tag{29.1}$$

where c stands for the expression

$$eu(S_1) > eu(S_2) \tag{29.2}$$

The execution of c in Expression (29.1) is such that $True(c)$ holds if and only if the inequality is true at the time of execution. The values of eu in Inequality (29.2) are simply statistics that have been recorded during past experience. Each time a substrategy S_i is executed, U is evaluated and $eu(S_i)$ is updated. When c is executed these numbers for the substrategies S_1, S_2 are simply retrieved and the appropriate arithmetic is carried out.

For a utility function U in which the cost of acting and, in particular, of computing is simply the time taken and in which the value of function U_0 is the utility of a static situation,[1] we have:

$$U(S) = time^\lambda - time^S + U_0^S \tag{29.3}$$

Using the notation of Russell and Wefald (1988), Q^S is the value of Q in the state resulting from the execution of strategy S. The term $time$ is just a function that returns the current time. For null strategy λ, $time^\lambda$ and $time^S$ are the times when S begins executing and finishes executing, respectively, and U_0^S is the utility of the state resulting from execution of S. The EU-strategy in Expression (29.1), written out in greater detail to show the evaluation of U and the updating of eu, is:

$$\left[\begin{array}{l} t := time; \\[2mm] \left[eu(S_1) > eu(S_2) \rightarrow \left[S_1; eu(S_1) := \dfrac{(U_0 + t - time + (\#(S_1) - 1)eu(S_1))}{\#(S_1)}\right]\right], \\[6mm] \left[S_2; eu(S_2) := \dfrac{(U_0 + t - time + (\#(S_2) - 1)eu(S_2))}{\#(S_2)}\right] \end{array}\right]$$

where $\#(T)$ is the number of times that substrategy T has been executed and ':=' is the usual assignment operator. The procedure *time* is just an implementation of the function mentioned above and returns the current time when it is executed. The value of $eu(S)$ at any time is just the sample mean of $U(S)$ over all past executions of S. For the sake of clarity these bookkeeping details about conditional actions are omitted from the presentation of EU-strategies below.

Unlike many search procedures such as A^* and planners such as Strips, EU-strategies involve no mechanism for backtracking. If a substrategy is executed that does not solve the current problem instance, there is no recourse. The containing strategy is not able to revise its decision to execute the first substrategy and execute an alternative substrategy in a further attempt to solve the current instance. For example in the EU-strategy

$$[c \to A; B] \tag{29.4}$$

it is not possible, once substrategy B has executed and failed to solve the current instance, then to execute substrategy A. Computations such as backtracking must be 'unwound' and encoded explicitly in the structure of an EU-strategy. One way to do this is to follow the last primitive action in a substrategy by a test for successfully solving the current problem instance. If the instance is not solved, another substrategy is executed that specifies the activity associated with further solution attempts. For example, after invoking substrategy B, a test q might indicate that the current problem instance has been solved and no further activity should occur. The test might also indicate that B failed to solve the current problem instance and that an alternate substrategy A should be invoked as a recourse. In an EU-strategy such 'backtracking' must be specified explicitly as in

$$[c \to \quad A,$$
$$[B; [q \to stop; A]]]$$

As before, if A is invoked and fails, the containing EU-strategy generates no further activity and fails. The point of defining strategies in this way is to make all the agent's activity (in particular, all its computation) explicit. This property of strategies permits a decision-theoretic evaluation of computations that is accurate and takes into account the precise conditions under which the computation is executed.

PROPERTIES OF EU-STRATEGIES

Selecting between two alternative substrategies in order to optimize overall performance is similar to the two-armed bandit problem (Goldberg, 1989). A two-armed bandit is composed of two one-armed bandits each with unknown pay-off probabilities. An agent playing a two-armed bandit attempts to maximize its cumulative pay-off and has to decide at each point in time whether to play the left or the right lever. The two-armed bandit points out a trade-off between the value of information and the cost associated with obtaining it. Information about the pay-off probability of the levers is, of course, valuable in that it helps the agent determine which lever to play. Such information, however, can only be obtained by playing both levers. While the agent plays the observed worst lever, it does not obtain the highest pay-off rate. While the agent plays only the observed best lever, it runs the risk of being misled by anomalous observations that do not reflect the real pay-off probabilities.

In general, since the outcome of a computation is unknown in advance of actually carrying it out, the selection of a substrategy occurs under the same uncertainty as that involved in the two-armed bandit problem. Whether a substrategy solves the part of the current problem it is intended to solve, what cost is incurred in obtaining the solution and what cost is involved in executing the solution cannot be determined in less time than is required to execute the substrategy. Thus, there is uncertainty about the outcome of executing each of the substrategies and only trying out both substrategies will reduce this uncertainty and provide information about their relative performance.

Any planning system operates under such uncertainty. For example, if a system includes two distinct procedures for generating solutions to the same problem instances, it is desirable to choose the procedure with the best performance on the problems in the current domain. In the absence of detailed knowledge supplied by the system's designer about the types of problems in the domain and the performance of these procedures on these problems, the system's only source of information is the actual measured performance of the procedures.

Experiments presented by Hartman (1990) incorporated a simple (non-optimal) solution to the two-armed bandit problem for substrategy selection. In a small fraction of the evaluations of substrategies such as Expression (29.1), the result of c was reversed and the observed worst substrategy was selected for execution. This small fraction of problem instances is sacrificed for the sake of keeping the performance statistics for each substrategy up to date.

The mean of U evaluated on the outcomes of executing S_i is a good approximation to $E[U, S_i]$ and the probability that the result of c in Inequality (29.2) disagrees with the actual inequality is a negative exponential function of the sample size n (Hoeffding, 1963). That is, given that problem instances are independent, if

$$E[U, S_1] > E[U, S_2] \tag{29.5}$$

and

$$\delta = E[U, S_1] - E[U, S_2] \tag{29.6}$$

then from Theorem 1 in Hoeffding (1963) it can be shown that

$$P(eu(S_1) < eu(S_2)) \leqslant \exp\left(-\frac{n}{2} \frac{(\delta + U_{max})^2}{U_{max}^2} \right) \tag{29.7}$$

where n is the minimum of the number of times that S_1 and S_2 have been executed and U_{max} is an upper bound on the range of U (which is assumed to exist). We use γ below to denote the bound $\exp[-\frac{1}{2}n(\delta + U_{max})^2/U_{max}^2]$. Implicitly it is a function of U, S_1 and S_2. As a greater number of trials is observed, the probability of an error due to anomalous observations of U becomes arbitrarily small.

Any planning system adopts some solution to the two-armed bandit problem associated with its choices. Suppose that, as mentioned above, an EU-strategy simply reverses the choice of the observed best option for some small randomly chosen fraction, say $1/k$, of each decision. Then given Inequality (29.5) and that the execution of the containing EU-strategy has come to the point of selecting one of S_1 or S_2, we have with high probability, namely,

$$1 - \max\{1/k, \gamma\} \tag{29.8}$$

that $eu(S_1) > eu(S_2)$ and that S_1 is executed. Similarly, if S_2 has the highest expected utility, the substrategy S_2 is selected for execution with high probability. Thus, the conditional action $[c \rightarrow S_1, S_2]$ approximates S_1 or approximates S_2 depending on which has the higher expected utility. This property of EU-strategies is discussed in more detail in Hartman (1991).

It is possible for EU-strategies to make accurate utility-based decisions about how to attempt to solve a problem and, in particular, about how to allocate resources. Resources are allocated by selecting from a set of substrategies that expend resources in different ways. This scheme allows for arbitrary resource allocation.

Consider the following EU-strategy:

$$S: [eu(S_1) > eu(S_2) \rightarrow S_1: [p_1 := C_1; R(p_1)]$$
$$S_2: [p_2 := C_2; R(p_2)]]$$

Each of the computations C_1 and C_2 produces information used by substrategy R in solving the current problem instance.

The utility of executing S_1 is:

$$U(S_1) = C([p_1 := C_1; R(p_1)]) + U_0^{[p_1 := C_1; R(p_1)]}$$

The historical average of $U(S_1)$ is stored in $eu(S_1)$. The case is similar for S_2. The only computational cost not recorded in $eu(S_1)$ or $eu(S_2)$ is the cost of deciding which of S_1 and S_2 to execute, namely,

$$time^\lambda - time^{eu(S_1) > eu(S_2)} \tag{29.9}$$

Taking this cost into account, i.e. determining that it yields an improvement in expected utility, can only be determined by agents using EU-strategies of a different form:

$$S_1': [eu(S_1) > eu(S) \rightarrow S_1, S]$$
$$S_2': [eu(S_2) > eu(S) \rightarrow S_2, S]$$

S_1' and S_2' measure the performance of S and an alternative and, with a probability given by Expression (29.8), select for execution the one with the highest expectation. That is, with high probability, they infer whether the comparison of expected utility estimates in S is worth doing. Of course, in carrying out such an inference they incur additional computational costs of their own. Two facts are important here: first, with a utility function in which computational cost is proportional to time, the cost of this potentially mistaken allocation of computational resources is a small constant; and second, for a small constant amount of effort per problem instance S can be compared with another EU-strategy.

In deciding to use EU-strategy S or S_1, we can look at their expected utilities. For EU-strategy S:

$$E[U, S] = E[C, eu(S_1) > eu(S_2)] + P(S_1)E[U, S_1] + (1 - P(S_1))E[U, S_2] \qquad (29.10)$$

where $P(S_1)$ is the probability that S_1 is executed during the execution of S. $P(S_2)$ is defined similarly and $P(S_2) = 1 - P(S_1)$. Given that $E[U, S_1] > E[U, S_2]$,

$$E[U, S_1] - E[U, S] = (1 - P(S_1))(E[U, S_1] - E[U, S_2])$$
$$- E[C, eu(S_1) > eu(S_2)]$$

With probability at most γ (from Inequality (29.7)) S_2 is the observed best substrategy and $P(S_1) = 1/k$. With probability at least $1 - \gamma$, S_1 is the observed best and $P(S_1) = 1 - 1/k$. Therefore, the expected value of $E[U, S_1] - E[U, S]$ is bounded by

$$(\gamma(k-1) + 1)\frac{\delta}{k} \qquad (29.11)$$

which is small for large values of k and small values of γ. Recall that γ is a negative exponential function of the number of problem instances. The value in Expression (29.11) is an upper bound on the expected loss of utility per problem instance when EU-strategy S is used to determine the optimal of S_1 and S_2 instead of simply using substrategy S_1 (which by assumption is the best of the two). Although a measure such as expected loss of utility is not in general meaningful, in the case of the simple utility model of computational cost used here, this measure indicates that only a small amount of computation is at stake in deciding between S and S_1.

An EU-strategy is able to compute a partial description of a plan and to execute external actions contained in this partial description before the plan is complete. In this way an agent is able to interleave the execution of internal and external actions, i.e. to interleave planning and 'execution'. This interleaving is possible since no distinction is made between the execution of external and internal actions. One benefit of such early execution of external actions is to simplify the process of completing a partial plan. By executing a sequence of external actions S in a partial plan R and examining the resulting state of the world, an agent is not required to complete R so as to account for all of the possible external states that result from executing S. It needs only to complete R so as to account for the actual external state obtained by executing S. This shortcut may amount to significant savings of computational effort and, hence, improvement in the expected utility of the remainder of the EU-strategy's problem-solving effort. Such early execution is a form of information gathering with the improvement in expected utility due to resulting information.

Interpreting computation as information gathering directly addresses a difficulty raised by Russell and Wefald (1988): that of defining the utility of a **partial computation**, i.e. a computation that does not commit the agent to executing an external action. In the formal system they describe there is no connection between a partial computation and the future actions or decisions of an agent. In any concrete realization of a problem-solving strategy there is some external action that follows a partial computation and to which there is an assignment of utility. For example, in EU-strategies the connection between computation and resulting external action is explicit. The external actions simply follow the computation in the specification of the containing substrategy.

The only activity of an EU-strategy that is not accounted for in the expected utility estimates is the activity up to and including the evaluation of the first expected utility comparison in each substrategy. Also, strategies do not involve any off-line and, hence, unevaluated computation as do the proposals of Russell and Wefald (1988), D'Ambrosio and Fehling (1989) and Etzioni (1989). The off-line computation in this proposals involves precomputing, in a learning or set-up phase, the statistical estimates of domain properties such as the expected utility of carrying out a particular computation. This off-line computation may involve exposing the agent to many problem instances in order to 'calibrate' it for the domain. By making use of off-line computation and not taking it into account in the utility appraisal of an agent's activity, it is possible to provide the agent with very high-quality information, equivalent to a great deal of computational effort. It is difficult under such circumstances to determine the quality of the agent's resource allocation decisions. In addition, one benefit of not depending on these off-line computations is that an EU-strategy is robust to some degree against changes in the statistical properties of the environment. If the agent has to calibrate for the domain while it is solving problems, then it simply continues to calibrate as the properties of the domain change.

Estimates of expected utility that strategies compute are conditioned by the internal state at the time of the decision. Internal states and, in particular, the state of execution of the strategy allow for the estimates to be conditioned by arbitrary propositions. Consider the following strategy:

$$
\begin{aligned}
&S \qquad\quad :[q \to S_1, S_2] \\
&S_1 :[eu(S_3) > eu(S_4) \to S_3{:}A, S_4{:}B] \\
&S_2 :[eu(S_5) > eu(S_6) \to S_5{:}A, S_6{:}B]
\end{aligned}
$$

S carries out an expected utility comparison in each of the two possible outcomes of q. The intuitive notion is that as the EU-strategy is executed, new information becomes available and this information is relevant to remaining choices. In particular, it is reasonable to expect that this new information bears on important properties of a substrategy such as the probability that it succeeds in solving the current problem instance.

While this scheme for conditioning statistical information and for conditioning decisions is very expressive it does not commit the agent to a large computational effort. The agent computes the actual decision in constant time. Although it is possible to construct EU-strategies in which arbitrarily long computational sequences are used in order to generate distinct execution states, such computation is not obligatory.

The expected utility evaluation that decisions are based on does not require any computation other than the constant time and space used to compare expected utilities and update statistics. In contrast, the proposal of Russell and Wefald (1988) involves an 'inner-loop' computation that takes time proportional to the branching factor in the search tree being explored.

Conditioning on execution states thus affords a great deal of expressive power in determining what information is brought to bear on a decision. At the same time, each decision commits the agent to only a constant time computational expenditure: none of the expressive power involves a 'built-in' cost to the agent. The low complexity of the expected utility comparison allows a strategy to make decisions about even small increments in its resource investments. In short the agent's capacity to decide about its resource expenditure is very fine-grained.

In addition, since only a small investment is required in order to be able to make a decision, it is possible for strategies to be very sensitive to the expenditure involved in a long computational sequence. For example, suppose there is a computational sequence $A_1, A_2, \ldots, A_n$ such that the more of the sequence that is executed, the better the outcome it yields. This computational sequence has the **anytime** property discussed by Dean and Boddy (1988). It is possible to introduce into the sequence decisions that determine, on the basis of expected utility, whether to continue the sequence. This additional investment involves of the order of no more time than is involved in part of the sequence that gets executed. One way to do this is with strategy S:

$$
\begin{aligned}
S &\quad : [A_1; [eu(S_1) > eu(S_2) \rightarrow S_1 : B, \\
S_2 &\quad : [A_2; [eu(S_3) > eu(S_4) \rightarrow S_3 : B, \\
&\qquad \cdots \\
S_{2i-2} &\quad : [A_i; [eu(S_{2i-1}) > eu(S_{2i}) \rightarrow S_{2i-1} : B, \\
&\qquad \cdots \\
S_{2n-2} &\qquad\qquad : [A_n; B] \ldots]
\end{aligned}
$$

where B is the substrategy that is executed after the A_is are executed. At each point it is possible for S to opt out of the computational sequence. The decision to opt out may be determined by more elaborate conditions than simply the expected utility estimate. For example, the agent may opt out when some preset time limit is exceeded. In any case a strategy structured in this way is very sensitive to its ongoing expenditure of its resources.

RELATED WORK

Feldman and Sproull (1974) used the rule of maximizing expected utility from the standard Bayesian decision theory to evaluate plans. The domain they use to present the evaluation is the **monkey and bananas** problem. For each action that an agent (the monkey) is able to execute there is an associated cost and set of outcomes. Associated with each possible outcome for an action is a probability of its occurrence given that the action is executed. The actions they considered included conditional branches that could make use of sensory information. A conditional action of this kind allows plans to determine explicitly when an action sequence has failed and to take remedial actions. Using a tree representation of plans, the evaluation involves computing the expected utility of leaves of the plan and backing up the maximum expected utility to all ancestors up to the root. The innovation in their work is that they avoid the combinatoric expense associated with the standard decision rule to obtain the optimal action sequence by computing the optimal plan only within a given set of plans. This computation is merely linear in the length of the description of the set of given plans as opposed to, in the standard decision rule, exponential in the length of the plan. EU-strategies extend this plan evaluation scheme to include the cost of generating the plans. In treating the internal computational actions in the same way as external physical actions with probabilistic outcomes we can use the same techniques to evaluate the utility of the computations that an EU-strategy carries out. With suitable care it is possible to embed this low time-complexity evaluation of planning costs in EU-strategies themselves.

Russell and Wefald (1988) have used their theory of value of computation in two search applications: the eight puzzle and a path planning problem. Application of their theory consists of approximating the utility values associated with states by values inferred from statistics. The applications involve an A^*-like search in which search costs are included in the cost estimates and have exhibited good performance. An important point to note is that the computational effort involved in making control decisions is not accounted for in their applications.

In related work (Hansson and Mayer, 1989a, 1989b), expected utility evaluations of node expansions are used to guide a heuristic search. The expected utility evaluations are obtained from estimates of the probability of an outcome given a particular value of an admissible heuristic function. The approach takes heuristic values as pieces of evidence and can make use of heuristic estimates from an unlimited number of nodes. In contrast, traditional heuristic search, as in A^*, is based on the evaluation of nodes only at the frontier of the search tree. These estimates are obtained off-line from a sample of the search space and are fixed during the operation of the problem-solver. Adequate performance is likely to depend on obtaining the probability estimates from a representative sample of the problem space and on the correlation between heuristic values and search outcomes not varying throughout the search space. Their approach does not account for the off-line computational cost and does not take into account the global computational cost of search in the node expansion decisions.

Etzioni (1989) discusses a control problem involving multiple goals and a time bound in which a set of solution methods is available for each of the goals. The solution to the control problem is a sequence of the available methods that is optimal with respect to a measure of expected utility. Within the sequence the selected methods for a goal are executed one at a time until the goal is solved or all methods have been tried. Such an attempt is made to solve each goal in turn. The methods in the sequence are executed until the time bound is exceeded. Since the solution methods for each goal are not guaranteed in fact to solve the goal, it is beneficial to incorporate more than one of these methods in the sequence for a particular goal.

The decision procedure Etzioni proposes orders methods by a greedy heuristic based on marginal utility, i.e. the ratio of utility to cost. The marginal utility of a method is a measure of the method's performance and represents the 'return' obtained per unit investment of, in this case, time. The decision rule orders the methods by estimates of their marginal utility. These estimates are obtained from statistical information conditioned by world state. That is, in a particular world state a method's performance is predicted from statistics about previous observations of the method's execution in that world state. The statistics collected for each method are success frequency and mean cost and are updated with each execution of the method.

Horvitz *et al.* (1989) and Horvitz (1986) study a particular trade-off in the medical diagnosis domain between the utility of further inference and the utility of immediate action. Using a sophisticated model of the costs, utilities and constraints they present a simple and efficient technique for deciding when to act. In addition, Horvitz (1987) provides a broad survey of the field of reasoning under computational resource constraints.

CONCLUSION

By connecting the choice of internal action, i.e. computation and planning, to resource expenditure and other aspects of performance, the approach presented here brings traditional planning within the scope of decision theory. Through standard decision-theoretic techniques it is possible within the limits imposed by the available knowledge for the strategy to approximate optimal decisions about what to do next. With high probability, choosing the substrategy from a set of alternatives that maximizes the approximated expected utility is optimal with respect to an external evaluation of utility. Error is introduced into this decision in virtue of the approximation errors inherent in estimating parameters such as the expected utility of executing a substrategy. These parameters determine the expected utility of a substrategy and approximations to them are used to approximate the substrategy's expected utility. Bounds on this error can be obtained from results presented by Hoeffding (1963). By considering precisely what knowledge is explictly encoded by a strategy and what knowledge is available to it for explicit reasoning, it is possible to incorporate decision-theoretic techniques without incurring the computational expense of the complete decision procedure which requires time exponential in the size of the strategy.

EU-strategies extend the techniques of decision-theoretic evaluation of conditional plans by interleaving planning and acting and by using an accurate estimate of expected utility that is efficiently computed. By removing the distinction between internal planning actions and external actions, EU-strategies permit both kinds of actions and action sequences to be evaluated in the same way. Thus, it is possible for EU-strategies to interleave planning and acting in such a way that they make appropriate decisions about when to act. The estimate of expected utility is accurate in the sense that it accounts for all of the computation carried out by the agent, including the computation involved doing the evaluation of expected utility. The hierarchical structure of EU-strategies allows an agent's behaviour to be related directly to the resulting outcomes and permits an accurate evaluation of computational costs.

By computing expected utility estimates incrementally, the effort associated with the search for successful substrategies is amortized over the lifetime of the agent. Until the estimates converge to stable and accurate values, a strategy makes varying choices of substrategies. These 'experiments' contribute to the computation of the utility estimates and, thus, with high probability to better decisions in the future. Such amortization of effort over the problem-solving history is necessary in order to accommodate the exponential time searches that are involved in solving hard problems (see, for example, Plaisted, 1984).

ACKNOWLEDGEMENTS

This work was supported by the National Science Foundation under grant DCR-8405720, DCR-8602958 and CCR-8320136 and by the National Institutes of Health under Public Health Service Research Grant 1 R01 NS22407-01.

NOTE

1. Although a utility function of this form is not critical in what follows, it is assumed throughout for the sake of concreteness.

REFERENCES

Brooks, R.A. (1991) Intelligence without reason, in *Proceedings of the International Joint Conference on Artificial Intelligence*, 569–595.

Canny, J. (1988) *The Complexity of Robot Motion Planning*, Cambridge, MA: MIT Press.

Chapman, D. (1987) Planning for conjunctive goals. *Artificial Intelligence*, **32**, 333–377.

Chapman, D. (1990) Vision, instruction and action. MIT AI TR-1085, June.

D'Ambrosio, B. and Fehling, M. (1989) Resource bounded agents in an uncertain world, *AI and Limited Rationality*, Working notes, AAAI 1989 Spring Symposium Series, Stanford University, March.

Dean, T. and Boddy, M. (1988) An analysis of time-dependent planning, in *Proceedings of the Seventh National Conference on Artificial Intelligence*.

Etzioni, O. (1989) Tractable decision-analytic control, in *Proceedings of the First International Conference on Principles of Knowledge Representation and Reasoning*, Brachman, R.J. and Levesque, H.J. (eds).

Feldman, J.A. and Sproull, R.F. (1974) Decision theory and artificial intelligence II: The hungry monkey. Technical report, Computer Science Department, University of Rochester. Also in *Cognitive Science*, **1**, 1977, 159–192.

Fikes, R.E. and Nilsson, N.J. (1971) STRIPS: A new approach to the application of theorem proving to problem solving. *Artificial Intelligence*, **2**, 189–208.

Fischer, M. and Rabin, M. (1974) Super-exponential complexity of Presburger arithmetic, *SIAM–AMS Proceedings*, **7**.

Georgeff, M.P. and Lansky, A.L. (1987) Reactive reasoning and planning. *AAAI*, 677–682.

Goldberg, D.E. (1989) *Genetic Algorithms in Search, Optimization and Machine Learning,* Boston, MA: Addison-Wesley.

Green, C. (1969) Application of theorem proving to problem solving, in *Proceedings of the International Joint Conference on Artificial Intelligence,* pp. 211–239.

Hansson, O. and Mayer, A. (1989a) Probabilistic heuristic estimates. Preliminary papers of the *Second International Workshop on Artificial Intelligence and Statistics.*

Hansson, O. and Mayer, A. (1989b) Heuristic search as evidential reasoning, in *Proceedings of the Fifth Workshop on Uncertainty in Artificial Intelligence.*

Hartman, L.B. (1990) Decision theory and the cost of planning. Technical Report 355, Computer Science, University of Rochester.

Hartman, L.B. (1991) Uncertainty and the cost of planning. Technical Report 372, Computer Science, University of Rochester.

Hayes, P.J. (1979) The naive physics manifesto, in *Expert Systems in the Microelectronic Age,* Michie, D. (ed.), Edinburgh: Edinburgh University Press.

Hayes, P.J. (1985) The revised naive physics manifesto, in *Formal Theories of the Common Sense World,* Hobbs, J.R. and Moore, R.C. (eds), Norwood, NJ: Ablex.

Hoeffding, W. (1963) Probability inequalities for sums of bounded random variables, *Journal of the American Statistical Association,* 13–30.

Hopcroft, J. and Wilfong, G. (1986) Motion of objects in contact. *International Journal of Robotics Research,* **4** (4), 32–46.

Horvitz, E.J. (1986) Reasoning under varying and uncertain resource constraints, *AAAI-86,* August.

Horvitz, E.J. (1987) Reasoning about beliefs and actions under computational resource constraints, in *Proceedings of the Third Workshop on Uncertainty in Artificial Intelligence,* July.

Horvitz, E.J., Cooper, G.F. and Heckerman, D.E. (1989) Reflection and action under scarce resources: Theoretical principles and empirical study, in *Proceedings of the International Joint Conference on Artificial Intelligence.*

Korf, R. (1985) *Learning to Solve Problems by Searching for Macro-Operators,* Boston, MA: Pitman Advanced Publishing Program.

Plaisted, D.A. (1984) Complete problems in the first-order predicate calculus. *JCSS,* **29,** 3–35.

Raiffa, H. (1968) *Decision Analysis: Introductory Lectures on Choices Under Uncertainty,* Reading, MA: Addison-Wesley.

Russell, S.J. and Wefald, E.H. (1988) Decision theoretic control of reasoning: General theory and an application to game-playing. UCB/CSD 88/435, University of California, Berkeley, Computer Science Divison (EECS).

Stefik, M. (1981a) Planning with constraints (Molgen: Part 1). *Artificial Intelligence,* **16,** 111–140.

Stefik, M. (1981b) Planning with constraints (Molgen: Part 2). *Artificial Intelligence,* **16,** 141–170.

Vere, S. (1983) Planning in time: Windows and durations for activities and goals. *IEEE Transactions on Pattern Analysis and Machine Intelligence,* **5**(3), 246–267.

Index

Accuracy
  admissible stochastic complexity
    models 340–1
  learning systems 168–9, 174–6,
    177–80
  neural networks 274–5
Action engine 42–3
Actions 42
Acyclic directed graphs *see* Directed
  acyclic graphs
Adaptive importance sampling 90–105
Adherence values 377
Admissible stochastic complexity
  models 338–46
Admissible transformations 56, 59
Aggregating modules 34
Aggregation, resolution by 349–51
Akaike information criterion (AIC) 251–3
ALARM network 147–8
AMIA 31–8
Analytics 313
Anytime property 396
Apparent error rate 203
Application areas 3–4, 11–12
Application interface model 41
APPROXIMATE tests 7
Approximation theory 221
Arc addition 110–12, 118
Architecture *see* System architecture
ARCO1 68–9, 72–9
Arc reversal procedure 96–8
Assumption-based Truth Maintenance
  System (ATMS) 296
Automatic planners 388
Auxiliary networks 95–104
Averaging of trees 184, 188–9, 194–5,
  197–9

Axiomatic aggregation of
  probabilities 350–1

Back-end manager (BEM) 41–3, 44
Back-propagation (BP) learning
  algorithm 217–40
Back-propagation procedures 265, 267–8,
  269–71, 274–6
Backtracking
  EU strategies 391
  hypertext 51, 52
Backward chaining 88
Bayesian
  aggregation procedures 353–4
  analysis 67–8
  averaging 194–5, 197–9
  decision problems 119–36
  models 82, 83–4
  networks *see* Belief networks
  probability revision 266, 269, 273–6
  smoothing 191–4, 198
  tree learning methods 184–99
Bayes
  nets 68
  risk 340–1
Bayes's rule
  belief networks 77
  geometrical interpretation 384–5
  learning classification trees 185
Beam searches 190
Bearings 124
Belief
  function model 82–4
  functions 82–8
  networks
    Bayesian decision problems 119–36
    causal structure inference 141, 144–5

Belief (contd).
    design 67–80
    filtering 90–105
    graphical 82–9
    probabilistic
      inference engines 106–18
      text understanding 300–6
    propagation 112–16, 118
    resolution 348–60
Best linear unbiased estimator (BLUE) 225
Bilingual corpora 281–94
Binary semantic trees 370–1
Bivariate statistic 23, 24
Blocked experiments 6
Bounds
    admissible models 339–40, 345
    neural networks 244, 246, 249–50,
      252–5
Box–Behnken designs 10
Branch-and-bound searches
    admissible stochastic complexity
      models 341
    belief networks 88–9
Breadth-first searches 108–10

C4 196, 197, 198
C4.5 321
Calculator problem 205–12
Canonical decision problem 129–30
CART 202–13
    learning classification trees 196, 197,
      198
Categorical variables 60
Causal
    explanations 213
    graphs 144–6
    models 327–34, 355–8
    probabilistic networks (CPNs) 106–18
    structure 141–55
    sufficiency 145
    systems 142
    theory 327–8
Causation 327–34
Central composite designs 10, 11–12
Central processing units (CPUs) 115
Channels 352
Classification 312–23
    accuracy 168–9, 174–6, 177–80
    problems 182, 335–46
    trees, learning 182–99
    CART 202–3, 213
    CONKAT 260
Classificators 312, 313, 317–23

Class probability trees 183
Clique-trees 106–7, 112–18
Clustering 315–16
Cognitive illusions 67
Coherence requirements 356, 357
Colliders 145
Combinations 120, 125–6, 133, 134
Common effect 332
Common-knowledge procedure 352–3,
    355
Competitive learning network 257–8, 260
Complete search procedures 387
Complicated influence 3, 11
Compressibility 341
Conceptual descriptors 171
Conceptual learning 178
CONCLUDE INSIGNIFICANT ONLY
    tests 7
CONCLUDE SIGNIFICANT ONLY
    tests 7
Conditional
    entropy 245–6, 249–50
    probabilities 119
Conditions 159
Confidence factors 269, 272–3
Configurations 124, 125
Confounding patterns 9
CONKAT 256–62
Connectionist networks
    back-propagation procedures 265,
      267–8, 269–71, 274–6
    CONKAT 256–62
Conservatism loss 85, 87
Consistency
    causal models 328
    probabilistic logic 370–85
    rules 35
Consistent regions 370–85
Constant-coefficient linear structures 153
Constraints
    non-monotonic reasoning 366
    precedence 124–5, 128
Constructive induction 172–4
CONSTRUCTOR 202, 203–6, 209–13
Constructors 172, 173
Context worlds 15
Contingency tables 204, 211
Continuity
    learning systems 169
    probability aggregation 350
Continuous variables 60
Contradictory explanation
    impropriety 161

Contributing properties   25
Control   82–9
Control factors   4, 5, 11
Correlation of variables   61
CPN-oriented topological ordering
    (COTO)   107–10, 118
CPNs   106–18
CPUs   115
Cross-validation   203, 210
Cut-points   185, 189–90
Cycling   269
Cyclops   91–105

Data
    analysis   13–15
    collection sheets   13
    transformation machine (DTM)   257
Databases   300, 304
Datum ratio   102
Decision
    analysis (DA)   67–80
    distance   85, 86–7
    problems   119–36
    theoretic control   387–99
    trees   68, 182–3
        classifiers   342–3
        ID3   266
        KAISER   156, 157, 158
        valuation-based systems   122, 129,
            131, 134
        variables   88
Defaults
    DEXPERT   5, 11
    non-monotonic reasoning   362, 365–6
Degress of freedom (df)   6–7, 9
Deletion sequences   133, 134–5
Delta rule   221–2
Dependence   69, 70, 74
Dependence relations   151
Description
    languages   168–9
    lengths   337, 344–5
Descriptors   171–2
Design
    belief networks   67–80
    generation   5–11, 15
    information screen   8, 9
    robust   4, 5, 11, 13
Desirable conditions (Dcond)   159, 160–1,
    163–5
Detectable difference   7–9, 10
Development tools   39–40
DEXPERT   3–15

Diabetes diagnosis problem   120–31
Diagnosis
    belief networks   76–7
    neural networks   258–9, 264–76
Dialectics   366
Dialogue control   41
Dimensional analysis   56–7, 58–9
Directed acyclic graphs (DAGs)
    causal modeling   327–8
    causal probabilistic networks   106–18
    causal structure inference   143–6
    probabilistic text understanding   300–1
Direct manipulation   34
DIRECT tests   7, 12
Dirichlet distribution   186, 187–8
Discriminational analysis   172–4, 178–80
Dispatches   115
Distance
    learning systems   175
    neural networks   246
    sentence alignment   285–9, 293
Documentation   34
Domain knowledge   156–7, 159–60,
    163–4
Domain knowledge base   157
D-optimal designs   10
DOX   43
d-separability   144–5, 149
Dwell period   103
Dynamic programming   285–9

Edge direction errors of
    commission   148–9
Edge existence errors of
    commission   147–9
Efficiency   63
Eight puzzle   397
Empirical information sets   355
Empirical learning   182
Energy demand modelling   31–2, 38
Energy policy   32
Epsilon-semantics   367
Equivalence   328
Errors
    admissible stochastic complexity
        models   340
    back-propagation procedures   221–3,
        270–1
    diagnosis   35
    of omission   148–9
    probabilistic model induction
        techniques   203, 204, 210, 212
    restricted   6

Errors (contd.)
  sentence alignment 290–2, 293
ε-semantics 367
Essential conditions (Econd) 159, 160–1, 164
ESTES 48–52
Evaluations
  belief networks 78–9
  BP learning algorithm 232–3
  sentence alignment 289–93
  statistical data analysis 173–4
Executions 394, 395–6
Expected mean squares (EMS), 6–7, 9–10
Expected utility (EU) strategies 388, 389–99
  decision problems 130–1
Expert systems 17–18, 28–9
  AMIA 31–8
  belief networks 67–80
  DEXPERT 3–15
  FOCUS 39–44
  hypertext 46–52
  metadata 54–63
  SPRINGEX 18–24
  STATISTICAL NAVIGATOR 24–8
Expirical symbolic learning system 168–80
Explantation-based learning (EBL) 314–16
Explanation capabilities
  hypertext 46–52
  SPRINGEX 18, 21
Explanatory power 206–10, 212
Explicable nil node impropriety 161
EXSYS 24, 25
Extreme vectors 372–4

Facet Attribution File (FAF) 316–17, 319
Faceted classification scheme 313
Facets 313, 316–23
Fact bases 355
FACTEX 9
Factor information 4–5
Facts 362
Faithfulness 144
Falsity-preservation 315
FAST 43
Faulty calculator data 187, 205–12
Feature constructions 172–4
Feedback 321
Filtering 90–105
Fisher's scoring method 246–9
Flexible matching process 174–6, 177–8
Focal elements 83

FOCUS 39–44
Forecasting
  AMIA 31–8
  belief networks 76–7
Forward chaining 88
Forward-propagation 222, 229
Fourier series approximation 221
Fractional factorial designs 9–10, 11, 13
Frames
  probabilistic text understanding 299–300
  valuation-based system 123–4
Front-end harness 40–1, 42, 44
Full factorial designs 5–9, 11, 13
Full-text documents 312
Function approximations 217–40
Functions 299
Fusion algorithm 119, 120, 131–5

Gain criterion 269, 272–3
Gain ratio 269, 272–3
Gauss–Markov theorem 225–6
Gauss–Seidel iterative method 247
Generalization
  back-propagation learning algorithm 233
  machine learning techniques 317–19, 323
Generation
  AMIA 36
  design 5–11, 15
  hypothesis 346
Genuine causes 331, 332
GLIMPSE 43
Graphical belief networks 82–9
Graphics
  FOCUS 44
  hypertext 47, 48, 49, 51–2
Graphs 13
Growth of trees 183, 189–90

Half–Brier score 196–7
Harness programming 44
Heading hierarchy 316
Hebbian learning law 221, 239
Help facilities
  DEXPERT 4
  hypertext 46–52
Heuristic searches 397
Hidden layers 244, 249–52
Holdout method 267
Homogeneity 57

HUGIN   112
Hybrid ML systems   315–16
Hybrid rule-based networks   344–5
Hypertext   46–52
Hypothesis
    generation   346
    node   75–6
    tests   29
    STATISTICAL NAVIGATOR   24–8

IC (Inductive Causation) algorithm   331,
    332
ID3   156, 157, 158
    evaluation   266, 268–9, 272–6
    SCIS   323
IDEAL system   305
IERS   43
Importance sampling
    adaptive   90–105
    function   93, 100
Improprieties   156, 158–9, 160–2, 167
Impropriety
    detector   157, 161, 167
    interpreter   157, 161–2, 167
    knowledge base   157–8
Incremental learning   346
Independence   362, 364–5
Indeterministic systems   142–4
Indexing module   313
INDUBI   169, 174, 177, 179
INDUCE   169
Induced causation   328, 329, 331, 332
Induction
    learning   182
    techniques   202–14
Inductive learner   157, 158
Inference
    belief networks   76–8
    causal structure   141–55
    non-monotonic   362–9
    probabilistic   106–18
    probabilistic text understanding   299
    valuation-based systems   120
Infinite norms   372–4
Influence
    belief networks   69, 70, 74
    complicated   3, 11
    diagrams   68, 90
    knowledge-based belief
        resolution   358–9
    valuation-based systems   119–20, 131,
        134

see also Belief networks
Information processing   76–8
Inseparable example impropriety   159
Inserted paragraphs   293
Integrated learning   179
Intelligent
    arc addition algorithm (IAAA)   110–12,
        118
    belief propagation   112–16, 118
INTELLIPATH   70
Interaction plots   13, 14
Inter-attribute relationships   159
Interfaces   see User-interfaces
Interpolation   218, 223–4
Interpretation   35
Interval scales   56
Irrelevance   367
Isometries   374, 378, 383–4, 385
Iterative communication of
    probabilities   352–3

Join-trees   see Clique-trees
Joint valuation   120, 130
Judgements probability   348–60

KAFTS   43
KAISER   156–67
Kalman filters   90
k-monotonicity   82
Knowledge acquisition
    belief networks   68–75
    CONKAT   256–62
    KAISER   156–67
    see also Learning
Knowledge-based
    belief resolution   354–9
    front-ends (KBFEs)   39–44
    modules   41
    systems   see Expert systems
Knowledge bases
    AMIA   34–5, 36–7
    belief networks   75
    CONKAT   258–9
    KAISER   157–8
    SPRINGEX   18, 22–4, 29
    STATISTICAL NAVIGATOR   25, 28,
        29
Knowledge Engineering Environment
    (KEE)   15
Knowledge-intensive, domain-specific
    learning (KDL)   314

Knowledge representation
belief networks   75–6
DEXPERT   15
Kohonen network   257–8, 259–60, 261

L_1 norm   374–7, 385
L_2 norm   379–84, 385
Lack-of-information based control   82–9
Lagrange multiplier method   228–9
Latent structures   328, 330–1
Learning
back-propagation   217–40
classification trees   182–99
CONKAT   257–8
inferring causal structures   141–55
KAISER   156–67
machine   312–23
module   317–21
probabilistic model induction
techniques   202–14
statistical techniques   168–80
see also Knowledge acquisition
Least Mean Squares (LMS) algorithm   217, 221
Least squares (LS) method   225–7
LED problem   187, 205–12
Lexicons   50, 51, 52
Linear causal
models   143–4
systems   144
Local computation   131–5
Logical independence   365
Logic sampling   77
Long-term energy demand modelling
31–2, 38
Lookahead heuristic
deletion sequences   134–5
learning classification trees   183,
189–90, 195, 197, 198–9
Lorenezen algorithm   6

Machine learning (ML) techniques   312–23
Main effect plots   13, 14
Mainpulation, direct   34
Mapping   160
Marginalization   120, 126–8, 133, 135–6
Marginals   126, 128
Marker-passers   304, 305
Markov
condition   144
random field (MRF) models   342
sequences   102–3
Mass functions   82–3

Matching process   174–6, 177–8, 319–20,
323
Maximum
a posteriori (MAP) techniques   336
likelihood (ML) estimators   225–6
likelihood training of neural
networks   241–55
mutual information (MMI)   241, 245–6
Meaningful statements   56–7
Meaningless statements   56
Means   57–8
Mean square error (MSE)   254
Mean sum of squared residuals
(MSSR)   230–1, 232–7
Measurement scales   54, 56–63
Measurement theory   56–7
MEDEE models   31
Medical diagnosis problem   132–4
MEDICL   269, 274
MERADIS   43
M-estimators   226–7
Metadata   54–63
Metadescriptors   172, 174
Metrical variables   60
Minimal detectable difference   7–8, 10
Minimal models   328–9
Minimality condition   144
Minimum
descriptive length (MDL)   251–3,
335–46
encoding techniques   184, 190, 198–9
fill-in algorithms   110, 111
Missing values   210–12, 213
Mixed causal structures   150–2
Model response   4, 5, 11
Modes of variables   60
Modularity   34
Monkey and bananas problem   396–7
Monotonicity   350
Monte Carlo analysis   77
adaptive importance sampling   91, 93,
95, 104
morphological analyser   298
MULREG   17
Multilayer feedforward networks   343–4
Multiple explanation impropriety   161
Multiple parents   116
Multi-ply lookahead   195, 197, 198–9
Multivariate statistic   22–3, 24
Mutual information   241, 245–6

Natural language understanding   295
Nested factors   6

Network constructors 303–4
Network evaluators 303–4
Neural knowledge acquisition module
    257
Neural networks 263–76
    admissible models 343–4
    back-propagation learning
        algorithm 217–40
    CONKAT 256–62
    learning 314
    maximum likelihood training 241–55
Neuro-computing 218–25
New ratio 102
Newton–Raphson algorithm 246–7
Nil node impropriety 158–9
No explanation impropriety 161
Noise
    BP algorithm 218, 224–5, 226–8,
        237–9
    DEXPERT 4, 5, 11, 13
    learning systems 169, 174–5
Noisy node impropriety 158
Non-linearity 62
Non-monotonic reasoning 362–9
Non-parametric statistic 23, 24
Non-serial dynamic programming 119
Non-smooth interpolation 223–4
Normalized Evidence Word Tokens
    (NEWTs) 316, 318–19, 321–2
Norms 370–85
Numerical descriptors 171

Object-oriented programming (OOP)
    35–6
Observable leaves 88
Occam's razor 328–9, 333
Off-line computations 395
One-ply lookahead heuristic
    deletion sequences 134–5
    learning classification trees 183,
        189–90, 197
Optimal strategies 131
Optimization, sequential 4, 5, 11–12
Options 194
Option trees 194–5, 197–9
Ordered variables 60
Orders 147
Ordinal relationships 159–60
Ordinal scales 56, 57–8
Overfitting 183, 218, 231
    back-propagation procedures 217
    ID3 273

Page layout recognition 176–80
Paragraph alignment 284–5, 286, 288,
    292
Parallel processors, utilization of 115
Parent nodes
    belief networks 90–1
    multiple 116
Parsers 296, 298, 303–4, 306
Partial
    computations 394–5
    correlations 149
    resolution 34
Partitioning classifiers 182
PATHFINDER 68–71, 74–9
PC algorithm 142, 145–9, 151
Perfect recall condition 125
Perfect representations 144
Permissible conditions (Pcond) 159,
    160–1, 164–5
Piecewise smooth functions 237–9
Plackett–Burman designs 10
Planning costs 387–99
Plan recognition 295, 300
Plausibility functions 82–4
Posterior probability distribution 184–5,
    187–9, 191, 199
    learning classification trees 184–5,
        187–9, 191, 199
    probabilistic text understanding 305
    sampling 102–3, 104
Potential causes 331–2
Potentials 124
Precedence constraints 124–5, 128
Precision 343, 378
Precomputing 395
Predecessors 128
Predictive accuracy 168–9, 174–6,
    177–80
Predictive power 210–12
Preferences 328
Pregrouping 237
Presentation 35
Presentation layer 41
Prior model entropy 345
Prior probability distribution 184–7,
    191–4, 196–9, 357–8
Probabilistic
    flexible matching process 174–6,
        177–8
    inference engines 106–18
    logic 370–85
    model induction techniques 202–14
    text understanding 295–310

Probability
  judgements 348–60
  score 287, 291
  valuations 124
Problem classes 388
Processing
  belief networks 76–8
  probabilistic text 296
Processors, parallel 115
Projections 330
  of configurations 125
Propagation 88, 112–16, 118
Prototypes 43–4
Pruning of trees 183, 190–4, 198–9
Pseudo-indeterministic systems 142–4
Pseudo-paragraphs 284

Qualitative structures 203–4
Quality heuristic 189, 190
Questionnaire generator 158, 162

Random-coefficient linear structures 153–5
Randomization 6, 9, 12
Randomness 362, 363–8
Ratio scales 56, 59–60
Recursive partitioning 183, 189
Recursive structural equation
    models 143–4
Redesign 12–13
Redundant chords 110
Redundant cliques, usage of 117
Referees 170–1
References 171
REGS 43
Reiter's default inferences 365–6
Rejection rate 102–3
Relationships 159–60
Replication problem 185
Representational measurement
    theory 56–7, 60
Representation language 170–2
RES 169–80
Resolution by aggregation 349–51
Response, model 4, 5 11
Response surface designs 10–12, 13
Restriction errors 6
Restructuring 190
Result-based mapping (RBM) 160, 165
Results
  presentation and interpretation 35
  reuse of 34

Robust
  back-propagation learning
      algorithm 217–40
  designs 4, 5, 11, 13
  objective function 227–8
Rule-based networks 344–5
Rule bases
  DEXPERT 15
  knowledge-based belief resolution 355
  SPRINGEX 18, 22–4
  STATISTICAL NAVIGATOR 25
Rule extractor module 257
Rules, network-building 304–5
Run-time tools 39
R-vagueness 84–5, 87

Sampling, adaptive importance 90–105
SAMSS 33, 36, 38
Scales, measurement 54, 56–63
Scenarios 33, 35, 36
Scoring method *see* Fisher's scoring
    method
Screening 3, 11
Seeheim model 40
Segmentation 237
Selectors 170–1
Semantic improprieties 160–1, 162
Seminormal defaults 366
Sensitivity 274–5
Sentence alignment 281–94
Separable user interfaces 40
SEPSOL 43
Sequential optimization 4, 5, 11–12
Servants 115
Set-up time 69
Shared causal models 356
Shootout project 79
Similar class impropriety 159
Similarity 175
Similarity-based learning (SBL) 314–16
Similarity networks 70, 74
Similar node impropriety 159
SIMPR 312–14, 316–23
Single-blind testing 296, 306–7
Slots of frames 299
SMECI-MASAI package 36, 38
Smooth functions, piecewise 237–9
Smoothing of trees 183, 184, 185–6,
    190–4, 197–8
Solutions 128–35
Specificity 274–5
Spline approximation 221

SPRINGEX   17, 18–24, 29
SPRING-STAT system   18
Stability   330
Star methodology   169
Statistical
    expert systems   *see* Expert systems
    information sets   355
    learning   179
    NAVIGATOR   17, 24–9
    techniques   168–80
Statistic rule   18, 22–4
Stochastic complexity models   335–46
Stochastic simulation   77
Strategies   128, 131
Strong implications   149
Structural improprieties   158–9
Subject classification   312–23
Subject Classification Intelligent System
    (SCIS)   312, 316–23
Subject Classification Maintenance System
    (SCMS)   316
Subsetting   269, 272–3
Substrategies   390–6, 398–9
Sucess rates   388
Sufficiency, causal   145
Supervised learning   182, 202
Support factors   160
Surrogate splits   211
Switches   152
Symbolic concept acquisition (SCA)   314
Symbolic learning systems   168–80
Synonyms   321–2
System architecture
    AMIA   37
    FOCUS   40–4
    KAISER   157–62
System capabilities   5–12

Target networks   94–5, 96, 98, 101–2,
    104
Task manager   42
Tasks   42
Teams   348–9
Template engine   43
TETRAD II program   147, 149
TETRIS   162–7
Text manipulation
    machine learning techniques   312–23
    probabilistic text understanding   295–
    310
    sentence alignment   281–94
Theorist's default axioms   366–7

Theory-based mapping (TBM)   160, 165–6
Time-warp applications   285–6
Topogical sorting algorithm   107, 108
TPL (TemPlate programming
    Language)   43, 44
Trace applications   285–6
Training
    auxiliary networks   98–100
    Bp algorithm   217–18, 232–7
    CONKAT   258
    learning systems   177–80, 315, 321
    maximum likelihood   241–55
    neural networks   267
    probabilistic model induction
        techniques   204–6, 209
    probabilistic text understanding   296,
    308
Training samples   182
Transformations, admissible   56, 59
Transition distances   289
Transmitted variation   5, 11
Tree algorithms   182–3
Treks   146
Triangulation methods   106–7
Truth-preservation   315
Tuning   218
Two-armed bandit problem   391–3

Unanimity   350–1
Understanding texts   295–310
Undirected paths   145
Uniqueness   57
Universal approximators   217
Unshielded colliders   145, 146
User-friendliness   34
User-interfaces
    AMIA   34, 35, 36–7
    DEXPERT   4
    FOCUS   40–1
Utilities   124
Utility valuations   120–1, 124, 128–34
Utilization of parallel processors   115

Vagueness   84–5, 87–9
Valuation-based system (VBS)   119–36
Valuation networks   119, 123–4, 129,
    132–3
Valuations   120, 124–5
Values   69, 70, 74
Variables
    belief networks   69, 70, 74
    valuation-based system   123, 128–9

Versions 33, 36
VL_{21} 169–72, 174–7
Voluntary control 332

Weight adaptation 223, 229–30
Weights 379, 382
Widrow–Hoff delta rule 221–2
Wilcoxon two-sample test 62

WIMP2 296
WIMP3 296–8, 300, 303–10
Windowing 269, 272–3

XOR problem 249–54
XTL (eXTraction programming
  Language) 43, 44

Milton Keynes UK
Ingram Content Group UK Ltd.
UKHW031533071024
449327UK00005B/77